Ordinary Differential Equations and Calculus of Variations

ORDINARY DIFFERENTIAL EQUATIONS AND CALCULUS OF VARIATIONS

Book of Problems

M. V. Makarets
Kiev T. Shevchenko University, Ukraine

V. Yu. Reshetnyak
Institute of Surface Chemistry, Ukraine

World Scientific
Singapore • New Jersey • London • Hong Kong

Published by

World Scientific Publishing Co. Pte. Ltd.

P O Box 128, Farrer Road, Singapore 9128

USA office: Suite 1B, 1060 Main Street, River Edge, NJ 07661

UK office: 57 Shelton Street, Covent Garden, London WC2H 9HE

ORDINARY DIFFERENTIAL EQUATIONS AND CALCULUS OF VARIATIONS

ISBN 981-02-2191-6

This book is printed on acid-free paper.

Printed in Singapore by Uto-Print

Contents

Preface

This problem book contains exercises for courses in differential equations and calculus of variations at universities and technical institutes.

It is designed for non-mathematics students and also for scientists and practicing engineers who feel a need to refresh their knowledge of such an important area of higher mathematics as differential equations and calculus of variations. Each section of the text begins with a summary of basic facts. This is followed by detailed solutions of examples and problems.

The book contains more than 260 examples and about 1400 problems to be solved by the students, a considerable part of which have been composed by the authors themselves.

Numerous references are given at the end of the book. These furnish sources for detailed theoretical approaches, and expanded treatment of applications.

In preparing this book for publication, Mr. Y.-S. Kim rendered a great help to us.

Chapter 1

FIRST ORDER DIFFERENTIAL EQUATIONS

1.1 Separable equations

A differential equation which can be written in the form

$$M(x)dx + N(y)dy = 0, \tag{1}$$

where M is a function of x alone and N is a function of y alone, is said to be *separable*. The solution is

$$\int M(x)dx + \int N(y)dy = C, \tag{2}$$

where C is an arbitrary constant. The problem is then reduced to the problem of evaluating the two integrals in (2). In Eq.(1) we say that the variables are separated.

Example 1. Find the solution of the equation

$$y' = e^{x+y}$$

which is such that $y = 0$ when $x = 0$.

The equation may be written as

$$y' = e^x e^y,$$

from which it is seen that the separated form is

$$e^{-y}dy = e^x dx.$$

Integrating now gives the general solution

$$-e^{-y} = e^x + C,$$

and we have to find the value of the constant C such that x and y vanish simultaneously. On putting $x = y = 0$, we have $-1 = 1 + C$ whence $C = -2$.

The appropriate solution is given by

$$e^{-y} = 2 - e^x.$$

Example 2. Solve the equation

$$xydx + (x+1)dy = 0. \tag{3}$$

If $y \neq 0$ and $x+1 \neq 0$, we can divide by y and $x+1$ and put the equation in the form

$$\frac{dy}{y} + \frac{xdx}{x+1} = 0.$$

Integrating,

$$\int \frac{dy}{y} + \int \frac{xdx}{x+1} = C,$$

or

$$\ln|y| + x - \ln|x+1| = C.$$

Taking exponential of both sides yields

$$y = C_1(x+1)e^{-x}, \ C_1 = \ln|C|.$$

Equation (3) has also solutions $y = 0$ and $x = -1$. The first one can be obtained from the general solution when arbitrary constant $C_1 = 0$ and therefore $y = 0$ is the particular solution. The second solution $x = -1$ can't be obtained from the general solution and therefore $x = -1$ is the *singular solution.* Then the solution of the problem (3) is

$$y = C_1(x+1)e^{-x} \quad \text{if } x \neq -1; \text{ also } x = -1.$$

Example 3. Solve the initial value problem

$$y' \cot x + y = 2; \ y\left(\frac{\pi}{3}\right) = 0. \tag{4}$$

If $y \neq 2$ and $\cot x \neq 0$ the differential equation can be written as

$$\frac{dy}{y-2} + \tan xdx = 0.$$

Integrating,

$$\int \frac{dy}{y-2} + \int \frac{\sin xdx}{\cos x} = C,$$

or

$$\ln|y-2| - \ln|\cos x| = C.$$

Whence

$$y = 2 + C_1 \cos x, \tag{5}$$

where $C_1 = \ln|C|$ is an arbitrary constant. To determine the particular solution satisfying the prescribed initial condition we substitute $x = \pi/3$ and $y = 0$ into

Eq.(5), obtaining $C_1 = -4$. Hence the desired particular solution is given explicitly by

$$y = 2 - 4\cos x.$$

Example 4. Solve the equation

$$y^2\left(x^3+1\right)dx + \left(x^3 - 5x^2 + 6x\right)dy = 0.$$

If $y \neq 0$ and $x \neq 0,\ 2,\ 3$ then the separated form is

$$-\frac{dy}{y^2} = \frac{(x^3+1)dx}{x^3 - 5x^2 + 6x},$$

or

$$-\frac{dy}{y^2} = \left(1 + \frac{5x^2 - 6x + 1}{x^3 - 5x^2 + 6x}\right).$$

Using partial fractions we can write

$$\frac{5x^2-6x+1}{x^3-5x^2+6x} = \frac{5x^2-6x+1}{x(x-2)(x-3)} = \frac{A}{x} + \frac{B}{x-2} + \frac{C}{x-3}.$$

Multiplying this by x we find

$$\frac{5x^2-6x+1}{(x-2)(x-3)} = A + \frac{Bx}{x-2} + \frac{Cx}{x-3}.$$

Then letting $x \to 0$, we have

$$A = \left[\frac{5x^2-6x+1}{(x-2)(x-3)}\right]_{x\to 0} = \frac{1}{6}.$$

Similarly, multiplying by $x-2$ and letting $x \to 2$ yields

$$B = \left[\frac{5x^2-6x+1}{x(x-3)}\right]_{x\to 2} = -\frac{9}{2}.$$

and multiplying by $x-3$ and letting $x \to 3$ yields $C = 28/3$. Putting these values, we get

$$-\frac{dy}{y^2} = \left(1 + \frac{1}{6}\frac{1}{x} - \frac{9}{2}\frac{1}{x-2} + \frac{28}{3}\frac{1}{x-3}\right)dx$$

which, on integration, gives

$$\frac{1}{y} = x + \frac{1}{6}\ln|x| - \frac{9}{2}\ln|x-2| + \frac{28}{3}\ln|x-3| + C,$$

where C is an arbitrary constant. The given equation has also the singular solutions $y = 0,\ (x \neq 0,\ x \neq 2,\ x \neq 3),\ x = 0\ (y \neq 0),\ x = 2\ (y \neq 0)$ and $x = 3\ (y \neq 0)$.

Example 5. Solve the equation

$$(x^3+1)dy - ydx = 0.$$

If $x \neq -1$ and $y \neq 0$ we can write this equation in the form

$$\frac{dy}{y} = \frac{dx}{x^3+1}.$$

Since $x^3+1 = (x+1)(x^2-x+1)$, we have using partial fractions

$$\frac{1}{x^3+1} = \frac{A}{x+1} + \frac{Bx+C}{x^2-x+1},$$

where A, B, C are undetermined constants. Multiplying by $(x+1)(x^2-x+1)$ we obtain

$$1 = A\left(x^2-x+1\right) + (Bx+C)(x+1) = x^2(A+B) + x(-A+B+C) + (A+C).$$

Since this is an identity we have on equating coefficients of like powers of x,

$$A+B=0,$$

$$-A+B+C=0,$$

$$A+C=1.$$

Solving these we find $A=1/3$, $B=-1/3$, $C=2/3$. On substituting these values and integrating we have

$$\int \frac{dy}{y} = \frac{1}{3}\int \frac{dx}{x+1} - \frac{1}{3}\int \frac{x-2}{x^2-x+1}dx$$

$$= \frac{1}{3}\ln|x+1| - \frac{1}{3}\int \frac{(x-1/2)d(x-1/2)}{(x-1/2)^2+3/4} + \frac{1}{2}\int \frac{dx}{(x-1/2)^2+3/4}$$

$$= \frac{1}{3}\ln|x+1| - \frac{1}{6}\ln\left(x^2-x+1\right) + \frac{1}{\sqrt{3}}\arctan\frac{2x-1}{\sqrt{3}} + C,$$

or

$$\ln|y| = \frac{1}{6}\ln\frac{(x+1)^2}{x^2-x+1} + \frac{1}{\sqrt{3}}\arctan\frac{2x-1}{\sqrt{3}} + C,$$

where C is an arbitrary constant. The original equation has also the singular solutions $x=-1$ $(y \neq 0)$ and $y=0$ $(x \neq -1)$.

If a differential equation can be written in the form

$$y' = f(ax+by) \tag{6}$$

then we put $z = ax + by$ and have

$$\frac{dz}{dx} = a + bf(z),$$

or

$$\frac{dz}{a + bf(z)} - dx = 0$$

in which the variables are separable.

Example 6. Solve the equation

$$(x + y + 1)dx + (2x + 2y - 1)dy = 0.$$

If $2(x + y) - 1 \neq 0$ the differential equation may be written as

$$y' = -\frac{(x + y) + 1}{2(x + y) - 1}.$$

Put $x + y = z$, so that

$$\frac{dz}{dx} = 1 + \frac{dy}{dx}.$$

Thus z is determined as a function of x by an equation of the form

$$\frac{dz}{dx} = 1 - \frac{z + 1}{2z - 1} = \frac{z - 2}{2z - 1},$$

which is separable

$$\frac{2z - 1}{z - 2} dz = dx, \quad z \neq 2,$$

or

$$\left(2 + \frac{3}{z - 2}\right) dz = dx.$$

Integrating,

$$\int \left(2 + \frac{3}{z - 2}\right) dz = \int dx,$$

or

$$2z + 3 \ln |z - 2| = x + C$$

from which, on putting $z = x + y$, we have the general solution

$$2(x + y) + 3 \ln |x + y - 2| = x + C,$$

or

$$x + 2y + 3 \ln |x + y - 2| = C,$$

where C is an arbitrary constant.

The original equation has also the singular solution $z = 2$ or $y = 2 - x$.

Example 7. Solve

$$\frac{dy}{dx} = \sqrt{4x + 2y - 1}.$$

By putting $z = 4x + 2y$, we have

$$z' = 4 + 2y' = 4 + 2\sqrt{z - 1}$$

whence

$$\frac{dz}{4 + 2\sqrt{z - 1}} = dx.$$

Integrating,

$$\int \frac{dz}{4 + 2\sqrt{z - 1}} = x + C.$$

Putting $z - 1 = u^2$ we have

$$\int \frac{dz}{4 + 2\sqrt{z - 1}} = \int \frac{udu}{u + 2} = u - 2\ln|u + 2|,$$

or

$$\sqrt{z - 1} - 2\ln\left(\sqrt{z - 1} + 2\right) = x + C$$

from which on substituting $z = x + y$, we have the general solution

$$\sqrt{4x + 2y - 1} - 2\ln\left(\sqrt{4x + 2y - 1} + 2\right) = x + C.$$

A system of differential equations of the form

$$\sum_{j=1}^{n+1} P_{ij}(y_j)dy_j = 0, \quad i = 1,\ 2,\ \ldots,\ n$$

is called *separable system.* These systems are integrable, and on integration, give

$$\sum_{j=1}^{n+1} \int P_{ij}(y_j)dy_j = C_i, \quad i = 1,\ 2,\ \ldots,\ n$$

where C_i are arbitrary constants.

Example 8. Solve the system of the equations

$$2y_1dy_1 + 3y_2^2dy_2 + dy_3 = 0,$$

$$\sin y_1dy_1 + \frac{dy_2}{y_2} + e^{y_3}dy_3 = 0.$$

From the preceding we can write the solution in the form

$$\int 2y_1dy_1 + \int 3y_2^2dy_2 + \int dy_3 = C_1,$$

$$\int \sin y_1 dy_1 + \int \frac{dy_2}{y_2} + \int e^{y_3}\, dy_3 = C_2,$$

or

$$y_1^2 + y_2^3 + y_3 = C_1,$$

$$-\cos y_1 + \ln|y_2| + e^{y_3} = C_2.$$

Sometimes system of equations can be reduced to the separable form on dividing by some function.

Example 9. Solve the system of the equations

$$y_3 y_2 \cos y_1 dy_1 + y_2^2 y_3 dy_2 + y_2 dy_3 = 0,$$

$$y_1^2 \frac{y_3}{y_2} dy_1 + 2y_3 dy_2 - \frac{1}{y_2 y_3} dy_3 = 0.$$

On dividing the first equation by the $y_2 y_3$ and the second equation by the y_3/y_2 we have

$$\cos y_1 dy_1 + y_2 dy_2 + \frac{1}{y_3} dy_3 = 0,$$

$$y_1^2 dy_1 + 2y_2 dy_2 - \frac{1}{y_3^2} dy_3 = 0.$$

Integrating, we get

$$\int \cos y_1 dy_1 + \int y_2 dy_2 + \int \frac{dy_3}{y_3} = C_1,$$

$$\int y_1^2 dy_1 + 2\int y_2 dy_2 - \int \frac{dy_3}{y_3^2} = C_2,$$

or

$$\sin y_1 + \frac{1}{2} y_2^2 + \ln|y_3| = C_1,$$

$$\frac{1}{3} y_1^3 + y_2^2 + \frac{1}{y_3} = C_2.$$

PROBLEMS

Solve each of the equation in Problem 1 through 40.

1. $y' = \dfrac{x^2}{y}$.
2. $y' = \dfrac{x^2}{y(1+x^3)}$.
3. $y' = y^2 \sin x$.
4. $xy' = \sqrt{1-y^2}$.
5. $y' = \dfrac{x^2}{1+y^2}$.
6. $xydy = \sqrt{y^2+1}dx$.
7. $(x^2-1)\,y' + 2xy^2 = 0;\ y(0) = 1$.

8. $y' = 3y^{2/3};\ y(2) = 0.$
9. $xy' + y = y^2;\ y(1) = 0.5.$
10. $2x^2yy' + y^2 = 2.$
11. $y' - xy^2 = 2xy.$
12. $\left(1 + \dfrac{dz}{dt}\right) e^{-z} = 1.$
13. $\dfrac{dy}{dx} = \dfrac{3x^2 + 4x + 2}{2(y-1)};\ y(0) = -1.$
14. $e^x dx - (1 + e^x) y dy = 0;\ y(0) = 1.$
15. $\dfrac{ydx}{x-1} + \dfrac{xdy}{y+1} = 0.$
16. $(x + 2x^3)\, dx + (y + 2y^3)\, dy = 0.$
17. $\dfrac{dx}{\sqrt{x}} + \dfrac{dy}{\sqrt{y}} = 0.$
18. $\dfrac{dx}{\sqrt{1-x^2}} + \dfrac{dy}{\sqrt{1-y^2}} = 0.$
19. $2x\sqrt{1-y^2}dx + ydy = 0.$
20. $y' = (y-1)(x+1).$
21. $y' = e^{x-y}.$
22. $y' = \dfrac{\sqrt{y}}{\sqrt{x}}.$
23. $y' = \dfrac{\sqrt{y}}{x}.$
24. $z' = 10^{x+z}.$
25. $\dfrac{dx}{dt} + t = 1.$
26. $y' = \cos(y - x).$
27. $y' - y = 2x - 3.$
28. $(x + 2y)y' = 1;\ y(0) = -1.$
29. $y' + y = 2x + 1.$
30. $y' = \cos(x - y - 1).$
31. $y' + \sin^2(x + y) = 0.$
32. $y' = 2\sqrt{2x + y + 1}.$
33. $y' = (x + y + 1)^2.$
34. $(y^2 + xy^2)\, dx + (x^2 - x^2y)\, dy = 0.$
35. $(1 + y^2)\,(e^{2x}dx - e^y dy) - (1 + y)dy = 0.$
36. $y_1dy_1 + y_2dy_2 + y_3dy_3 = 0,$
$y_1^2dy_1 + (y_2^3 + 1)\, dy_2 + (2 - y_3)dy_3 = 0.$
37. $y_1^2dy_1 - 3y_2^3dy_2 + 4y_3dy_3 + 5y_4^4dy_4 = 0,$
$\cos y_1 dy_1 + \arctan y_2 dy_2 + \tan y_3 dy_3 + y_4 e^{-y_4^2}\, dy_4 = 0.$
38. $y_1y_2y_3dy_1 - y_2^2y_3(1 + y_1)dy_2 + (1 + y_1)(1 + y_3)y_2dy_3 = 0,$
$y_1^2\,(1 + yt_1^2)\, y_3dy_1 + y_3 \arctan y_2 dy_2 + (1 + y_2^2)\, dy_3 = 0.$
39. $y_1y_2dy_1 + y_1y_2dy_2 = 0,$

$(dy_1 + dy_2)y_2 = 0.$

40. $(y_1 + y_2)(dy_1 + dy_2) + (y_2 - y_3)(dy_2 - dy_3) + (y_3 - y_4)(dy_3 - dy_4) = 0,$
$dy_1 + 5dy_2 + 6dy_3 + 7dy_4 = 0,$
$y_3 dy_4 + y_4 dy_3 = 0.$

1.2 Homogeneous equations

An equation of the form

$$\frac{dy}{dx} = f(x, y) \tag{1}$$

is said to be *homogeneous* whenever the function f does not depend on x and y separately, but only on their ratio y/x or x/y. Thus homogeneous equations are of the form

$$\frac{dy}{dx} = f\left(\frac{y}{x}\right). \tag{2}$$

Function $f(x, y)$ is called homogeneous of degree n if

$$f(tx, ty) = t^n f(x, y)$$

that is, if x and y be replaced by tx and ty, t^n factors out of the resulting function, and the remaining factor is the original function. For example,

$$x^2 + 2y^2, \quad \sin\frac{x}{y}, \quad \ln\frac{x+y}{x},$$

$$x + y + \sqrt{xy}, \quad \frac{1}{x+y}$$

are homogeneous of degrees 2, 0, 0, 1 and -1, respectively, since

$$(tx)^2 + 2(ty)^2 = t^2\left(x^2 + 2y^2\right),$$

$$\sin\frac{tx}{ty} = \sin\frac{x}{y},$$

$$\ln\frac{tx+ty}{tx} = \ln\frac{x+y}{x},$$

$$tx + ty + \sqrt{txty} = t(x + y + \sqrt{xy}),$$

$$\frac{1}{tx+ty} = t^{-1}\frac{1}{x+y}.$$

A differential equation

$$M(x, y)dx + N(x, y)dy = 0 \tag{3}$$

is called homogeneous if $M(x, y)$ and $N(x, y)$ are homogeneous functions of the same degree. Writing the equation (3) in the form

$$\frac{dy}{dx} = -\frac{M(x,y)}{N(x,y)}$$

the second member is now homogeneous of degree zero and so is a function of y/x,

$$\frac{dy}{dx} = F\left(\frac{y}{x}\right).$$

The form of a homogeneous equation suggests that it may be simplified by introducing a new variable, which we denote by v, to represent ratio of y to x. Thus

$$y = xv \tag{4}$$

and Eq. (2) becomes

$$\frac{dy}{dx} = F(v).$$

Looking on v as a new dependent variable, we must consider v as a function of x. Differentiating Eq.(4) gives

$$\frac{dy}{dx} = x\frac{dv}{dx} + v$$

and hence Eq.(2) becomes

$$x\frac{dv}{dx} + v = F(v),$$

or

$$\frac{dv}{F(v) - v} = \frac{dx}{x} \tag{5}$$

The variables x and v separated. Solving Eq.(5) and then replacing v by y/x gives the solution of the original equation.

Example 1. Solve the differential equation

$$y' = \frac{y}{x}(\ln y - \ln x + 1).$$

The expression on the right - hand side is clearly a function of y/x only.

$$y' = \frac{y}{x}\left(\ln\frac{y}{x} + 1\right).$$

The substitution $y = xv$ gives

$$v + x\frac{dv}{dx} = v(\ln v + 1),$$

or

$$\frac{dv}{v\ln v} = \frac{dx}{x}, \; v \neq 0, \; v \neq 1.$$

Integrating both sides yields

$$\int \frac{dv}{v \ln v} = \int \frac{d(\ln v)}{\ln v} = \int \frac{dx}{x},$$

or

$$\ln|\ln|v|\,| = \ln|x| + \ln|C|.$$

This is equivalent to

$$\ln|v| = Cx,$$

or

$$v = e^{Cx}.$$

Hence substituting for v in terms of y gives the general solution

$$y = xe^{Cx}.$$

$v = 1$ or $y = x$ is the particular solution corresponding $C = 0$.

Example 2. Solve the differential equation

$$(x + 2y)dx - xdy = 0.$$

The equation being homogeneous, we set

$$y = vx,$$

$$dy = vdx + xdv.$$

Then

$$(x + 2vx)dx - x(xdv + vdx) = 0,$$

if $x \neq 0$, the resulting equation is

$$(1 + 2v)dx - vdx - xdv = 0,$$

or

$$(1 + v)dx - xdv = 0$$

whence

$$\frac{dx}{x} = \frac{dv}{v + 1}.$$

Integrating both sides yields

$$\ln|x| + \ln|C| = \ln|v + 1|,$$

where C is an arbitrary constant. Hence we obtain

$$Cx = (v + 1).$$

Finally, substituting for v in terms of y gives the solution of the origin equation in the form

$$Cx = \left(\frac{y}{x} + 1\right), \text{ also } x = 0.$$

PROBLEMS

Show that the equations in Problems 1 through 30 are homogeneous, and find their solutions.

1. $(x - y)dx + (x + y)dy = 0.$
2. $(y - 2xy)dx + x^2 dy = 0.$
3. $2x^3 y' = y\left(2x^2 - y^2\right).$
4. $y^2 + x^2 y' = xyy'.$
5. $\left(x^2 + y^2\right) y' = 2xy.$
6. $xy' - y = x \tan \dfrac{y}{x}.$
7. $xy' = y - x \exp\left(\dfrac{y}{x}\right).$
8. $xy' - y = (x + y) \ln \dfrac{x + y}{x}.$
9. $xy' = y \cos\left(\ln \dfrac{y}{x}\right).$
10. $\left(y + \sqrt{xy}\right) dx - xdy = 0$
11. $xy' - \sqrt{x^2 - y^2} - y = 0.$
12. $(x + y)dx - (x - y)dy = 0$
13. $\left(x^2 + 2xy - y^2\right) dx + \left(y^2 + 2xy - x^2\right) dy = 0; \quad y(1) = -1.$
14. $xdy - ydx = ydy$
15. $y^2 dx + \left(x^2 - xy\right) dy = 0.$
16. $\left(x^2 + xy + y^2\right) dx = x^2 dy.$
17. $\dfrac{dx}{x^2 - xy + y^2} = \dfrac{dy}{2y^2 - xy}.$
18. $y' = \dfrac{2xy}{3x^2 - y^2}.$
19. $y' = \dfrac{x}{y} + \dfrac{y}{x}; \; y(-1) = 0.$
20. $xy' = y + \sqrt{y^2 - x^2}.$
21. $ydx + \left(2\sqrt{xy} - x\right) dy = 0.$
22. $xy' = y \ln \dfrac{y}{x}.$
23. $y'(y' + y) = x(x + y); \; y(0) = 0.$
24. $(xy' + y)^2 = y^2 y'.$
25. $x^2 y'^2 - 3xyy' + 2y^2 = 0.$
26. $xy' - y = \sqrt{x^2 + y^2}.$
27. $y\left(\dfrac{dy}{dx}\right)^2 + 2x\dfrac{dy}{dx} - y = 0.$

28. $y' + \dfrac{x+2y}{x} = 0.$

29. $y' = \dfrac{y}{x+y}.$

30. $xy' = x + \dfrac{1}{2}y;\ y(0) = 0.$

Now consider the equation

$$\frac{dy}{dx} = f\left(\frac{a_1x + b_1y + c_1}{a_2x + b_2y + c_2}\right) \tag{6}$$

which can be made homogeneous by change of variables. We put

$$x = x_1 + h,$$

$$y = y_1 + k$$

then

$$a_1x + b_1y + c_1 = a_1x_1 + b_1y_1 + a_1h + b_1k + c_1,$$
$$a_2x + b_2y + c_2 = a_2x_1 + b_2y_1 + a_2h + b_2k + c_2,$$
$$\frac{dy_1}{dx_1} = \frac{dy}{dx}.$$

If we choose h and k so that

$$a_1h + b_1k + c_1 = 0,$$

$$a_2h + b_2k + c_2 = 0, \tag{7}$$

then Eq.(6) takes the form

$$\frac{dy_1}{dx_1} = f\left(\frac{a_1x_1 + b_1y_1}{a_2x_1 + b_2y_1}\right)$$

which is homogeneous. This equation is called the *linear fractional equation.*

Equation (7) can be solved for k and h, provided

$$D = a_1b_2 - a_2b_1 \neq 0.$$

By translating the origin to the point (h, k) we have removed the constant terms in the linear expressions.

If $D = 0$, the first-degree terms of numerator and dominator are proportional and the equation falls into the form

$$\frac{dy}{dx} = f(ax + by)$$

in which the variables are separable (see Section 1.1, Eq.(6)).

Example 3. Solve the differential equation

$$y' = \frac{x+y-2}{y-x-4}.$$

Put $x = x_1 + h,\ y = y_1 + k$, and choose h and k so that

$$h + k - 2 = 0,$$

$$-h + k - 4 = 0.$$

These two equations give $h = -1,\ k = 3$ and therefore $x = x_1 - 1,\ y = y_1 + 3$. On substituting in the differential equation, we get

$$\frac{dy_1}{dx_1} = \frac{x_1 + y_1}{y_1 - x_1}$$

which is homogeneous and may be solve by substitution $y_1 = vx_1$. Then we have

$$v + x_1 \frac{dv}{dx_1} = \frac{1+v}{v-1},$$

or

$$x_1 \frac{dv}{dx_1} = \frac{1 - v^2 + 2v}{v-1}.$$

Separating the variable and integrating yield

$$\int \frac{v-1}{1+2v-v^2} dv = \int \frac{1}{x_1} dx_1 + \ln|C|,\ v \neq -1 \pm \sqrt{2}.$$

Using partial fractions, we have

$$-\int \frac{1}{2}\left(\frac{1}{v-(1+\sqrt{2})} + \frac{1}{v-(1-\sqrt{2})}\right) dv = \ln|x_1| + \ln|C|,$$

or

$$-\frac{1}{2}\left(\ln\left|v - \left(1+\sqrt{2}\right)\right| + \ln\left|v - \left(1-\sqrt{2}\right)\right|\right) = \ln|x_1| + \ln|C|$$

which is equivalent to

$$v^2 - 2v - 1 = Cx_1^{-2},$$

or

$$y_1^2 - 2y_1x_1 - x_1^2 = C.$$

Since $y_1 = y + 3,\ x_1 = x + 1$ the required general solution is

$$y^2 - 2xy - x^2 + 4x - 8y = C,$$

$v = -1 \pm \sqrt{2}$ are the particular solutions corresponding to $C = 0$.

Example 4. Solve the differential equation

$$(2x-4y+6)dx+(x+y-3)dy=0.$$

Equations (7) are

$$2h-4k+6=0, \quad h+k-3=0$$

whence $h=1,\ k=2$, and we put

$$x=x_1+1,\ y=y_1+2.$$

We have

$$(2x_1-4y_1)dx_1+(x_1+y_1)dy_1=0$$

which is homogeneous. Setting $y_1=vx_1$, we have

$$(2x_1-4vx_1)dx_1+(x_1+x_1v)(vdx_1+x_1dv)=0,$$

or

$$x_1\left(2-3v+v^2\right)dx_1+x_1^2(1+v)dv=0,$$

if $x_1\neq 0, v\neq 1, v\neq 2$ we have

$$-\frac{(1+v)dv}{(v-1)(v-2)}=\frac{dx_1}{x_1}.$$

Integrating both sides yields

$$\ln|x_1|+\ln|C|=-\int\frac{(1+v)dv}{(v-2)(v-1)}=\int\left(-\frac{3}{v-2}+\frac{2}{v-1}\right)dv$$

$$=-3\ln|v-2|+2\ln|v-1|=\ln\left|\frac{(v-1)^2}{(v-2)^3}\right|.$$

Where C is an arbitrary constant.

Returning to the original variables

$$C(y-2x)^3=(y-x-1)^2, \tag{8}$$

also $v=2$ or $y=2x$, ($v=1$ or $y=x+1$ is a particular solution and may be obtained from the general solution (8) when arbitrary constant $C=0$).

PROBLEMS

Using substitution of the form $x=x_1+h,\ y=y_1+k$ reduce the next equations to homogeneous, and find their solutions.

31. $y'=\dfrac{2y-x+5}{2x-y-4}.$

32. $y'=-\dfrac{4x+3y+15}{2x+y+7}.$

33. $y' = \dfrac{x+3y-5}{x-y-1}$.

34. $y' = 2\left(\dfrac{y+2}{x+y+1}\right)^2$.

35. $(2x+y+1)dx - (4x+2y-3)dy = 0$.

36. $(x-y-1)+(y-x+2)y' = 0$.

37. $(x+4y)y' = 2x+3y-5$.

38. $(y+2)dx = (2x+y-4)dy$.

39. $(y'+1)\ln\dfrac{y+x}{x+3} = \dfrac{y+x}{x+3}$.

40. $y' = \dfrac{x-2y+5}{y-2x-4}$.

41. $y' = \dfrac{3x-y+1}{2x+y+4}$.

1.2.1 Quasihomogeneous Equations

A function $f(x,y)$ is called quasihomogeneous of degree k if under the definite α and β

$$f(t^\alpha x, t^\beta y) = t^k f(x,y).$$

A differential equation

$$\frac{dy}{dx} = f(x,y) \tag{9}$$

is called *quasihomogeneous* if $f(x,y)$ is quasihomogeneous of degree $\beta - \alpha$ that is

$$f(t^\alpha x, t^\beta y) = t^{\beta-\alpha} f(x,y). \tag{10}$$

For example, for

$$f(x,y) = \frac{x^6 - y^4}{x^4 y}$$

we have $\alpha = 2,\ \beta = 3$, since

$$\frac{(t^2x)^6 - (t^3y)^4}{(t^2x)^4 t^3 y} = t^{3-2}\frac{x^6-y^4}{x^4y}.$$

The form of a quasihomogeneous equation suggests that it may be simplified by introducing a new variable, which we will denote by u

$$y = ux^{\beta/\alpha}$$

and equation (9) becomes separable.

Example 5. Show that the equation is quasihomogeneous, and solve it

$$2xdy + \left(x^2y^4+1\right)ydx = 0. \tag{11}$$

Writing this equation as

$$\frac{dy}{dx} = -\frac{y(x^2y^4+1)}{2x},\ x \neq 0$$

we have

$$f(x,y) = -\frac{y(x^2y^4+1)}{2x}.$$

Looking on $f(x,y)$ as the quasihomogeneous function we must find α and β in Eq.(10). Then

$$\frac{t^\beta y(t^{2\alpha}x^2t^{4\beta}y^4+1)}{2t^\alpha x} = t^{\beta-\alpha}\frac{x^2y^4+1}{2x},$$

or

$$\frac{y(x^2y^4t^{2\alpha+4\beta-\alpha+\beta}+t^{\beta-\alpha})}{2x} = t^{\beta-\alpha}\frac{x^2y^4+1}{2x}.$$

Whence $2\alpha+4\beta-\alpha+\beta = \beta-\alpha$ or $\alpha+2\beta=0$. The equation being quasihomogeneous ($\beta/\alpha = -1/2$), we set

$$y = x^{-1/2}u$$

and respectively

$$\frac{dy}{dx} = -\frac{1}{2}x^{-3/2}u + x^{-1/2}\frac{du}{dx}.$$

The resulting equation is

$$x^{-1/2}\frac{du}{dx} - \frac{1}{2}x^{-3/2}u = -\frac{x^{-1/2}u(u^4+1)}{2x},$$

or

$$\frac{du}{dx} = -\frac{u^5}{2x}.$$

Separating the variables we have

$$\frac{dx}{x} = -2\frac{du}{u^5},\ u \neq 0.$$

Integrating both sides yields

$$\ln|x| + \ln|C| = \frac{u^{-4}}{2}.$$

Taking the exponential of both sides, we obtain

$$Cx = \exp\left(\frac{u^{-4}}{2}\right).$$

Finally, substitution for u in terms of y gives the general solution of Eq.(11) in the form

$$Cx = \exp\left(\frac{x^{-2}y^{-4}}{2}\right).$$

Also $x = 0$ $(y \neq 0)$ and $y = 0$ $(x \neq 0)$.

A quasihomogeneous equation may be reduced to a homogeneous one by introducing a new variable z

$$y = z^\alpha$$

where α to be chosen such that all the terms in the equation are of the same degree.

Example 6. Solve the equation

$$2xy'\left(x - y^2\right) + y^3 = 0.$$

Put $y = z^\alpha$ we have

$$2x\alpha z^{\alpha-1}z'\left(x - z^{2\alpha}\right) + z^{3\alpha} = 0.$$

The first term $2x^2\alpha z^{\alpha-1}z'$ is of degree $(\alpha-1)+2 = \alpha+1$, the second term $-2x\alpha z^{3\alpha-1}z'$ is of degree $(3\alpha - 1 + 1) = 3\alpha$ and the last term $z^{3\alpha}$ is of degree 3α. All the terms are of the same degree if $\alpha + 1 = 3\alpha$ or $\alpha = 1/2$.

On making the substitution $y = z^{1/2}$, we have

$$xz^{-1/2}z'(x - z) + z^{3/2} = 0,$$

or

$$z' = \frac{z^2}{x(z - x)}, \; x \neq 0.$$

This equation being homogeneous, we set $z = ux$ and have

$$\frac{dz}{dx} = x\frac{du}{dx} + u = \frac{u^2}{u - 1}$$

whence

$$x\frac{du}{dx} = \frac{u}{u - 1},$$

which on separation, takes the form

$$\frac{u - 1}{u}du = \frac{dx}{x}, \; u \neq 0.$$

Integrating both sides yields

$$u - \ln|u| = \ln|x| + \ln|C|.$$

Putting y^2/x for u we have the general solution of the given equation

$$y^2 = x\ln|Cy^2|.$$

Also $y = 0$.

PROBLEMS

Show that the next equations are quasihomogeneous, and find their solutions.

42. $x^3(y' - x) = y^2$.
43. $2x^2y' = y^3 + xy$.
44. $ydx + x(2xy + 1)dy = 0$.
45. $2y' + x = 4\sqrt{y}$.
46. $y' = y^2 - \frac{2}{x^2}$.
47. $2xy' + y = y^2\sqrt{x - x^2y^2}$.
48. $\frac{2}{3}xyy' = \sqrt{x^6 - y^4} + y^2$.
49. $2y + (x^2y + 1)\,xy' = 0$.
50. $y(1 + xy)dx + (1 - xy)xdy = 0$.
51. $(x^2y^2 + 1)\,ydx + (x^2y^2 - 1)\,xdy = 0$.
52. $(x^2 - y^4)\,y' - xy = 0$.
53. $y\left(1 + \sqrt{x^2y^4 - 1}\right)dx + 2xdy = 0$.

1.3 Exact equations

A differential equation

$$M(x,y)dx + N(x,y)dy = 0 \tag{1}$$

is said to be *exact* if there exists a function, denote by $U(x,y)$, such that

$$dU = \frac{\partial U}{\partial x}dx + \frac{\partial U}{\partial y}dy = Mdx + Ndy.$$

That is, if the first member of the equation is the exact differential of a function of x and y. The differential equation then takes the form

$$dU = 0,$$

and its solutions are defined implicitly by

$$U(x,y) = C,$$

where C is an arbitrary constant.

Example 1. Solve the differential equation

$$\frac{1}{y}dx - \frac{x}{y^2}dy = 0.$$

This equation is exact, since it may be written

$$d\left(\frac{x}{y}\right) = 0$$

and its solution is

$$\frac{x}{y} = C.$$

On the other hand the same equation in the form

$$ydx - xdy = 0$$

is not exact.

THEOREM. *Equation (1) is exact if and only if*

$$\frac{\partial M}{\partial y} \equiv \frac{\partial N}{\partial x} \tag{2}$$

The solution of the equation (1) is

$$\int_a^x M(x,y)dx + \int_b^y N(a,y)dy = C \tag{3}$$

where a, b are any convenient constants.

Example 2. Solve the differential equation

$$2xydx + \left(x^2 - y^2\right) dy = 0.$$

It is clear that

$$\frac{\partial M(x,y)}{\partial y} = 2x = \frac{\partial N(x,y)}{\partial x},$$

and the given equation is exact. If we take $a = 0,\ b = 0$ several terms drop out and Eq.(3) gives

$$\int_0^x 2xydx + \int_0^y \left(-y^2\right) dy = C,$$

or

$$x^2y - \frac{y^3}{3} = C.$$

In solving exact equation (1) the student has various procedures available. The following example presents one of such methods.

Example 3. Solve the differential equation

$$e^{-y}dx - (2y + xe^{-y})dy = 0.$$

Since

$$\frac{\partial}{\partial y}(e^{-y}) = -e^{-y} = \frac{\partial}{\partial x}\left[-\left(2y + xe^{-y}\right)\right] = -e^{-y}$$

the equation is exact. Then there exists a function $U(x,y)$ such that

$$\frac{\partial U}{\partial x} = M = e^{-y}, \quad \frac{\partial U}{\partial y} = N = -\left(2y + xe^{-y}\right). \tag{4}$$

Integrating the first of Eq.(4) with respect to x, holding y constant, gives

$$U(x,y) = \int e^{-y}dx + h(y) = xe^{-y} + h(y). \tag{5}$$

The function h is an arbitrary function of y, playing the role of the arbitrary constant. Now choose $h(y)$ so that

$$\frac{\partial U}{\partial y} = N.$$

From Eq.(5)

$$\frac{\partial U}{\partial y} = -xe^{-y} + h'(y).$$

Setting $\dfrac{\partial U}{\partial y}$ equal to N and solving for $h'(y)$ gives

$$-xe^{-y} + h'(y) = -2y - xe^{-y}.$$

Whence

$$h'(y) = -2y.$$

Integrating

$$h(y) = -y^2.$$

The constant of integrating can be omitted since any solution of previous differential equation will suffice; we do not require the most general one. Substituting for $h(y)$ in Eq.(5) gives

$$U(x,y) = xe^{-y} - y^2,$$

hence the solution of the original equation is given by

$$xe^{-y} - y^2 = C.$$

Example 4. Solve the differential equation

$$x\left(2x^2 + y^2\right) + y\left(x^2 + 2y^2\right)y' = 0.$$

Here

$$\frac{\partial M}{\partial y} = 2xy = \frac{\partial N}{\partial x},$$

so the equation is exact. We have

$$U(x,y) = \int x\left(2x^2 + y^2\right)dx + f(y) = \frac{1}{2}\left(x^4 + y^2x^2\right) + f(y).$$

The partial derivative $\dfrac{\partial U}{\partial y}$ of the obtained function $U(x,y)$ must equal $y(x^2 + 2y^2)$, which yields

$$yx^2 + f' = y\left(x^2 + 2y^2\right)$$

whence

$$f' = 2y^3.$$

Hence

$$f = \frac{1}{2}y^4$$

and the solution is

$$(x^4 + y^2x^2 + y^4) = C.$$

It is sometimes easier in finding the function $U(x, y)$ to hold x constant and integrate N with respect to y.

Example 5. Solve

$$\left(x^3e^{x+y} + 3x^2e^{x+y} - x\right)dx + \left(x^3e^{x+y} + y\right)dy = 0.$$

This equation is exact since

$$\frac{\partial}{\partial y}\left(x^3e^{x+y} + 3x^2e^{x+y} - x\right) = \left(x^3e^{x+y} + 3x^2e^{x+y}\right) = \frac{\partial}{\partial x}\left(x^3e^{x+y} + y\right).$$

Since the integration of $(x^3e^{x+y} + 3x^2e^{x+y})$ with respect to x is more complicated, than $(x^3e^{x+y} + y)$ with respect to y, we use

$$\frac{\partial U}{\partial y} = x^3e^{x+y} + y$$

hold x constant. Integrating this with respect to y, gives

$$U = x^3e^{x+y} + \frac{y^2}{2} + h(x).$$

Here the function $h(x)$ is to be found from

$$\frac{\partial U}{\partial x} = h' + 3x^2e^{x+y} + x^3e^{x+y} = 3x^2e^{x+y} + x^3e^{x+y} - x,$$

or

$$h' = -x$$

whence

$$h(x) = -\frac{x^2}{2},$$

and the solution is

$$U(x, y) = x^3e^{x+y} + \frac{y^2}{2} - \frac{x^2}{2} = C.$$

Exact systems

A system of differential equations

$$\sum_{j=1}^{n+1} M_{ij}(y_1, y_2, \ldots, y_{n+1})dy_j = 0, \quad i = 1,\ 2,\ \ldots,\ n$$

is said to be *exact system* if

$$\frac{\partial M_{ij}}{\partial y_{j+k_j}} = \frac{\partial M_{ij+k_j}}{\partial y_j},\ i, j = 1, \ldots, n; \quad k_j = 1,\ \ldots,\ (n+1-j).$$

The solution of this system is

$$\sum_{j=1}^{n+1} \int_{y_{j0}}^{y_j} M_{ij}(y_1, \ldots, y_{j-1}, s, y_{j+1\ 0}, \ldots, y_{n+1\ 0})ds = C_i, \quad i = 1,\ \ldots,\ n,$$

where $y_{j\ 0}$ are any convenient constants.

Example 6. Solve

$$(1 + y_2 + y_3)dy_1 + (1 + y_1 + y_3)dy_2 + (1 + y_1 + y_2)dy_3 = 0,$$

$$y_2 y_3 dy_1 + y_1 y_3 dy_2 + y_1 y_2 dy_3 = 0.$$

We have $n = 2$

$$M_{11} = 1 + y_2 + y_3,\ M_{12} = 1 + y_1 + y_3,\ M_{13} = 1 + y_1 + y_2,$$

$$M_{21} = y_2 y_3,\ M_{22} = y_1 y_3,\ M_{23} = y_1 y_2.$$

Since

$$\frac{\partial M_{11}}{\partial y_2} = \frac{\partial M_{12}}{\partial y_1} = 1,$$

$$\frac{\partial M_{12}}{\partial y_3} = \frac{\partial M_{13}}{\partial y_2} = 1,$$

$$\frac{\partial M_{21}}{\partial y_2} = \frac{\partial M_{22}}{\partial y_1} = y_3,$$

$$\frac{\partial M_{22}}{\partial y_3} = \frac{\partial M_{23}}{\partial y_2} = y_1.$$

the given system is found to be exact, thus putting $y_{1\ 0} = y_{2\ 0} = y_{3\ 0} = 0$ we get

$$\int_0^{y_1} ds + \int_0^{y_2} (1 + y_1)ds + \int_0^{y_3} (1 + y_1 + y_2)ds$$

$$= y_1 + (1 + y_1)y_2 + (1 + y_1 + y_2)y_3 = C_1,$$

and

$$\int_0^{y_3} y_1 y_2 ds = y_1 y_2 y_3 = C_2.$$

Then the resulting solution is given by the following system

$$y_1 + y_2 + y_3 + y_1y_2 + y_1y_3 + y_2y_3 = C_1,$$

$$y_1y_2y_3 = C_2.$$

PROBLEMS

Determine whether or not each of the equation in Problems 1 through 24 is exact. If exact, find the solution.

1. $(2 - 9xy^2)\,x\,dx + (4y^2 - 6x^3)\,y\,dy = 0.$
2. $\frac{y}{x}dx + (y^3 + \ln x)\,dy = 0.$
3. $(2x + 3) + (2y - 2)\,y' = 0.$
4. $(2x + 4y) = (2x - 2y)\,y' = 0.$
5. $(9x^2 + y - 1) - (4y - x)\,y' = 0.$
6. $(2xy^2 + 2y) + (2x^2y + 2x)\,y' = 0.$
7. $\frac{3x^2 + y^2}{y^2}dx - \frac{2x^3 + 5y}{y^3}dy = 0.$
8. $2x\left(1 + \sqrt{x^2 - y}\right)dx - \sqrt{x^2 - y}dy = 0.$
9. $(1 + y^2 \sin 2x)\,dx - 2y\cos^2 x dy = 0.$
10. $3x^2(1 + \ln y)\,dx = \left(2y - \frac{x^3}{y}\right)dy.$
11. $\left(\frac{x}{\sin y} + 2\right)dx + \frac{(x^2+1)\cos y}{\cos 2y - 1}dy = 0.$
12. $(e^x \sin y - 2y \sin x)\,dx + (e^x \cos y + 2\cos x)\,dy = 0.$
13. $(e^x \sin y + 3y)\,dx - (3x - e^x \sin y)\,dy = 0.$
14. $(ye^{xy}\cos 2x - 2e^{xy}\sin 2x + 2x)\,dx + (xe^{xy}\cos 2x - 3)\,dy = 0.$
15. $\left(\frac{y}{x} + 6x\right)dx + (\ln x - 2)dy = 0.$
16. $(x\ln y + xy)dx + (y\ln x + xy)dy = 0.$
17. $\frac{xdx}{(x^2 + y^2)^{3/2}} + \frac{ydy}{(x^2 + y^2)^{3/2}} = 0.$
18. $(y - x)dx + xdy = 0.$
19. $(3x^2 + 6xy^2)\,dx + (6x^2y + 4y^3)\,dy = 0.$
20. $(x^3 - 3xy^2)\,dx + (y^3 - 3x^2y)\,dy = 0.$
21. $\frac{xdx + ydy}{\sqrt{x^2 + y^2}} + \frac{xdy - ydx}{x^2} = 0.$
22. $xdy + ydy + \frac{xdy - ydx}{x^2 + y^2} = 0.$
23. $\frac{2x(1 - e^y)}{(1 + x^2)^2}dx + \frac{e^y}{1 + x^2}dy = 0.$
24. $\left[1 + \exp\left(\frac{x}{y}\right)\right]dx + \exp\left(\frac{x}{y}\right)\left(1 - \frac{x}{y}\right)dy = 0.$

Determine whether or not each of the systems in Problems 25 through 27 is exact. If exact, find the solution.

25. $(y_2 + y_3)dy_1 + (y_1 + y_3)dy_2 + (y_1 + y_2)dy_3 = 0,$
$y_2 dy_1 + y_1 dy_2 + y_3 dy_3 = 0.$

26. $\frac{y_2}{y_3}dy_1 + \frac{y_1}{y_3}dy_2 - \frac{y_1 y_2}{y_2}dy_3 = 0,$
$y_1^2 dy_1 + y_2^2 dy_2 + y_3^2 dy_3 = 0.$

27. $y_2 y_4 \left(3y_1^2 + y_3 y_4\right) dy_1 + \left(y_1^3 y_4 + y_1 y_3 y_4^2\right) dy_2 + y_1 y_2 y_4^2 dy_3$
$+ \left(y_1^3 y_2 + 2y_1 y_2 y_3 y_4\right) dy_4 = 0,$
$(y_2 + y_3)dy_1 + (y_1 + y_3)dy_2 + (y_2 + y_1 + y_3 y_4)dy_3 + y_1 y_3 dy_4 = 0.$

1.3.1 Integrating Factors

We noted that the equation

$$ydx - xdy = 0$$

is not exact, but that if we multiply by $1/y^2$ we have an exact equation

$$\frac{dx}{y} - \frac{xdy}{y^2} = d\left(\frac{x}{y}\right) = 0,$$

or by x/y^3

$$\frac{xdx}{y^2} - \frac{x^2dy}{y^3} = d\left(\frac{1}{2}\frac{x^2}{y^2}\right) = 0,$$

or by $1/xy$

$$\frac{dx}{x} - \frac{dy}{y} = d\left(\ln|x| - \ln|y|\right) = 0.$$

The multipliers $1/y^2$, x/y^3, $1/xy$ are called *integrating factors* of the given equation.

DEFINITION. *A nonzero function $\mu(x,y)$ is called integrating factor of the equation*

$$Mdx + Ndy = 0$$

if the equation

$$\mu(Mdx + Ndy) = 0$$

is exact.

The Swiss mathematician Leonard Euler (1707-1783) introduced this useful concept in 1734.

THEOREM 1. *The differential equation (1) has an infinite number of integrating factors.*

THEOREM 2. *If μ_1 and μ_2 are integrating factors of (1) whose ratio is not constant, the general solution of the equation (1) is*

$$\mu_1 = C\mu_2.$$

Example 7. Solve

$$ydx - xdy = 0.$$

We find that $\mu_1 = 1/y^2$ and $\mu_2 = x/y^3$ are integrating factors. Then the general solution is

$$\frac{1}{y^2} = C\frac{x}{y^3},$$

or

$$\frac{y}{x} = C.$$

In general case it is very difficult to find an integrating factor. We have for some simple cases:

a. If, in the equation

$$Mdx + Ndy = 0$$

we have

$$\frac{1}{N}\left(\frac{\partial M}{\partial y} - \frac{\partial N}{\partial x}\right) = f(x)$$

a function of x alone, then

$$\mu = \exp\left(\int f(x)dx\right)$$

is an integrating factor.

Example 8. Solve

$$(xy - 1)dx + dy = 0.$$

Here

$$\frac{1}{N}\left(\frac{\partial M}{\partial y} - \frac{\partial N}{\partial x}\right) = x$$

is a function of x alone. So we have the integrating factor

$$\mu = \exp\left(\int xdx\right) = \exp\left(\frac{x^2}{2}\right).$$

Then we have

$$\exp\left(\frac{x^2}{2}\right)(xy - 1)dx + \exp\left(\frac{x^2}{2}\right)dy = 0, \tag{6}$$

$$\frac{\partial M}{\partial y} = x\exp\left(\frac{x^2}{2}\right) = \frac{\partial N}{\partial x} = x\exp\left(\frac{x^2}{2}\right)$$

and Eq.(6) is exact. Integrating $N = \exp\left(\frac{x^2}{2}\right)$ with respect to y, holding x constant gives

$$U(x, y) = y\exp\left(\frac{x^2}{2}\right) + h(x).$$

The arbitrary function $h(x)$ is so that

$$\frac{\partial U}{\partial x} = yx\exp\left(\frac{x^2}{2}\right) + h'(x) = M = \exp\left(\frac{x^2}{2}\right)(xy-1),$$

or

$$h'(x) = -\exp\left(\frac{x^2}{2}\right)$$

and

$$h(x) = -\int \exp\left(\frac{x^2}{2}\right) dx.$$

Hence the solution of the original equation is given by

$$U(x,y) = y\exp\left(\frac{x^2}{2}\right) - \int \exp\left(\frac{x^2}{2}\right) dx = C.$$

Similarly , if

$$\frac{1}{M}\left(\frac{\partial N}{\partial x} - \frac{\partial M}{\partial y}\right) = f(y)$$

is a function of y alone, then

$$\mu = \mu(y) = \exp\left(\int f(y)dy\right)$$

is an integrating factor.

Example 9. Solve the differential equation

$$\left(2xy^2 - y\right) dx + \left(y^2 + x + y\right) dy = 0.$$

Here

$$M = 2xy^2 - y,\ N = y^2 + x + y$$

$$\frac{1}{M}\left(\frac{\partial N}{\partial x} - \frac{\partial M}{\partial y}\right) = \frac{1 - 4xy + 1}{2xy^2 - y} = -\frac{2}{y}.$$

Consequently

$$\mu = \frac{1}{y^2}.$$

Multiplying the original equation by $1/y^2$ gives

$$\left(2x - \frac{1}{y}\right) dx + \left(1 + \frac{x}{y^2} + \frac{1}{y}\right) dy = 0$$

which is exact since

$$\frac{\partial}{\partial y}\left(2x - \frac{1}{y}\right) = \frac{1}{y^2} = \frac{\partial}{\partial x}\left(1 + \frac{x}{y^2} + \frac{1}{y}\right).$$

And the general integral is (see Eq.(3))

$$\int_0^x \left(2x - \frac{1}{y}\right) dx + \int_{y_0}^y \left(1 + \frac{1}{y}\right) dy = C,$$

or

$$x^2 - \frac{x}{y} + y + \ln y = C_1$$

where

$$C_1 = C + y_0 + \ln y_0.$$

b. More generally, let μ be a function of some function of x and y, $\omega(x, y)$, let we say. Then if

$$\frac{\dfrac{\partial M}{\partial y} - \dfrac{\partial N}{\partial x}}{N\dfrac{\partial \omega}{\partial x} - M\dfrac{\partial \omega}{\partial y}} = f(\omega)$$

$f(\omega)$ is a function of ω , then

$$\mu = \exp\left(\int f(\omega) d\omega\right).$$

As an example, letting $\omega = x^2 + y^2$ we have, if

$$\frac{\dfrac{\partial M}{\partial y} - \dfrac{\partial N}{\partial x}}{2Nx - 2My} = f\left(x^2 + y^2\right)$$

a function of the $x^2 + y^2$, then

$$\mu = \exp\left[\int f\left(x^2 + y^2\right) d\left(x^2 + y^2\right)\right].$$

Example 10. Solve the differential equation

$$\left(y - \frac{1}{x}\right) dx + \frac{dy}{y} = 0.$$

Here $M = y - \frac{1}{x}$, $N = \frac{1}{y}$. Assuming that μ is a function of $\omega(x, y) = x/y$ we have

$$\frac{\dfrac{\partial M}{\partial y} - \dfrac{\partial N}{\partial x}}{N\dfrac{\partial \omega}{\partial x} - M\dfrac{\partial \omega}{\partial y}} = \frac{1}{\dfrac{1}{y^2} - \left(y - \dfrac{1}{x}\right)\left(-\dfrac{x}{y^2}\right)} = \frac{1}{\dfrac{x}{y}}$$

then

$$\mu = \exp\left(\int \frac{1}{\frac{x}{y}} d\frac{x}{y}\right) = \exp\left(\ln\left|\frac{x}{y}\right|\right) = \frac{x}{y}.$$

Multiplying the original equation by x/y gives

$$\left(x - \frac{1}{y}\right)dx + \frac{x}{y^2}dy = 0$$

and we have an exact equation

$$\frac{\partial}{\partial y}\left(x - \frac{1}{y}\right) = \frac{1}{y^2} = \frac{\partial}{\partial x}\left(\frac{x}{y^2}\right).$$

If we take $a = 0$ the second term in Eq.(3) drops out and integrating gives the solution of the original equation

$$\int_0^x \left(x - \frac{1}{y}\right) dx = \frac{x^2}{2} - \frac{x}{y} = C.$$

Example 11. Solve the differential equation

$$\left(2y + \frac{1}{(x+y)^2}\right) dx + \left(3y + x + \frac{1}{(x+y)^2}\right) dy = 0.$$

Here

$$M = 2y + \frac{1}{(x+y)^2},\ N = 3y + x + \frac{1}{(x+y)^2}.$$

Assuming that μ is a function of $x + y$, we have

$$\frac{1}{N-M}\left(\frac{\partial M}{\partial y} - \frac{\partial N}{\partial x}\right) = \frac{1}{x+y}$$

then

$$\mu = \exp\left[\int \frac{1}{x+y} d(x+y)\right] = \exp\left[\ln(x+y)\right] = x + y.$$

Multiplying the original equation by $\mu = x + y$ we get

$$\left[2(x+y)y + \frac{1}{x+y}\right] dx + \left[(3y+x)(x+y) + \frac{1}{x+y}\right] dy = 0.$$

This equation is exact

$$\frac{\partial}{\partial y}\left[2(x+y)y + \frac{1}{x+y}\right] = 2x + 4y - \frac{1}{(x+y)^2}$$

$$= \frac{\partial}{\partial x}\left[(3y+x)(x+y) + \frac{1}{x+y}\right].$$

And the general integral is

$$\int_0^y \left[(3y+x)(x+y) + \frac{1}{x+y}\right] dy + \int_{x_0}^x \frac{1}{x} dx$$

$$= y^3 + x^2y + 2xy^2 + \ln|x+y| - \ln|x| + \ln|x| - \ln|x_0| = C,$$

or

$$y^3 + x^2y + 2xy^2 + \ln|x+y| = C_1$$

where $C_1 = C + \ln|x_0|$.

In simple equations it is often possible to group terms into combinations which are exact and to solve by inspection.

Example 12. Solve the differential equation

$$\left(x^2 - \sin^2 y\right) dx + x \sin 2y dy = 0.$$

Multiplying by $1/x^2$ gives

$$dx + \frac{x \sin 2y dy - \sin^2 y dx}{x^2} = dx + d\left(\frac{\sin^2 y}{x}\right) = 0, \ x \neq 0.$$

Here each part is an exact differential and we have immediately

$$x + \frac{\sin^2 y}{x} = C, \text{ also } x = 0.$$

Integrating factors are sometimes obtained by inspection. Certain combinations of differential suggests trial factors.

Example 13. Solve the differential equation

$$xy^2 dx + \left(x^2 y - x\right) dy = 0.$$

The equation may be written as

$$xy(ydx + xdy) - xdy = 0,$$

or

$$xyd(xy) - xdy = 0,$$

from which multiplying by $1/xy$ gives

$$d(xy) - \frac{dy}{y} = 0.$$

The general solution is therefore

$$xy - \ln|y| = C.$$

Example 14. Solve the differential equation

$$\left(x^2 + y^2 + x\right) dx + ydy = 0.$$

The combination

$$xdx + ydy = d\left(x^2 + y^2\right)$$

is suggestive, and we have

$$\left(x^2 + y^2\right) dx + \frac{1}{2} d\left(x^2 + y^2\right) = 0,$$

or multiplying by $1/(x^2 + y^2)$

$$dx + \frac{1}{2} d\left[\ln\left(x^2 + y^2\right)\right] = 0,$$

whence the general solution is

$$x + \frac{1}{2}\ln\left(x^2 + y^2\right) = C.$$

Again, $xdy - ydx$ reminds one of

$$d\left(\frac{y}{x}\right) = \frac{xdy - ydx}{x^2}.$$

Thus the differential equation

$$ydx - \left(x^2 y + x\right) dy = 0$$

may be written as

$$\frac{ydx - xdy}{x^2} + ydy = 0$$

whence

$$\frac{y^2}{2} + \frac{y}{x} = C.$$

The following formulas are often exploited in similar fashion:

$$d(xy) = xdy + ydx,$$

$$d\left(\frac{x}{y}\right) = \frac{ydx - xdy}{y^2},$$

$$d\left(\sqrt{x^2+y^2}\right) = \frac{xdx+ydy}{\sqrt{x^2+y^2}},$$

$$d\left(\sqrt{x^2-y^2}\right) = \frac{xdx-ydy}{\sqrt{x^2+y^2}},$$

$$d\left(\ln\frac{x}{y}\right) = \frac{dx}{x} - \frac{dy}{y}.$$

PROBLEMS

Show that the equations in Problems 28 through 37 are not exact, but become exact when multiplied by the given integrating factor. Then solve the equations.

28. $(y^2+xy)\,dx - x^2dy = 0;\ \mu = (xy^2)^{-1}$.
29. $(3xy+y^2) + (x^2+xy)\,y' = 0;\ \mu = x$.
30. $x^2y^3 + x\,(1+y^2)\,y' = 0;\ \mu = (xy^3)^{-1}$.
31. $\left(\frac{\sin y}{y} - 2e^{-x}\sin x\right)dx + \left(\frac{\cos y + 2e^{-x}\cos x}{y}\right) = 0;\ \mu = ye^x$.
32. $ydx + (2x - ye^y)dy = 0;\ \mu = y$.
33. $\left(1-\frac{x}{y}\right)dx + \left(2xy + \frac{x}{y} + \frac{x^2}{y^2}\right)dy = 0;\ \mu = \frac{1}{x}$.
34. $(x^3 - xy^2 - y)\,dx + (x^2y - y^3 + x)\,dy = 0;\ \mu_1 = \frac{1}{xy};\ \mu_2 = \frac{1}{x^2-y^2}$.
35. $(y-x)e^{-x}dx + xe^{-x}dy = 0;\ \mu = e^x$.
36. $(x-y)dx + (x+y)dy = 0;\ \mu = \mu\left(\frac{1}{x^2+y^2}\right)$.
37. $(1-x^2y)\,dx + x^2(y-x)dy = 0;\ \mu = \mu\left(\frac{1}{x^2}\right)$.

In each of Problems 38 through 72 find an integrating factor or change the variables, and solve the given equation.

38. $(y + x^3y^2)\,dx + xdy = 0$.
39. $(4x^2+y)\,dx - xdy = 0$.
40. $(x^2+y^2+y)\,dx - xdy = 0$.
41. $ydy = (xdy + ydx)\sqrt{1+y^2}$.
42. $xy^2\,(xy'+y) = 1$.
43. $y^2dx - (xy + x^3)\,dy = 0$.
44. $(x^2 + 3\ln y)\,ydx = xdy$.
45. $y^2dx + (xy + \tan xy)dy = 0$.
46. $y(x+y)dx + (xy+1)dy = 0$.
47. $y\,(y^2+1)\,dx + x\,(y^2 - x + 1)\,dy = 0$.
48. $(x^2 + 2x + y)\,dx = (x - 3x^2y)\,dy$.
49. $ydx - xdy = 2x^3\tan\left(\frac{y}{x}\right)dx$.
50. $y^2dx + (e^x - y)dy = 0$.
51. $xydx = (y^3 + x^2y + x^2)\,dy$.
52. $x^2y(ydx + xdy) = 2ydx + xdy$.

53. $(x^2 - y^2 + y)\,dx + x(2y-1)dy = 0.$
54. $(2x^2y^2 + y)\,dx + (x^3y - x)\,dy = 0.$
55. $y\,(x + y^2)\,dx + x^2(y-1)dy = 0.$
56. $x(\ln y + 2\ln x - 1)dy = 2ydx.$
57. $(x^2+1)\,(2xdx + \cos ydy) = 2x \sin ydx.$
58. $(2x^3y^2 - y)\,dx + (2x^2y^3 - x)\,dy = 0.$
59. $x^2y^3 + y + (x^3y^2 - x)\,y' = 0.$
60. $(x \sin\alpha + y\cos\alpha)dx + (y\sin\alpha - x\cos\alpha)dy = 0.$
61. $(x^2 - y)\,dx + x(y+1)dy = 0.$
62. $y^2(ydx - 2xdy) = x^3(xdy - 2ydx).$
63. $\left(\frac{x}{y} + 1\right)\,dx + \left(\frac{x}{y} - 1\right)\,dy = 0.$
64. $(x^2 + y)\,dx - xdy = 0.$
65. $(2xy^2 - y)\,dx + (y^2 + x + y)\,dy = 0.$
66. $x^2y^3 + y + (x^3y^2 - x)\,y' = 0,\ \mu = \mu(xy).$
67. $(y + x^2)\,dy + (x - xy)dx = 0,\ \mu = \mu(x^2 + y^2).$
68. $(x^2 + y)\,dx - xdy = 0.$
69. $(xy^2 + y)\,dx - xdy = 0.$
70. $(2x^3y^2 - y)\,dx + (2x^2y^3 - x)\,dy = 0.$
71. $xy^2dx + (x^2y - x)dy = 0.$
72. $(x^2 + y^2 + 1)\,dx - 2xydy = 0,\ \mu = \mu\,(x^2 \pm y^2)\,.$

1.4 Linear equations

The general form of a first order *linear* differential equation is

$$A(x)\frac{dy}{dx} + B(x)y + C(x) = 0.$$

On division by the first coefficient, it can be put in the form

$$\frac{dy}{dx} + P(x)y = Q(x). \tag{1}$$

If the second member is zero, $Q(x) = 0$, the equation can be solved by separating the variables

$$\frac{dy}{dx} + P(x)y = 0, \tag{2}$$

$$\frac{dy}{y} = -Pdx; \quad \ln|y| = -\int P(x)dx + C,$$

$$y = C_1 \exp\left[-\int P(x)dx\right], \tag{3}$$

where C_1 is an arbitrary constant.

In general case (1) we multiply Eq.(1) by $\exp\left[\int P(x)dx\right]$

$$\exp\left[\int P(x)dx\right]\left(\frac{dy}{dx}+P(x)y\right)=\exp\left[\int P(x)dx\right]Q(x) \tag{4}$$

the first member in (4) is now recognized as derivative of $\exp\left[\int P(x)dx\right]y$

$$\left(\exp\left[\int P(x)dx\right]y\right)'=\left(\exp\left[\int P(x)dx\right]\right)'y+\exp\left[\int P(x)dx\right]y'$$

$$=\exp\left[\int P(x)dx\right]P(x)y+\exp\left[\int P(x)dx\right]y'$$

and we can get the general solution by an integration of (4)

$$\exp\left[\int P(x)dx\right]y=\int\exp\left[\int P(x)dx\right]Q(x)dx+C,$$

or

$$y=C\exp\left[-\int P(x)dx\right]+\exp\left[-\int P(x)dx\right]\int\exp\left[\int P(x)dx\right]Q(x)dx,$$

where C is an arbitrary constant.

Example 1. Solve the differential equation

$$\frac{dy}{dx}+\frac{1}{x}y=x^2.$$

Here $P(x)=1/x$, and we multiply by $\exp\left[\int P(x)dx\right]=x$,

$$x\left(\frac{dy}{dx}+\frac{1}{x}\right)=x^3,$$

or

$$(xy)'=x^3$$

whence, integrating,

$$xy=\frac{1}{4}x^4+C,$$

or

$$y=\frac{1}{4}x^3+\frac{C}{x}.$$

Example 2. Find the solution of the initial value problem

$$y'-2xy=x;\quad y(0)=1.$$

Here $P(x)=-2x$,

$$\int P(x)dx=-\int 2xdx=-x^2.$$

Hence multiplying by e^{-x^2}, we have

$$e^{-x^2}(y' - 2xy) = xe^{-x^2}$$

so that

$$\left(ye^{-x^2}\right)' = xe^{-x^2}.$$

Therefore

$$ye^{-x^2} = \int xe^{-x^2}\,dx + C = -\frac{1}{2}e^{-x^2} + C$$

and finally

$$y = -\frac{1}{2} + Ce^{x^2}.$$

To satisfy the initial condition $y(0) = 1$ we must choose $C = 3/2$. Hence

$$y = -\frac{1}{2} + \frac{3}{2}e^{x^2}$$

is a solution of the given initial value problem.

THEOREM 1. *If $y = U(x)$ is a particular solution of (1) and if $y = V(x)$ is a particular solution of (2), then $y = CV(x) + U(x)$ is the general solution of (1).*

We see from this Theorem that if we notice or can find in any way a particular solution of (1), the problem is then reduced to the solution of the less complicated Eq.(2), the general solution of (1) can be written down at once.

Example 3. Solve the differential equation

$$y' + y\tan x = \tan x.$$

We note that $y = 1$ is a solution. The equation

$$y' + y\tan x = 0$$

can be solved by separating the variables

$$\frac{dy}{y} = -\frac{\sin x}{\cos x}dx,$$

or

$$\ln|y| = \ln|\cos x|.$$

Taking exponential of both sides yields

$$y = \cos x.$$

The general solution then is

$$y = C\cos x + 1.$$

THEOREM 2. *If $y = U(x)$ and $y = V(x)$ are different particular solutions of (1), then the general solution is*

$$y = C\,[U(x) - V(x)] + V(x).$$

This can be put in various forms. Since we have also

$$y = C_1\,[U(x) - V(x)] + U(x).$$

Example 4. Solve the differential equation

$$y' - \frac{1}{x-1}y = -\frac{1}{x-1}.$$

This equation is found to be satisfied by $y = 1$ and by $y = x$. The general solution is then

$$y = C(x-1) + 1.$$

Consider the following method of solving the general linear differential equation of the first order (1).

If $Q(x)$ is not identically zero, then we assume that the solution is of the form (3)

$$y = A(x)\exp\left[-\int P(x)dx\right] \tag{6}$$

where an arbitrary constant C_1 is now a function of x and $\exp\left[-\int P(x)dx\right]$ is a solution of corresponding homogeneous equation. By substituting for y in the given differential equation we have

$$A'(x)\exp\left[-\int P(x)dx\right] - A(x)P(x)\exp\left[-\int P(x)dx\right]$$

$$+P(x)A(x)\exp\left[-\int P(x)dx\right] = Q(x),$$

or

$$A'(x) = Q(x)\exp\left[\int P(x)dx\right].$$

Hence

$$A(x) = C + \int Q(x)\exp\left[\int P(x)dx\right]dx.$$

Then substituting for $A(x)$ in (6) one can determine y. This technique is known as the *method of variation of parameters.*

Example 5. Solve the differential equation

$$xy' - 2y = 2x^4. \tag{7}$$

In order to use the method of variation of parameters, it is necessary to know a solution of the corresponding homogeneous equation

$$xy' - 2y = 0.$$

This equation can be solved by separating the variables

$$\frac{dy}{y} = 2\frac{dx}{x}.$$

Integrating,

$$\int \frac{dy}{y} = 2\int \frac{dx}{x} + \ln|C|,$$

or

$$\ln|y| = 2\ln|x| + \ln|C|,$$

where C is an arbitrary constant. Hence taking the exponential of both sides, we obtain

$$y = Cx^2.$$

Then we assume that the general solution of the original equation is of the form

$$y = A(x)x^2 \tag{8}$$

Substituting (8) in Eq.(7)

$$x\left[2xA(x) + A'(x)x^2\right] - 2A(x)x^2 = 2x^4,$$

or

$$A'(x) = 2x.$$

Integrating both sides yields

$$A(x) = x^2 + C_1,$$

where C_1 is an arbitrary constant. Finally, substituting for $A(x)$ in (8) gives the general solution of Eq.(7) in the form

$$y = \left(x^2 + C_1\right)x^2.$$

By a change of variable many equations can be reduced to linear form. We illustrate this by some examples.

Example 6. Solve the differential equation

$$(2e^y - x)\,y' = 1.$$

This equation is not linear. But it becomes linear if we consider x as a function of y. Then

$$\frac{dx}{dy} = \frac{1}{\frac{dy}{dx}}$$

and

$$\frac{dx}{dy} + x = 2e^y.$$

The corresponding homogeneous equation is

$$\frac{dx}{dy} + x = 0,$$

or

$$\frac{dx}{x} = -dy.$$

Integrating

$$\ln|x| = -y + \ln|C|,$$

or

$$x = Ce^{-y}.$$

Then we assume that the general solution of the original equation is of the form

$$x = A(y)e^{-y}.$$

By substituting for x we have

$$-Ae^{-y} + e^{-y}\frac{dA}{dy} + Ae^{-y} = 2e^{y},$$

or

$$\frac{dA}{dy} = 2e^{2y}.$$

Integrating,

$$A = e^{2y} + C_1.$$

The general solution then is

$$x = \left(e^{2y} + C_1\right) e^{-y}.$$

Example 7. Solve the differential equation

$$2xy\frac{dy}{dx} + y^2 = e^x.$$

This equation is not linear. Putting $v = y^2$, it becomes linear

$$\frac{dv}{dx} = 2y\frac{dy}{dx},$$

$$x\frac{dv}{dx} + v = e^x \tag{9}$$

Solving corresponding homogeneous equation

$$x\frac{dv}{dx} + v = 0,$$

$$\frac{dv}{v} = -\frac{dx}{x},$$

$$v = \frac{C}{x}$$

and the general solution of (9) is of the form

$$v = \frac{A(x)}{x}.$$

Then substituting in (9) we have

$$x\left(\frac{A'}{x} - \frac{A}{x^2}\right) + \frac{A}{x} = e^x,$$

$$A' = e^x.$$

Integrating,

$$A = e^x + C_1,$$

where C_1 is an arbitrary constant.

The general solution of the original equation is then

$$y^2 = (e^x + C_1)\frac{1}{x}.$$

Similarly the equation

$$\frac{dy}{dx} + P(x) = Q(x)e^{my}$$

which may be written

$$e^{-my}\frac{dy}{dx} + P(x)e^{-my} = Q(x)$$

becomes linear after the change of variable $v = e^{-my}$

$$\frac{dv}{dx} = -me^{-my}\frac{dy}{dx}$$

and

$$-\frac{1}{m}\frac{dv}{dx} + P(x)v = Q(x).$$

The proper change of variable in each case was suggested by the presence of a function of y and its derivative.

PROBLEMS

In each of Problems 1 through 50 find the general solution of the given differential equation.

1. $xy' + 2y = 3x, y\,(0) = 0.$
2. $xy' + 3y = x^2.$

3. $y' + ay = e^{mx}$.
4. $(1 + x^2)\, y' - 2xy = (1 + x^2)^2$.
5. $y' + 2xy = 2xe^{-x^2}$.
6. $(2x + 1)y' = 4x + 2y$.
7. $y' + y \tan x = \sec x$.
8. $(xy + e^x)dx - xdy = 0$.
9. $x^2 y' + xy + 1 = 0$.
10. $y = x(y' - x \cos x)$.
11. $2x\,(x^2 + y)\, dx = dy$.
12. $(xy' - 1) \ln x = 2y$.
13. $xy' + (x + 1)y = 3x^2 e^{-x}$.
14. $(x + y^2)\, dy = ydx$.
15. $(\sin^2 y + x \cot y)\, y' = 1$.
16. $(2x + y)dy = ydx + 4 \ln y dy$.
17. $y' = \dfrac{y}{3x - y^2}$.
18. $(1 - 2xy)y' = y(y - 1)$.
19. $y' + y = x^2 + 2$.
20. $y' - 3y = e^{3x} + e^{-3x}$.
21. $y' - 2\dfrac{y}{x} = x^4$.
22. $y' = 2y + e^{2x} + 1$.
23. $xy' + 2y = (3x + 2)e^{3x}$.
24. $2xy' - y = x^3 - x$.
25. $y' + y \tan x = \sin 2x$.
26. $2xy' + y = \dfrac{2ax^2 + 1}{x}$.
27. $(x^2 - y^2 - 1)\, y' = 2xy$.
28. $(y^2 - 6x)\, y' + 2y = 0$.
29. $(x - 2xy - y^2)\, y' + y^2 = 0$.
30. $y' - y = 2x - x^2$.
31. $y' - y = x - 1$; $y(0) = 1$.
32. $y' + 3y = x^2 + 1$.
33. $y' + 2y = x^3$.
34. $y' - y \cot x = 2x - x^2 \cot x$.
35. $y' + y \cos x = \dfrac{1}{2} \sin 2x$.
36. $y' + y = \sin x + \cos x$.
37. $x \ln x y' + y = 2 \ln x$; $y(e) = 0$.
38. $y' - 2xy = \cos x - 2x \sin x$; $y(0) = 1$.
39. $y' - 2xy = 1$.
40. $x \ln x y' - y = x(\ln x - 1)$.
41. $y' - \dfrac{xy}{x^2 + 1} = \dfrac{x^2 - x + 1}{x^2 + 1} e^x$.

42. $y' + y\cos x = e^{-\sin x}$.
43. $y' - \dfrac{2y}{\sin 2x} = -\dfrac{\sin^2 x}{\cos x}$.
44. $y' + xy = x^3;\ y(0) = -2$.
45. $y' + \dfrac{y}{x} = \dfrac{\sin x}{x^2}$.
46. $y' + y\tan x = x\cos^2 x;\ y(0) = 1$.
47. $x^2 dy - 2xy dx = 3dx$.
48. $y'\sin x - y = 2\sin^2\dfrac{x}{2}$.
49. $e^{x^2} dy + \left(2xye^{x^2} - x\sin x\right) dx = 0$.
50. $y'x\cos x + y(x\sin x + \cos x) = 1$.

Find a particular solution by inspection; find a solution when the term not involving y is replaced by zero; and write down the general solution.

51. $y' - y = 2$.
52. $y' + y = 2e^x$.
53. $xy' - y = 1$.
54. $y' = y + 1$.
55. $y' + y = x + 1$.

1.4.1 Bernoulli's Equation

The nonlinear differential equation

$$\frac{dy}{dx} + P(x)y = Q(x)y^n \tag{10}$$

where n is a constant but not necessary an integer, known as *Bernoulli's equation*, was studied in 1695 by the Swiss mathematician Jacob Bernoulli (1654-1705).

We rule out cases $n = 0$ and $n = 1$, for which the equation is already linear. The substitution $v = y^{1-n}$ reduces Bernoulli's equation to a linear equation. This method of solution was found by the German mathematician Gotfrid Wilhelm Leibniz (1646-1716) in 1696.

Equation (10) may be written in the form

$$y^{-n}\frac{dy}{dx} + Py^{1-n} = Q(x).$$

Since

$$\frac{1}{1-n}\frac{dv}{dx} = y^{-n}\frac{dy}{dx}$$

we get the linear equation

$$\frac{1}{1-n}\frac{dv}{dx} + P(x)v = Q(x).$$

Example 8. Solve the differential equation

$$x^2y' + 2xy - y^3 = 0.$$

We have

$$x^2\frac{1}{y^3}\frac{dy}{dx} + 2xy^{-2} = 1$$

and, setting $v = y^{-2}$, we have

$$-\frac{1}{2}\frac{dv}{dx} = y^{-3}\frac{dy}{dx}$$

hence we obtain the linear equation

$$-\frac{x^2}{2}\frac{dv}{dx} + 2xv = 1. \tag{11}$$

Solving corresponding homogeneous equation

$$-\frac{x^2}{2}\frac{dv}{dx} + 2xv = 0$$

we have

$$\frac{dv}{v} = 4\frac{dx}{x},$$

or

$$v = Cx^4$$

Then the general solution of (10) is of the form

$$v = A(x)x^4$$

Substituting for v in (11) gives

$$\frac{dv}{dx} = 4Ax^3 + A'x^4,$$

$$-\frac{x^2}{2}\left(4Ax^3 + A'x^4\right) + 2xAx^4 = 1,$$

or

$$A' = -\frac{2}{x^6}.$$

Integrating,

$$A = \frac{2}{5}x^{-5} + C_1,$$

where C_1 is an arbitrary constant. Hence the solution of (11) is

$$v = \left(C_1 + \frac{2}{5}x^{-5}\right)x^4.$$

Finally, substituting for v in terms of y gives the general solution of the original equation in the form

$$y = \left(\frac{2}{5}x^{-1} + C_1x^4\right)^{-1/2}.$$

Practically to solve Bernoulli's equation it is helpful to use the substitution $y = u(x)v(x)$ where $v(x)$ is a particular solution of (10) if $Q(x) \equiv 0$.

Example 9. Solve the differential equation

$$x^2y' + 2xy - y^3 = 0.$$

Writing this equation in the form

$$y' + \frac{2}{x}y = \frac{y^3}{x^2}$$

we have $P(x) = 2/x$, $Q(x) = 1/x^2$. Putting $y = u(x)v(x)$ we get

$$vu' + u\left(v' + \frac{2v}{x}\right) = \frac{(uv)^3}{x^2} \tag{12}$$

Let $v(x)$ is a particular solution of

$$v' + \frac{2}{x}v = 0$$

then separating the variables we find

$$\frac{dv}{v} = -\frac{2}{x}dx, \quad v(x) = x^{-2}$$

On substituting $v(x)$ in (12) we have

$$x^{-2}u' = u^3x^{-8},$$

or

$$\frac{du}{u^3} = \frac{dx}{x^6}.$$

Integrating both sides yields

$$u^{-2} = \frac{2}{5}x^{-5} + C,$$

or

$$u = \left(C + \frac{2}{5}x^{-5}\right)^{-1/2}.$$

Hence the general solution of the given equation is

$$y = \left(Cx^4 + \frac{2}{5}x^{-1}\right)^{-1/2}.$$

PROBLEMS

Solve each of the following equations.

56. $y' + 2y = y^2 e^x$.
57. $(x+1)(y' + y^2) = -y$.
58. $y' = y^4 \cos x + y \tan x$.
59. $xy' - 2x^2\sqrt{y} = 4y$.
60. $xy' + 2y + x^5 y^3 e^x = 0$.
61. $2y' - \dfrac{x}{y} = \dfrac{xy}{x^2-1}$.
62. $y'x^3 \sin y = xy' - 2y$.
63. $(2x^2 y \ln y - x)\, y' = y$.
64. $y' - 4y = 2e^x\sqrt{y}$.
65. $y' - y + y^2(x^2 + x + 1) = 0$.
66. $y' + y = xy^2$.
67. $y' + 2y = 2\sqrt{y}$.
68. $xy' + y = y^2 \ln x$.
69. $y' + 2xy = 2x^3 y^3$.
70. $3y^2 y' - ay^3 = x + 1$.
71. $y^{n-1}(ay' + y) = x$.
72. $dx + (x + y^2)\, dy = 0$.
73. $(xy + x^2 y^3)\, y' = 1$.
74. $ydx + (2x - 6y^4)\, dy = 0$.
75. $y' = \dfrac{y}{x + y^2}$.

1.4.2 Darboux's Equation

The equation

$$M(x,y)dx + N(x,y)dy + R(x,y)(xdy - ydx) = 0$$

where $M(x,y)$ and $N(x,y)$ are homogeneous functions of degree m and $R(x,y)$ is homogeneous function of degree n becomes Bernoulli's equation after the change of variable $y = xu(x)$. This equation is known as *Darboux's equation*, after the French mathematician Gaston Darboux (1842-1917).

Example 10. Solve the differential equation

$$ydx + xdy + y^2(xdy - ydx) = 0.$$

Here $M = y$, $N = x$ are homogeneous functions of degree 1 and $R = y^2$ is homogeneous function of degree 2. The equation being Darboux's equation, we set

$$y = xu(x), \quad dy = xdu + udx.$$

Then

$$xudx + x(xdu + udx) + x^2u^2\left[x(xdu + udx) - xudx\right] = 0,$$

or

$$2xudx + \left(x^2 + x^4u^2\right) du = 0.$$

Looking on x as the function of u we have

$$\frac{dx}{du} + x\frac{1}{2u} = -\frac{u}{2}x^3, \ x \neq 0.$$

This is Bernoulli's equation. Putting $x = z(u)v(u)$ where $v(u)$ is a particular solution of equation

$$\frac{dv}{du} + v\frac{1}{2u} = 0,$$

or

$$v = u^{-1/2}.$$

We get

$$vz' + z\left(v' + \frac{v}{2u}\right) = -\frac{u}{2}(zv)^3$$

on substituting $v = u^{-1/2}$ we have

$$u^{-1/2}z' = -\frac{u}{2}z^3u^{-3/2},$$

or

$$\frac{dz}{z^3} = -\frac{1}{2}du.$$

Integrating both sides yields

$$z^{-2} = u + C$$

whence

$$x = u^{-1/2}(u + C)^{-1/2}$$

and on substitution $u = y/x$ the solution of the original equation is given by

$$x^{-2} = \frac{y}{x}\left(C + \frac{y}{x}\right),$$

or

$$Cxy + y^2 = 1, \text{ also } x = 0.$$

PROBLEMS

Find the solution of the following Darboux's equations.

76. $(x^2 + y^2 + y)\,dx - xdy = 0.$
77. $(y^3 + 2xy^2)\,dy - 2y^3dx + (x + y)(xdy - ydx) = 0.$
78. $x^2y^3dx + x^3y^2dy + ydx - xdy = 0.$
79. $(x^2y + y^3)\,dx - xydx + x^2dy = 0.$
80. $(x^3 - xy^2)\,dx + 2x^2ydy - (xdy - ydx) = 0.$

1.4.3 Riccati's Equation

The equation

$$y' = P(x)y^2 + Q(x)y + R(x) \qquad (13)$$

is called a *Riccati's equation.* Certain special cases of this equation were studied by the Italian mathematician Count Jacopo Francesco Riccati (1676 - 1754). In general, this equation cannot be solved by quadratures (that is, a finite number of integrations). If, however, some particular solution $y_1(x)$ of this equation is known, we can obtain a more general solution containing one arbitrary constant through the substitution

$$y = y_1(x) + \frac{1}{v(x)}.$$

Example 11. Solve the differential equation

$$y' = 1 + x^2 - 2xy + y^2.$$

We find that $y = x$ is a solution. Substituting

$$y = x + \frac{1}{v}$$

we have

$$\frac{dy}{dx} = 1 - \frac{1}{v^2}\frac{dv}{dx} = 1 + x^2 - 2x\left(x + \frac{1}{v}\right) + \left(x + \frac{1}{v}\right)^2.$$

Simplifying, we have

$$\frac{dv}{dx} = -1.$$

Integrating,

$$v = C - x,$$

where C is an arbitrary constant. Whence the general solution of the original equation is

$$y = x + \frac{1}{C - x}.$$

PROBLEMS

Find a particular solution by inspection; find the general solution of the following Riccati's equations.

81. $x^2y' + xy + x^2y^2 = 4.$
82. $3y' + y^2 + \dfrac{2}{x^2} = 0.$
83. $xy' - (2x + 1)y + y^2 = -x^2.$
84. $y' - 2xy + y^2 = 5 - x^2.$
85. $y' - 2ye^x - y^2 = e^{2x} + e^x.$
86. $y' + y + y^2 = 2.$

87. $y' = -\frac{1}{x^2} - \frac{y}{x} + y^2.$

88. $y' = \frac{2\cos^2 x - \sin^2 x + y^2}{2\cos x}.$

89. $y' + y^2 = -\frac{1}{4}x^2.$

90. $xy' = y^2 - 2xy + x^2 + 2x - y.$

91. $x^2y' = x^2y^2 + xy + 1.$

92. $x^2y' + (xy-2)^2 = 0.$ If we know two or three special solutions $y = y_i(x)$ of (13), the general solution is represented or follows. When $y_1(x)$ and $y_2(x)$ are the known solutions,

$$\frac{y - y_1(x)}{y - y_2(x)} = C \exp\left[\int P(x)(y_1(x) - y_2(x))dx\right].$$

When $y_1(x)$, $y_2(x)$, $y_3(x)$ are the known solutions,

$$\frac{y - y_1(x)}{y - y_2(x)} = C\frac{y_3(x) - y_1(x)}{y_3(x) - y_2(x)}.$$

Example 12. Solve

$$y' + y + y^2 = 2.$$

We find that $y_1 = 1$ and $y_2 = -2$ are the particular solutions. Then the general solution is

$$\frac{y-1}{y+2} = C\exp\left[\int(-3)dx\right] = Ce^{-3x}$$

or

$$y = -2 + \left(C_1e^{-3x} + \frac{1}{3}\right)^{-1}, \text{ also } y = -2.$$

Riccati's differential equation

$$y' + ay^2 = bx^m$$

if $m = 0, -2, \frac{4k}{(1-2k)}$ where k is integer is solved by quadrature. (In general, it is reduced to *Bessel's* differential equation by $ay = u'/u$.)

a. If $m = 0$ we have

$$y' + ay^2 = b$$

or

$$y' = b - ay^2$$

this equation can be solved by separating the variables

$$\frac{dy}{b - ay^2} = dx, \qquad x + C = \int \frac{dy}{b - ay^2}.$$

b. If $m = -2$ we have

$$x^2y' + ax^2y^2 = b$$

Putting $y = u/x$ we get the equation for u

$$y' = \frac{u'}{x} - \frac{u}{x^2},$$

or

$$x^2y' + ax^2y = xu' - u + au^2 = b.$$

Whence

$$xu' = b + u - au^2$$

or

$$\frac{du}{b + u - au^2} = \frac{dx}{x}.$$

The variables u and x separated.

c. If $m = \frac{4k}{1-2k}$, $k > 0$ then the substitution

$$v = x^{m+3}, \ y = \frac{1}{x^2u(v)} + \frac{1}{ax} \tag{14}$$

gives

$$y' = -\frac{1}{ax^2} - \frac{2}{x^3u} - \frac{\frac{du}{dv}}{x^2u^2}(m+3)x^{m+2},$$

or

$$y' + ay^2 = \left[-u' + \frac{a}{(m+3)x^{m+4}}\right]\frac{m+3}{u^2}x^m = bx^m$$

whence

$$\frac{du}{dv} + \frac{b}{m+3}u^2 = \frac{a}{m+3}v^{-(m+4)/(m+3)} \equiv \frac{a}{m+3}v^{m_1} \tag{15}$$

where

$$m_1 = \frac{4(k-1)}{1-2(k-1)}.$$

By proceeding to do this substitution over and over again $m_k = 0$ can be obtained and we have the case $m = 0$.

If $m = \frac{4k}{1-2k}$, $k < 0$ then the substitution

$$v = x^{-(m+1)}, \quad y = \left[v^2u(v) + \frac{m+1}{b}v\right]^{-1} \tag{16}$$

gives

$$\frac{du}{dv} - \frac{b}{m+1}u^2 = -\frac{a}{m+1}v^{m_1}$$

where $m_1 = \dfrac{4(k+1)}{1-2(k+1)}$. And the original equation is reduced to the case $m = 0$ again.

Example 13. Solve the differential equation

$$y' - y^2 = x^{-8/3}.$$

Here $a = -1$, $b = 1$ and $m = -\dfrac{8}{3}$. From the equality $-\dfrac{8}{3} = \dfrac{4k}{1-2k}$ we have $k = 2$ hence the substitution (14) we must make twice.

The substitution (14) gives

$$y = \frac{1}{x^2 u} - \frac{1}{x}, \qquad v = x^{1/3}$$

and

$$y' = \frac{1}{x^2} - \frac{2}{x^3 u} - \frac{1}{x^2 u^2}\frac{du}{dv}\frac{1}{3}x^{-2/3}.$$

Then the original equation takes the form

$$\frac{du}{dv} + 3u^2 = -3v^{-4} \tag{17}$$

Here $a = 3$, $b = -3$ and $m_1 = -4$ and from the equality $-4 = \dfrac{4k}{1-2k}$ we have now $k = 1$.

On making the substitution

$$z = v^{-1}, \quad u = \frac{1}{v^2 w} + \frac{1}{3v}$$

we have

$$u'_v = -\frac{2}{v^3 w} - \frac{1}{3v^2} + \frac{1}{v^2 w^2}\frac{dw}{dz}\frac{1}{v^2}$$

and (17) takes the form

$$u'_v + 3u^2 = \frac{1}{v^4 w^2}\frac{dw}{dz} - \frac{1}{3v^2} - \frac{2}{v^3 w} +$$

$$+3\left(\frac{1}{v^4 w^2} + \frac{2}{3}\frac{1}{v^3 w} + \frac{1}{9v^2}\right) = \frac{1}{v^4 w^2}\frac{dw}{dz} + \frac{3}{v^4 w^2} = -3v^{-4},$$

or

$$\frac{dw}{dz} = -3\left(1 + w^2\right).$$

Separating the variables and integrating both sides yields

$$-3z + C = \arctan w,$$

where C is an arbitrary constant.

Finally, substitution for z and w in terms of y and x gives the solution of the original equation

$$-\frac{3}{x^{-1/3}}+C=\arctan\left[\frac{(xy+1)x^{2/3}}{x^{1/3}-\frac{1}{3}x(xy+1)}\right].$$

PROBLEMS

Find the general solution of the Riccati's equations

93. $y'+y^2=2x^{-2}$.
94. $4y'+y^2=-4x^{-2}$.
95. $y'+y^2=2x^{-4}$.
96. $y'+y^2=x^{-4/3}$.
97. $y'+y^2=-x^{-4}$.
98. $y'-y^2=2x^{-2/3}$.

1.4.4 Bool's Equation

The equation

$$y'-\frac{a}{x}y+\frac{b}{x}y^2=Cx^{n-1}$$

is known as *Bool's equation*, after the English mathematician George Boole (1815 - 1864). The substitution $v=x^a$, $y=uv$ where $u=u(v)$ reduces Bool's equation to the form

$$au'+bu^2=Cv^m$$

where $m=\frac{n}{a}-2$ and therefore in the case when $n=\frac{2a}{1-2k}$, $k=0,\pm1,\cdots$ is solved by quadrature.

Example 14. Solve

$$xy'+\frac{1}{2}y+\frac{1}{2}y^2=-\frac{1}{2}x.$$

It is Bool's equation, and we rewrite it in the form

$$y'+\frac{1}{2}\frac{y}{x}+\frac{1}{2}\frac{y^2}{x}=-\frac{1}{2}.$$

Here $a=-1/2$, $b=1/2$, $n=1$. Putting $v=x^{-1/2}$, $y=ux^{-1/2}$, where $u=u(v)$ we obtain

$$\frac{dy}{dx}=\frac{d}{dx}\left(ux^{-1/2}\right)=\frac{du}{dx}x^{-1/2}-\frac{1}{2}ux^{-3/2}=x^{-1/2}\frac{du}{dv}\frac{dv}{dx}-\frac{1}{2}ux^{-3/2}$$

$$=-\frac{1}{2}x^{-2}u'_v-\frac{1}{2}x^{-3/2}u.$$

Substituting this into the given equation we get

$$-\frac{1}{2}u'_v x^{-2} - \frac{1}{2}ux^{-3/2} + \frac{1}{2}ux^{-3/2} + \frac{1}{2}u^2x^{-2} = -\frac{1}{2},$$

or

$$\frac{du}{dv} - u^2 = v^{-4} \tag{18}$$

We obtain the Riccati equation where $a = -1$, $b = 1$, $m = -4$ and $k = 1$ then substitution (14) gives

$$z = v^{-1}, \; u = \frac{1}{v^2 w} - \frac{1}{v},$$

$$u'_v = \frac{1}{v^2w^2}\frac{dw}{dv} - \frac{2}{v^3w} + \frac{1}{v^2} = \frac{1}{v^2w^2}\frac{dw}{dz}\frac{dz}{dv} - \frac{2}{v^3w} + \frac{1}{v^2}$$

$$= \frac{1}{v^2w^2}\frac{dw}{dz}\frac{1}{v^2} - \frac{2}{v^3w} + \frac{1}{v^2}.$$

and (18) takes the form

$$u'_v - u^2 = \frac{1}{v^4w^2}\frac{dw}{dz} - \frac{2}{v^3w} + \frac{1}{v^2} - \left(\frac{1}{v^2w} - \frac{1}{v}\right)^2$$

$$= \left(\frac{dw}{dz} - 1\right)v^{-4}w^{-2} = v^{-4},$$

or

$$w' = \left(1 + w^2\right).$$

Separating the variables and integrating both sides gives

$$z + C = \arctan w,$$

or

$$\tan(C + z) = w.$$

By substitution for z and w in terms of y and x we obtain the general solution of the original equation

$$y = -1 + x^{1/2}\cot\left(C + x^{1/2}\right).$$

PROBLEMS

Find the solution of the following Bool's equations

99. $xy' - 5y - y^2 = x^2$.

100. $3xy' - 9y - y^2 = x^{2/3}$.

101. $y' + \frac{5}{x}y + \frac{3}{x}y^2 = x$.

102. $y' + \frac{3}{x}y + \frac{y^2}{x} = x$.

By differentiating it is possible sometimes to reduce integral equation to differential one. We illustrate this by the following example.

Example 15. Solve the integral equation

$$\int_0^x (x-t)y\,dt = 2x + \int_0^x y(t)dt.$$

Differentiating both sides yields

$$2 + y(x) = \int_0^x y(t)dt.$$

Putting $x = 0$ we find the initial condition $y(0) = -2$. Differentiating once again gives

$$y' = y.$$

Separating the variables we have

$$\frac{dy}{y} = dx$$

Integrating now gives the general solution

$$y = Ce^x$$

and we have to find the value of the constant C such that $y(0) = -2$. On putting $x = 0$, $y = -2$, we have $-2 = C$. The appropriate solution is given by

$$y = -2e^x.$$

1.5 Nonlinear equations

There is a first order differential nonlinear equation of the form

$$F(x, y, p) = 0, \tag{1}$$

where $p \equiv y'(x)$ and F is a function of three variables.

THEOREM *There exists a unique solution $y = y(x)$, $x_0 - h \le x \le x_0 + h$ (where h is sufficiently small) of Eq.(1) that satisfies the condition $y(x_0) = y_0$ for which $y'(x_0) = y_1$, where y_1 is one of the real roots of the equation $F(x_0, y_0, y_1) = 0$, if in a closed neighborhood of a point (x_0, y_0, y_1) the function F possesses the properties:*

1) $F(x, y, p)$ is continuous in all the arguments;

2) the derivative $F'_p(x, y, p)$ exists and is nonzero;

3) there exists the derivative $|F'_y(x, y, p)| < M$.

Any solution of Eq.(1) of the form $y = \varphi(x, C)$ in which C is an arbitrary constant is called a *general solution* of the equation. Any solution that may be obtained from the general solution of Eq.(1) by assigning particular value to the constant C is called a *particular solution* of that equation. A solution of Eq.(1) that cannot be obtained by assigning specific value to the arbitrary constant in the general solution is called a *singular solution* of that equation.

In order to solve of Eq.(1) one should first investigate

1) whether it is solvable for a variable (for example: $F(x, y, p) = y - \ln(xy') + x^2y' = 0$, and then $y = \ln(xy') - x^2y'$);

2) or it is unsolvable for every variable (for example $F(x, y, p) = xy - \ln(xy') + \sin(yy') = 0$).

Then in the first case one should investigate

1) whether all obtained functions are real;

2) or some obtained functions are complex.

Finally, in the case 1.1 the following cases are possible:

1) Eq.(1) is solvable for $y'(x)$;

2) Eq.(1) is solvable for $y(x)$;

3) Eq.(1) is solvable for x.

In general, the cases 2) and 1.2) are beyond the scope of our text.

1.5.1 *Solvable Equations. General Solution*

1. Eq.(1) is solvable for $y'(x)$. Solving Eq.(1) for p leads to:

$$y'(x) = f_i(x, y), \quad i = 1, 2, \ldots, l, \tag{2}$$

where l is a number of solutions. Further we must solve each of the l equations. A general solution of Eq.(2) at any fixed i can be found in any of the following forms, namely: an explicit function $y = y(x, C)$; an inverse function $x = x(y, C)$; a parametric function $x = x(t, C)$, $y = y(t, C)$, where t is a parameter; an implicit relation $\varphi(x, y, C) = 0$.

Example 1. Find the general solution of the differential equation

$$y'^2 - 4xy = 0$$

at $xy > 0$.

Proceeding as stated above we obtain $y' = 2\sqrt{xy}$ and $y' = -2\sqrt{xy}$. Upon integrating we find

$$y(x) = \left(\frac{2}{3}x^{3/2} + C_1\right)^2,$$

and

$$y(x) = \left(-\frac{2}{3}x^{3/2} + C_2\right)^2,$$

where C_i are arbitrary constants. Putting $C_2 = -C_1 \equiv C$ we obtain the general solution of the given equation

$$y = \left(\frac{2}{3}x^{3/2} + C\right)^2.$$

The given equation also has the singular solution $y = 0$ (see below).

a) At $f_i(x, y) \equiv f_i(y)$ Eq.(2) becomes separable and then

$$x = \int \frac{dy}{f_i(y)} + C_i, \tag{2a}$$

where C_i are arbitrary constants.

Example 2. Find the general solution of the differential equation

$$y'^2 - 4y = 0$$

at $y > 0$.

According to Eq.(2) we find $y' = 2\sqrt{y}$, $y' = -2\sqrt{y}$. Hence solutions of the given equation are inverse functions $x = \sqrt{y} + C_1$ and $x = -\sqrt{y} + C_2$, where C_i are arbitrary constants. Putting $C_2 = -C_1 \equiv C$ we obtain the general solution of the given equation

$$y = (x + C)^2.$$

Here $y = 0$ is the singular solution.

b) At $f_i(x, y) \equiv f_i(x)$ Eq.(2) becomes separable and then:

$$y = \int f_i(x)dx + C_i, \tag{2b}$$

where C_i are arbitrary constants.

Example 3. Find the general solution of the differential equation

$$y'^2 - 4x = 0$$

at $x > 0$.

According to Eq.(2) we obtain $y' = 2\sqrt{x}$, $y' = -2\sqrt{x}$. Hence solutions of the given equation are

$$y = \frac{4}{3}x^{3/2} + C_1$$

and

$$y = -\frac{4}{3}x^{3/2} + C_2,$$

where C_i are arbitrary constants. These solutions can be written as a general solution in the form

$$(y + C)^2 = \frac{16}{9}x^3,$$

where C is an arbitrary constant.

2. Eq.(1) is solvable for y. Solving Eq.(1) for y leads to

$$y = \varphi_i(x, y'), \quad i = 1, 2, \ldots, l, \tag{3}$$

where l is a number of solutions. Further we must solve each of the l equations. Let $p(x) = y'(x)$ is a new function of x alone. Differentiating Eq.(3) with respect to x then yields

$$p(x) = \varphi'_{ix}(x,p) + \varphi'_{ip}(x,p)p'(x).$$

Hence

$$p'(x) = \frac{p - \varphi'_{ix}(x,p)}{\varphi'_{ip}(x,p)}, \quad i = 1, 2, \ldots, l. \tag{4}$$

This differential equation is completely analogous to Eq.(2). If an explicit function $p = p(x, C)$ is a general solution of Eq.(4), then by substituting this function in Eq.(3) one find a general solution of Eq.(1) as an explicit function $y = \varphi_i(x, p(x, C))$. If an inverse function $x = x(p, C)$ is a general solution of Eq.(4), then this function together with Eq.(3) give a general solution of Eq.(1) in the parametric form $x = x_i(p, C)$, $y = \varphi_i(x_i(p, C), p)$, where p is a parameter. There is a similar case when Eq.(4) has a general solution in the parametric form $x = x_i(t, C)$, $p = p_i(t, C)$, where t is a parameter. If an implicit expression $\psi_i(x, p, C) = 0$ is a general solution of Eq.(4), then this expression with Eq.(3) give a general solution of Eq.(1) as an implicit parametric relation $y = \varphi_i(x, p)$, $\psi_i(x, p, C) = 0$.

Example 4. Find the general solution of the differential equation

$$\frac{1}{x}y'^2 + xy' - y = 0.$$

According to Eq-s (3) and (4) we find

$$p'(x) = \frac{p^2}{x\,(x^2 + 2p)}.$$

By substituting $p = x^2/z$, where z is a new unknown function, we obtain, after some minor calculations,

$$xz' = \frac{z(2z+3)}{z+2}.$$

Using partial fraction and integrating gives

$$\frac{2}{3}\ln|z| - \frac{1}{6}\ln|2z+3| = \ln|Cx|,$$

where C is an arbitrary constant and $2z + 3 \neq 0$, $z \neq 0$. Now we obtain, recalling the substitution for p,

$$x^2 = \frac{3C^6p^4}{1 - 2C^6p^3}.$$

Hence the general solution of the given equation has the parametric form

$$x^2 = \frac{3C^6p^4}{1 - 2C^6p^3},$$

$$y = \frac{p^2}{x} + xp,$$

where p is a parameter. The singular solutions are $y = 0$ and $y = -2x^3/9$.

a) At $\varphi_i(x,p) \equiv \varphi_i(p)$ Eq.(3) takes the form

$$y = \varphi_i(p), \tag{3a}$$

Eq.(4) becomes separable and then

$$x = \int \varphi'_{ip}(p)\frac{dp}{p} + C_i \equiv g_i(p) + C_i, \tag{4a}$$

where C_i are arbitrary constants. Eq-s (3a) and (4a) give general solutions of Eq.(1) in the parametric form $y = \varphi_i(p)$, $x = g_i(p) + C_i$, where p is a parameter.

Example 5. Find the general solution of the differential equation

$$y'^2 - y' - y = 0.$$

From Eq-s(3a) and (4a) we find, after some minor calculations,

$$y = p\,(p-1)\,, \quad x = 2p - \ln|p| + C, \quad \text{if } p \neq 0,\ p \neq 1,$$

where C is an arbitrary constant. The conditions $p = 0$ and $p = 1$ then lead to the singular solution $y = 0$.

b) **Lagrange's equation**. An equation of the form

$$y = x\psi(y') + f(y')$$

is known as a *Lagrange's equation*, after the French mathematician Joseph-Louis Lagrange (1736-1813). In this case Eq.(4) becomes linear with respect to the inverse function $x(p)$

$$\frac{dx}{dp} - x\frac{\psi'(p)}{p - \psi(p)} = \frac{f'(p)}{p - \psi(p)}, \tag{4b}$$

when $p - \psi(p) \neq 0$. A solution of Eq.(4b) can be found by the method of variation of parameter (see Sec. 1.4). Hence a general solution of the Lagrange equation can be always written in the parametric form

$$x = x(p, C), \quad y = x\psi(p) + f(p),$$

where p is a parameter and C is an arbitrary constant. In those points p_i^*, where $p_i^* - \psi(p_i^*) = 0$, we have the singular solutions as the straight lines

$$y = x\psi(p_i^*) + f(p_i^*),$$

where $i = 1, 2, \ldots, k$ and k is a number of the singular solutions.

Example 6. Find the general solution of the Lagrange equation

$$xy'^2 - y' - y = 0.$$

According to Eq.(4b) we obtain

$$\frac{dx}{dp} + \frac{2x}{p-1} = \frac{1}{p(p-1)},$$

if $p \neq 0$ and $p \neq 1$. Solving it (see Sec.1.4) we find

$$x(p) = \frac{C + p + \ln|p|}{(p-1)^2},$$

where C is an arbitrary constant. Hence

$$x = \frac{C + p + \ln|p|}{(p-1)^2},$$

$$y = xp^2 - p$$

is the general solution of the given equation. If $p = 0$ or $p = 1$ we find two singular solutions $y = 0$ and $y = x - 1$, respectively.

c) **Clairaut's equation**. An equation of the form

$$y = xy' + f(y') \tag{4c}$$

is known as a *Clairaut's equation*, after the French mathematician Alexis Clairaut (1713-1765). Notice that this is a particular case of Lagrange's equation, if $\psi(y') = y'$. Hence the condition $p - \psi(p) = 0$ is always fulfilled. In this case Eq.(4) can be written in the form

$$(x + f'(p))\frac{dp}{dx} = 0.$$

Hence the general solutions of the Clairaut equation is

$$p = C, \qquad y = Cx + f(C),$$

where C is an arbitrary constant. Moreover there exists a singular solution of the form

$$x = -f'(p), \qquad y = xp + f(p),$$

where p is a parameter. Notice that the singular solution is the *envelope* of the family of integral curves defined by the general solution.

Example 7. Find the general solution of the Clairaut equation

$$y = xy' + e^{y'}.$$

Since $f(p) = \exp(p)$, we have the general solution

$$y = Cx + e^C,$$

where C is an arbitrary constant. The singular solution is

$$x = -e^p, \quad y = xp + e^p,$$

where p is a parameter.

3. Eq.(1) is solvable for x. Solving Eq.(1) for x leads to

$$x = \psi_i(y, y'), \quad i = 1, 2, \ldots, l, \tag{5}$$

where l is a number of solutions. Further we must solve each of the l equations. Let $p(y) = y'(x)$ is a new function of y alone. A differentiation of Eq.(5) with respect to x then yields, taking $dp(y)/dx = pp'(y)$ into account,

$$1 = \psi'_{iy}(y, p)p + \psi'_{ip}(y, p)pp'(y).$$

Then

$$p'(y) = \frac{1 - p\psi'_{iy}(y, p)}{p\psi'_{ip}(y, p)}, \quad i = 1, 2, \ldots, l. \tag{6}$$

This differential equation is completely analogous to Eq.(2). If an explicit function $p(y, C)$ is a general solution of Eq.(6), then by substituting this function in Eq.(5) one finds a general solution of Eq.(1) as an inverse function $x = \psi_i(y, p_i(y, C))$. If an inverse function $y = y_i(p, C)$ is the general solution of Eq.(6), then this solution together with Eq.(5) give a general solution of Eq.(1) in the parametric form $y = y_i(p, C)$, $x = \psi_i(y_i(p, C), p)$, where p is a parameter. There is a similar case when a general solution of Eq.(6) has the parametric form $y = y_i(t, C)$, $p = p_i(t, C)$, where t is a parameter. If an implicit relation $\varphi_i(y, p, C) = 0$ is a general solution of Eq.(6), then this function together with Eq.(5) give a general solution of Eq.(1) as an implicit parametric relation $x = \psi_i(y, p)$, $\varphi_i(y, p, C) = 0$, where p is a parameter.

Notice that the Lagrange's and Clairaut's equations can be written as

$$x = y\varphi(y') + g(y')$$

and

$$x = \frac{y}{y'} + g(y'),$$

respectively, that is, as particular cases of Eq.(5).

Example 8. Find the general solution of the differential equation

$$y'^3 - y - x = 0.$$

Proceeding as stated above we obtain

$$x = y'^3 - y \equiv p^3 - y$$

and

$$p'(y) = \frac{1+p}{3p^3}.$$

Further by integrating we find $y(p)$ as well as the general solution of the given equation in the parametric form

$$y = (p+1)^3 - \frac{9}{2}(p+1)^2 + 9(p+1) - 3\ln|p+1| + C,$$

$$x = p^3 - y, \quad \text{if} \quad p \neq -1,$$

where C is an arbitrary constant. At $p = -1$ we find the singular solution $y = -x - 1$.

a) If $\psi_i(y, p) \equiv \psi_i(p)$ then Eq.(5) takes the form

$$x = \psi_i(p), \tag{5a}$$

Eq.(6) becomes separable and then

$$x = \psi_i(p), \quad y = \int p\psi'_{ip}(p)dp + C_i,$$

where C_i are arbitrary constants.

Example 9. Find the general solution of the differential equation

$$y'^3 - x = 0.$$

Likewise as in Ex.8 we obtain, after some minor calculations,

$$x = p^3, \quad y = \frac{3}{4}p^4 + C,$$

where C is an arbitrary constant.

1.5.2 Solvable Equations. Singular Solution

A singular solution may exist if even one of the conditions of the uniqueness theorem is violated. There is considered case $F'_p = 0$ only. Thus there is a system of equations

$$\begin{aligned} F(x, y, p) &= 0, \\ F'_p(x, y, p) &= 0, \end{aligned} \tag{7}$$

which determines so called *singular set*. Eliminating p from system (7) yields

$$y = \varphi_i(x), \quad i = 1, 2, \ldots, l,$$

where l is a number of solutions of Eq.(7), φ_i are functions of x alone. Substituting each of these functions in Eq.(1) we find some of its solutions

$$y = \varphi_j(x), \quad j = 1, 2, \ldots, k \leq l,$$

where k is a number of solutions of Eq.(1), φ_j are some functions from the functions φ_i. A solution of Eq.(1) of the form $y = \varphi(x)$, where φ is a function of x alone, is a singular solution of that equation if it is tangent in every point to a general solution of that equation. This singular solution is the envelope of the family of general solutions of Eq.(1). Hence we substitute each of the solutions φ_j in the conditions of tangency of two curves

$$\begin{aligned} \varphi_j(x) &= g(x, C), \\ \varphi_j'(x) &= g'(x, C), \end{aligned} \tag{8}$$

where $g(x, C)$ is a general solution of Eq.(1). If system (8) is compatible at any function $\varphi_j(x)$ and it has a solution for x and C, then this $\varphi_j(x)$ is the singular solution of Eq.(1).

Example 10. Find the singular solution of the differential equation

$$xy'^2 - 2y'y + 4x = 0,$$

if $y = (4x^2 + C^2)/2C$ is its general integral.

According to Eq.(7) we have

$$\begin{aligned} xp^2 - 2py + 4x &= 0, \\ 2px - 2y &= 0. \end{aligned}$$

Eliminating p then yields

$$y = 2x \equiv \varphi_1(x), \qquad y = -2x \equiv \varphi_2(x).$$

Substituting $\varphi_1(x)$ and $\varphi_2(x)$ in the given equation we assure ourselves that they are its solutions. Hence $k = l = 2$. Now according to Eq.(8) we obtain, after some minor calculations,

$$\begin{aligned} \pm 2x &= (4x^2 + c^2)/2c, \\ \pm 2 &= 4x/c. \end{aligned}$$

This system is compatible and has two solutions $x = \pm C/2$. Hence $y = \pm 2x$ are the singular solutions of the given equation as well as the envelopes of the family of general solutions.

Example 11. Find the singular solution of the differential equation

$$y'^3 - y - x = 0.$$

According to Eq.(7) we have

$$\begin{aligned} p^3 - y - x &= 0, \\ 3p^2 &= 0. \end{aligned}$$

Eliminating p gives $y = -x$. It is easy to verify that this function does not satisfy the given equation. Hence the singular solution $y = -x - 1$, which has been found in Ex. 8, is not an envelope of the family of general solutions.

a) A singular solution may exist in some other cases. They can be found if a general solution of Eq.(1) is written as the implicit relation $\Phi(x, y, C) = 0$. In this case a singular solution is the envelope of the single family of curves $\Phi(x, y, C) = 0$. First by eliminating C from the system

$$\begin{aligned} &\Phi(x, y, C) = 0, \\ &\Phi'_C(x, y, C) = 0, \end{aligned} \tag{9}$$

we obtain $y = \varphi_i(x)$, $\quad i = 1, 2, \ldots, l$, where $\varphi_i(x)$ are functions of x alone. Further, proceeding as stated above we find all singular solutions of Eq.(1).

Example 12. Find the singular solution of the differential equation

$$y = x + y' + \ln y'$$

if $y = C + \exp(x - C)$ is its general solution.

According to Eq.(9) we obtain

$$\begin{aligned} &y - C - \exp(x - C) = 0, \\ &-1 + \exp(x - C) = 0. \end{aligned}$$

Eliminating C then yields $y = x + 1$. It is easy to verify that this function satisfies the given equation. Now we write the system (8)

$$\begin{aligned} &x + 1 = C + \exp(x - C), \\ &1 = \exp(x - C). \end{aligned}$$

This system is compatible and it has the solution $x = C$. Hence $y = x + 1$ is the singular solution of the given equation.

Complex solutions of Eq.(1)

In this case we must investigate Eq.(1) by the methods of analytic theory of differential equations. This problem is beyond the scope of our text.

Example 13. Find the solution of the differential equation

$$y'^3 + (x - y)^2 = 0.$$

Proceeding as in Ex. 1 we obtain

$$y_1'(x) = e^{i\pi/3}(x - y)^{2/3}, \quad y_2'(x) = e^{i\pi}(x - y)^{2/3}, \quad y_3'(x) = e^{i5\pi/3}(x - y)^{2/3}.$$

There exists on the real axis only solution $y_2(x)$. A general solution of the given cquation exists in the complex plane.

1.5.3 Unsolvable Equations

In order to solve these equations we must use some suitable methods. There does not exist a general method for solving of these equations .

Example 14. Find the solution of the differential equation

$$x + y + y' + \exp(x + y + y') = 0.$$

Let $\alpha = -x - y - y'$ is so real number that $\alpha = e^{-\alpha}$. It is obvious that $0 < \alpha < 1$. Now we obtain the linear nonhomogeneous equation

$$y' + y = -x - \alpha.$$

Further proceeding as has been stated in Sec.1.4 we find the general solution of this equation

$$y = Ce^{-x} + 1 - x - \alpha,$$

where C is an arbitrary constant. It is also a general solution of the given equation. Using Eq.(9) for this general solution, we convince ourself that the envelope of this single family of curves does not exist.

Notice that a general parametric representation of Eq.(1) can be written in the form

$$x = \varphi(u, v), \qquad y = \psi(u, v), \qquad p = \vartheta(u, v),$$

where φ, ψ, ϑ are such functions of u and v alone, that Eq. (1) is satisfied identically. Then, taking into account that $dy = y'dx$ and putting $v = v(u)$, we obtain

$$\frac{dv}{du} = \frac{\vartheta\varphi'_u - \psi'_u}{\psi'_v - \vartheta\varphi'_v},$$

that is, the first order differential equation which is solved for the derivative. However, this equation is integrable by quadratures only in the case 1, as has been shown above, and it is not integrable by quadratures in the case 2.

PROBLEMS

Find the solutions of the following equations, first solving them for $y'(x)$.

1. $y'^2 - y^2 = 0.$
2. $y'^2 - 4x^2 = 0.$
3. $y'^2 - 4y^3(1 - y) = 0.$
4. $y'^2 - 4|y| = 0.$
5. $y'^3 + y^2 - yy'(y' + 1) = 0.$
6. $8y'^3 - 27y = 0.$
7. $y'^3 - \dfrac{y'}{4x} = 0.$
8. $yy'^3 + x - 1 = 0.$
9. $y'^2 - xy' = 0.$
10. $y^2(y'^2 + 1) - 1 = 0.$

11. $\sqrt{y'^2+1} - xy' = 0.$
12. $y'^2 - 2yy' - y^2(e^x - 1) = 0.$
13. $yy'^2 - (xy+1)\,y' + x = 0.$
14. $yy'\,(yy' - 2x) - x^2 + 2y^2 = 0.$
15. $y^2y'^2 - 2xyy' + 2y^2 - x^2 = 0.$
16. $y'^2 + y^2 - 1 = 0.$
17. $y'^2 - 2xy' - 8x^2 = 0.$
18. $y'^2 - y^3 + y^2 = 0.$
19. $(xy' + 3y)^2 - 7x = 0.$
20. $y'^3 - xy'^2 - 4yy' + 4xy = 0.$

Find the solutions of the following equations, first solving them for $y(x)$.

21. $8y'^3 - 27y = 0.$
22. $2y'^3 + y'^2 - y = 0.$
23. $y'^2 - xy' - y + \frac{1}{2}x^2 = 0.$
24. $y'^2 - ye^{y'} = 0.$
25. $y'^2 - e^y + 1 = 0.$
26. $xy'^2 - xyy' - 1 = 0.$
27. $y' \sin y' + \cos y' - y = 0.$
28. $y'^2 + \frac{1}{2}x^3y' - x^2y = 0.$
29. $y'^2 + (x+1)y' - y = 0.$
30. $x^2y'^3 - xy' + y = 0.$

Find the solutions of the following equations, first solving them for x.

31. $yy'^3 + x - 1 = 0.$
32. $y'^3 + y' - x = 0.$
33. $\ln y' + \sin y' - x = 0.$
34. $y'^3 - 4xyy' + 8y^2 = 0.$
35. $xy'^2 + xy' - y = 0.$
36. $xy'^2 - 2y' - x = 0.$
37. $y'^3 + xy^2y' - y^3 \ln y = 0.$
38. $xy'^3 - yy'^2 + 1 = 0.$
39. $\sin y' + y' - x = 0.$
40. $y'^2 + \frac{2}{x}y' + e^{2y} = 0.$

Solve each of the following Clairaut's equations.

41. $y = xy' - y' - 2.$
42. $y = xy' + y'^2 + x.$
43. $y = xy' + 2\sqrt{y'^2+1}.$
44. $y = xy' + \frac{1}{4}y'^2.$
45. $y = xy' - y'^2.$
46. $y = xy' + y'^2.$

Solve each of the following Lagrange's equations.

47. $y = 3xy' - y'^3$.
48. $y = 2xy' - y'^2$.
49. $y = 2xy' - 4y'^3$.
50. $y = -xy' + y'^2$.
51. $y = xy'^2 - 2y'^3$.
52. $y = \frac{1}{2}xy' + 2\frac{x}{y'}$.

Find the singular solutions of the following differential equations using the given general integrals.

53. $y'^2 + y^2 = 1, \qquad y - 2\sin\left(\frac{x}{2} + C\right) = 0$.
54. $x^2y'^2 + (2x - y)\,yy' + y^2 = 0, \qquad xy - Cy + C^2 = 0$.
55. $y'^3 - 4xyy' + 8y^2 = 0, \qquad y - C(x - C)^2 = 0$.
56. $y'^2 - xy' - y + \frac{x^2}{2} = 0, \qquad -y + \frac{x^2}{2} + Cx + C^2 = 0$.

Find the solutions of the following differential equations.

57. $y'^2 - e^{y'} + x = 0$.
58. $8y'^3 - 12y'^2 - 27(y - x) = 0$.
59. $xy' - \ln y' + y = 0$.
60. $2y' - \ln y' - x = 0$.
61. $x^2y'^5 - xyy'^3 - xyy'^2 + y^2 = 0$.
62. $x^2y'^2 - 2xyy' - 3y^2 - x^2 = 0$.
63. $y'^2 - \left(\frac{y}{y'}\right)^2 + 1 = 0$.
64. $3y'^4 - y' - y = 0$.
65. $y'^2 + 2xy' + y = 0$.
66. $2xy' - \sin y' - y = 0$.

1.6 Applications in physics

Here we consider some simple examples of how differential equations may arise in the analysis of problems in physics.

1.6.1 Mechanics

The second Newton's law can be written in the form

$$m\mathbf{a}(t) = \mathbf{F}(\mathbf{r}, \mathbf{v}, t), \tag{1}$$

where m is the constant mass of a particle, $\mathbf{r}(t)$ is its space coordinate at time t, $\mathbf{v}(t) = \dot{\mathbf{r}}(t)$ is its velocity at time t, $\mathbf{a}(t) = \dot{\mathbf{v}}(t) = \ddot{\mathbf{r}}(t)$ is its acceleration at time t, and $\mathbf{F}$ is an external force action on the particle with coordinates $\mathbf{r}$ and velocity $\mathbf{v}$ at time t. If the initial conditions

$$\mathbf{r}(t_0) = \mathbf{r}_0, \qquad \dot{\mathbf{r}}(t_0) = \mathbf{v}_0 \tag{2}$$

are prescribed, then coordinates $\mathbf{r}(t)$ are those solution of Eq.(1), which satisfy the initial conditions (2) at $t = t_0$.

Example 1. A body of mass m moves along of a straight line with no force upon it except a resistance proportional to the speed, that is, $F = -kv$, where k is a constant. Determine the total path of body up to stop, if its initial coordinate is zero and initial velocity is $v_0 > 0$.

According to Eq.(1) we have

$$m\ddot{x} = -k\dot{x}.$$

Upon integrating both sides of this equation we obtain

$$m\dot{x} = -kx + C_1,$$

where C_1 is an arbitrary constant. Dividing of this equation by the term on the right hand side and integrating then yields

$$-\frac{m}{k}\ln|-kx + C_1| = t + C_2,$$

where C_2 is an arbitrary constant. Hence

$$x = \frac{1}{k}\left(C_1 - \exp\left(-\frac{k}{m}(t + C_2)\right)\right).$$

To determine the constants C_1 and C_2 we substitute this solution into the prescribed initial conditions, obtaining

$$0 = C_1 - \exp(-\frac{k}{m}C_2),$$
$$v_0 = \frac{1}{m}\exp(-\frac{k}{m}C_2).$$

Solution of this system is $C_1 = mv_0$, $C_2 = -(m/k)\ln(mv_0)$. Hence

$$x = \frac{mv_0}{k}\left(1 - e^{-kt/m}\right).$$

Since

$$v = \dot{x}(t) = v_0 e^{-kt/m},$$

then $v = 0$ at $t \to \infty$ and the total path of body up to stop we find approaching the limit as $t \to \infty$

$$s = \lim_{t\to\infty} x(t) = \frac{mv_0}{k}.$$

Example 2. A body of mass m falls from rest under the gravitational force which is given by Newton's law, that is,

$$F = -\frac{Gm}{x^2},$$

where G is the gravitational constant and x is the distance from the origin to the position of body at time t. Find the relation between the distance x and the time t. Determine the falling time of body on the origin from an initial distance H.

According to Eq.(1) we have

$$m\ddot{x} = -\frac{Gm}{x^2}.$$

Multiplying the equation by $\dot{x}$ and integrating the obtained product gives, after some simplification,

$$\frac{\dot{x}^2}{2} = \frac{G}{x} + C_1,$$

where C_1 is an arbitrary constant. Further we find

$$\dot{x} = -\sqrt{2\left(C_1 + \frac{G}{x}\right)}.$$

The negative sign is required because x decreases when t increases. The initial condition $v(0) = 0$ requires that $C_1 = -G/H$. Dividing this equation by the term on the right hand side and then integrating by making change of variable $x = H\sin^2 s$, we obtain

$$H\sqrt{\frac{H}{2G}}\left(\arcsin\sqrt{\frac{x}{H}} - \sqrt{\frac{x}{H}\left(1 - \frac{x}{H}\right)}\right) = -t + C_2,$$

where C_2 is an arbitrary constant. The initial condition $x(0) = H$ requires that

$$C_2 = \frac{\pi H}{2}\sqrt{\frac{H}{2G}}.$$

Further we find, recalling that $\pi/2 - \arcsin x = \arccos x$,

$$t = H\sqrt{\frac{H}{2G}}\left(\sqrt{\frac{x}{H}\left(1 - \frac{x}{H}\right)} + \arccos\sqrt{\frac{x}{H}}\right),$$

where $x \geq 0$. Setting $x = 0$ gives the falling time of body on the origin

$$T = \frac{\pi H}{2}\sqrt{\frac{H}{2G}}.$$

1.6.2 Hydrodynamics

It is known that the viscous incompressible liquid flows from an orifice in the bottom of a vessel with the speed

$$v(h) = k\sqrt{2gh(t)}, \tag{3}$$

where k is a constant, g is the acceleration of gravity, $h(t)$ is called the head of the liquid at time t. The continuity equation takes the form in these conditions

$$v(h)S(h) = \text{const}, \tag{4}$$

where $S(h)$ is the area of the upper surface of the liquid when the head is equal h. Eq-s (3) and (4) and expression for the speed $v(h) = -\dot{h}$ make a system equations of motion of viscous incompressible liquid.

Example 3. Find the time required to empty a cylinder vessel full of water through an orifice of radius r in the bottom. The radius of the cylinder is R and the height is H.

Taking into account Eq-s (3) and (4) and recalling that $S(h) = \pi R^2$ and $S(0) = \pi r^2$, we obtain

$$\begin{gathered} v(h) = k\sqrt{2gh}, \\ -\pi R^2 \dot{h}(t) = \pi r^2 v(h). \end{gathered}$$

The initial condition is $h(0) = H$. Hence

$$\frac{dh(t)}{dt} = -k\left(\frac{r}{R}\right)^2 \sqrt{2gh}, \qquad h(0) = H.$$

Dividing this equation by $h^{1/2}$ and integrating both sides gives, after some minor calculations,

$$h(t) = \left(C_1 - k\left(\frac{r}{R}\right)^2 \sqrt{\frac{g}{2}}t\right)^2,$$

where C_1 is an arbitrary constant. The initial condition requires that $C_1 = \sqrt{H}$. Hence

$$h(t) = H\left(1 - k\left(\frac{r}{R}\right)^2 \sqrt{\frac{g}{2H}}t\right)^2.$$

The time required to empty the vessel is determined by the condition $h(T) = 0$. This condition then yields

$$T = \frac{1}{k}\left(\frac{R}{r}\right)^2 \sqrt{\frac{2H}{g}}.$$

1.6.3 Electrical Networks

According to Kirchhoff's second law: In a closed circuit the impressed voltage is equal to the sum of the voltage drops in the rest of circuit, that is,

$$U_1 + U_2 + \ldots + U_n = \mathcal{E}, \tag{5}$$

where U_i are the voltages drop across the conductors in the circuit, $\mathcal{E}$ is the impressed voltage. The voltages drop across a resistance, a capacitor and an inductance are equal to

$$U_R = IR, \qquad U_C = \frac{Q}{C} = \frac{1}{C}\int_{t_0}^{t} I(t)\,dt, \qquad U_L = L\,\frac{dI(t)}{dt}, \tag{6}$$

respectively, where $I(t)$ is the current at time t, R is the resistance, C is the capacitance, L is the inductance and Q is the total charge on the capacitor.

Example 4. Determine the temporal variation of the current in a series circuit consisting of a resistance R, a capacitance C and an impressed voltage $\mathcal{E}$ after the switch is closed. Find the current at the initial time.

According to Eq-s (5) and (6) we have

$$RI(t) + \frac{Q(t)}{C} = \mathcal{E}, \quad \text{at} \quad t > 0.$$

The initial condition is $Q(0) = 0$. Since $I = \dot{Q}$ we obtain

$$R\dot{Q}(t) = \mathcal{E} - \frac{Q(t)}{C}.$$

Dividing this equation by the term on the right hand side and integrating both sides give

$$-RC\ln\left|\mathcal{E} - \frac{Q}{C}\right| = t + C_1,$$

where C_1 is an arbitrary constant. The initial condition $Q(0) = 0$ requires that

$$C_1 = -RC\ln|\mathcal{E}|.$$

Putting $\mathcal{E} > 0$ we obtain

$$Q(t) = \mathcal{E}C\left(1 - e^{-t/RC}\right).$$

and

$$I(t) = \frac{\mathcal{E}}{R}e^{-t/RC}.$$

The current at $t = 0$ is $I(0) = \mathcal{E}/R$.

Example 5. Determine the temporal variation of the current in a closed series circuit consisting of a capacitor C and an inductance L if at $t = 0$ the initial current is $I_0 > 0$ and the initial charge on the capacitor is zero.

According to Eq-s (5) and (6) we have

$$\frac{Q(t)}{C} + L\frac{dI(t)}{dt} = 0.$$

The initial conditions are $I(0) = I_0$ and $Q(0) = 0$. Since $I(t) = \dot{Q}(t)$ we obtain

$$\frac{Q(t)}{C} + L\ddot{Q}(t) = 0.$$

In a way analogous to that used in Ex.2 we find

$$\dot{Q}^2 + \frac{Q^2}{CL} = C_1,$$

where C_1 is an arbitrary constant. Hence

$$\dot{Q} = \pm\left(C_1 - \frac{Q^2}{CL}\right)^{1/2}.$$

The initial conditions require that sign is positive and $C_1 = I_0^2$. Then dividing this equation by the term in the right hand side and integrating both sides gives

$$\sqrt{LC}\arcsin\left(\frac{Q}{I_0\sqrt{LC}}\right) = t + C_2,$$

where C_2 is an arbitrary constant. Hence

$$Q(t) = I_0\sqrt{LC}\sin\left(\frac{t + C_2}{\sqrt{LC}}\right)$$

and

$$I(t) = I_0\cos\left(\frac{t + C_2}{\sqrt{LC}}\right).$$

The initial conditions require that $C_2 = 0$. Hence

$$I(t) = I_0\cos\left(\frac{t}{\sqrt{LC}}\right).$$

1.6.4 Kinetic Theory

1. It is known that liquid evaporates in the air with the rate

$$\frac{dM(t)}{dt} = kS\left(p(t) - p_s\right), \tag{7}$$

where $M(t)$ is mass of the liquid at time t, k is a constant, S is the area of the evaporating surface, $p(t)$ is the vapor pressure at time t, p_s is the saturated vapor pressure at constant temperature. The Eq.(7) is correct when the vapor distributes spatially uniform into the surrounding air, that is,

$$p(t) = p_0 + \frac{Am(t)}{V}, \tag{8}$$

where p_0 is the vapor pressure at the initial time t_0 , A is a constant, $m(t) = M_0 - M(t)$ is the mass of the vapor at the time t, M_0 is the mass of the liquid at the initial time t_0 , V is the volume of the surrounding air.

Example 6. Determine the temporal variation of the mass of water during its evaporation from a cylinder vessel of volume v_0, if $v_0 << V$ $p(0) = p_0$, $M(0) = M_0$.

According to Eq-s (7) and (8) we have

$$\dot{M} = kS\left(A\frac{M_0 - M(t)}{V - v(t)} + p_0 - p_s\right),$$
$$v(t) = \frac{M(t)}{\rho},$$

where $v(t)$ is the volume of water at time t, ρ is the density of water. Since $V >> v_0 \geq v(t)$ then neglecting of $v(t)$ in the first equation gives

$$\frac{dM(t)}{dt} = kS\left(A\frac{M_0 - M(t)}{V} + p_0 - p_s\right),$$

Integration of this equation (see Ex.4) then yields

$$\ln\left|A\frac{M - M(t)}{V} + p_0 - p_s\right| = -\frac{kASt}{V} + C$$

where C is an arbitrary constant. By the initial conditions we find $C = \ln|p_0 - p_s|$. Putting $p_s > p_0$ we obtain

$$M(t) = M_0 - \frac{V}{A}(p_s - p_0)\left(1 - e^{-kASt/V}\right).$$

2. It is known that the pressure of the gas at an altitude z can be determined as

$$p(z) = \int_z^\infty \rho(z)g(z)\,dz, \tag{9}$$

where $\rho(z)$ is the density of gas at the altitude z, $g(z)$ is the acceleration of gravity at this altitude. Then the Clapeyron ideal gas law

$$\rho(z) = \frac{p(z)}{RT(z)}, \tag{10}$$

where R is the gas constant, $T(z)$ is the temperature of gas at the altitude z, together with Eq.(9) make a system of equations for ρ, p and T.

Example 7. Assuming that the gravitational force is constant, determine the behavior of the pressure of the isothermal ideal gas. The gas pressure at $z = 0$ is equal p_0.

According to Eq-s (9) and (10) we have

$$p(z) = \int_z^\infty \frac{p(z)g\,dz}{RT},$$

that is, an integral equation for $p(z)$. After differentiation with respect to z we find

$$\frac{dp}{dz} = -\frac{pg}{RT}.$$

Integrating (see Ex. 4) then yields

$$\ln p(z) = -\frac{zg}{RT} + C,$$

where C is an arbitrary constant. Notice that p is intrinsically positive so that absolute value is not needed. The initial condition gives $C = \ln p_0$. Thus we obtain

$$p = p_0 e^{-zg/RT},$$

so called Boltzman's formula.

3. According to Newton's law of cooling, the cooling rate of a body is given by the equation

$$\frac{dT(t)}{dt} = -k\left(T(t) - T_{ext}\right), \tag{11}$$

where k is a constant, $T(t)$ is the temperature of the body at time t, T_{ext} is the temperature of the surrounding medium.

Example 8. Determine the temporal variation of temperature of a body into the surrounding medium with the constant temperature T_{ext}, if the initial temperature of the body is T_0.

According to Eq.(11) we have

$$\frac{dT(t)}{dt} = -k\left(T(t) - T_{ext}\right).$$

Integrating (see Ex. 4) leads to

$$\ln|T(t) - T_0| = -kt + C,$$

where C is an arbitrary constant. The initial condition requires that

$$C = \ln|T_0 - T_{ext}|.$$

Putting $T_0 > T_{ext}$ we obtain

$$T(t) = T_{ext} + (T_0 - T_{ext})\,e^{-kt}.$$

1.6.5 Nuclear Physics

It is known that the rate of the radioactive decay is given by the equation

$$\frac{dm(t)}{dt} = -km(t), \tag{12}$$

where k is a decay constant, $m(t)$ is mass of radioactive material at time t.

Example 9. Determine the decay constant of Radii if its half decay time is 1600 years.

According to Eq.(12) we have

$$m'(t) = -km(t).$$

An integration then yields

$$\ln m(t) = -kt + C,$$

where C is an arbitrary constant. Notice that m is intrinsically positive so that absolute value is not needed. The initial condition $m(0) = m_0$ requires that $C = \ln m_0$. Then

$$m(t) = m_0 \exp(-kt),$$

where m_0 is the mass of Radii at the initial time. The half decay time $T_{1/2}$ is determined by the condition

$$\frac{m_0}{2} = m_0 e^{-kT_{1/2}}.$$

Hence we obtain $k = (\ln 2)/T_{1/2} = (\ln 2)/1600$ (measured in $1/\text{year}$).

1.6.6 Optics

It is known that the intensity of light into absorbing medium at the point $z + \delta z$ is determined as

$$I(z + \delta z) = I(z) - kI(z)\delta z,$$

where $I(z)$ is the intensity of light at the point z, k is the constant of absorption. Subtracting $I(z)$, dividing by δz and approaching the limit $\delta z \to 0$ leads to the equation for the intensity of light

$$\frac{dI(z)}{dz} = -kI(z), \tag{13}$$

which is completely analogous to Eq.(12).

PROBLEMS

1. A body of mass m falls from rest in air under a constant gravitational force and a resistance proportional to each of the following powers of the speed: a) first; b) second. Determine the falling time of body in each case if body's initial altitude is equal to H.

2. Suppose everything is as in Problem 1, except that resistance is $F_r = -k_1 v - k_2 v^2$. Determine the falling time of body. Find the limiting velocity in each of the following cases: $k_1 \to 0$; $k_2 \to 0$.

3. A body of mass m is thrown vertically upward from the surface of the earth under a constant gravitational force and a resistance proportional to each of the following powers of the speed : a) first; b) second. Determine the maximum height attained by the body, and the time at which the maximum height is reached, if bodies initial speed is v_0.

4. Suppose everything is as in Problem 3. Determine the time at which the body returns to its starting point. Approach each of the following limits: the resistance is very large; the resistance is very small.

5. A body of mass m moves along of a straight line under a resistance force proportional to $(1+\alpha)$-th power of the speed. Determine the temporal variation of the coordinate and velocity if body's initial speed and coordinate are v_0 and zero, respectively. Approach each of the following limits: $\alpha \to -0.5$; $\alpha \to 0$; $\alpha \to 1$; $\alpha \to 2$.

6. A body of mass m moves along of a straight line under the Hook's force $F = -kx$. Determine the temporal variation of its coordinate, if at $t = 0$ the initial speed is zero and the initial coordinate is x_0.

7. A body of mass m falls from the rest under a force $F = -k/x^3$, where k is a constant and x is the distance from the origin to the position of body at time t. Determine the temporal variation of bodies coordinate, if the initial distance is h.

8. Find the time required to empty each of the following vertical vessels full of water: a) the cone of height H and top base of radius R; b) the cone of height H and bottom of radius R; c) the sphere of radius R; d) the circular paraboloid $z = (x^2 + y^2)/H$, $0 \le z \le H$; e) the circular paraboloid $z = H - (x^2 + y^2)/H$, $0 \le z \le H$. Assume that the radius of an orifice in the bottom is equal to r. Note that the spherical vessel has also a small hole at the top.

9. Find the time required to empty a horizontal cylinder full of water through an orifice of radius r in the lower point of cylinder. The radius and the length of the cylinder are R and H, respectively.

10. Determine the temporal variation of the current in a series circuit consisting of an inductance L, a resistance R and an impressed voltage $\mathcal{E}$, if at $t = 0$ the switch is closed.

11. Determine the temporal variation of the current in a series circuit consisting of a capacitance C, a resistance R, if at $t = 0$ the capacitor has a total charge equal to Q, and the switch is closed.

12. Determine the temporal variation of the current in a closed series circuit consisting of an inductance L, a resistance R, if at $t = 0$ the current is equal to I_0.

13. Determine the temporal variation of the current in a closed series circuit consisting of an inductance L and a capacitance C, if at $t = 0$ the capacitor has the maximum total charge $Q_0 > 0$.

14. Determine the temporal variation of the mass of water which evaporates from

a full cylinder vessel of the volume v_0 inside another closed vessel of the volume $V \geq v_0$. Suppose that at $t = 0$ the mass of water is equal to m_0, the vapor pressure $p_0 < p_s$. Approach the limit as $(p_s - p_0) \to 0$ at $V >> v_0$.

15. Determine the temporal variation of the mass of water which evaporates from the full cone vessel of a height H and a radius of the top base R. Assume that $v_0 << V$ and $p(0) = p_0 < p_s$.

16. Determine the air pressure above earth level, taking into account that the acceleration of gravity on an altitude h is $g(h) = g_0/(1 + h/R)^2$, where g_0 is the acceleration of gravity at $h = 0$ and R is the radius of earth. Assume that the air is the ideal gas at the constant temperature T.

17. Find the differential equation for the pressure of the "ideal" isothermal self-gravitate gas into a spherically cloud.

18. The body in air at 20 °C cools from 100 °C to 60 °C in 10 minutes. When will its temperature be 25 °C?

19. Find the time required to evaporate a spherical liquid drop of a radius R. Assume that $V >> R^3$ and $p(0) = p_0 < p_s$.

20. Determine the temporal variation of temperature of a body suppose that the temperature of the surrounding medium is a function of the time $T_{ext}(t)$. Assume that $T(0) = T_0$.

21. It is known that 206 gram of Plumb is produced owing to the radioactive decay of 238 gram of Uran. Determine the age of such rock, where the relation of Uran mass to Plumb mass is equal to 50:7. The half decay time of Uran is equal to 4.5×10^9 years. Assume that at $t = 0$ the mass of Plumb is equal to zero.

22. A ray of light gets into a layer of water bounded by two parallel mirrors. Determine the profile of temperature into the layer of water, if the distance between mirrors is d, the intensity of the light is I_0, the duration of expose is T and the coefficient of absorption is k. Assume that the ray of light is normal to the surface of mirrors.

23. What's the share of light that has been absorbed into the layer of water thickness of 2 m, if half of the light has been absorbed into the layer of thickness 0.35 m.

1.7 Miscellaneous problems

This section consists of a list of problems, which can be solved by the methods of the previous sections. They are presented so that the reader may have some practice in identifying the methods applicable to a given equation.

PROBLEMS

1. $(6x + y - 1)dx + (4x + y - 2)dy = 0$.
2. $y\,(x^2y^2 + 1)\,dx + (x^2y^2 - 1)\,xdy = 0$.
3. $(x - 2)dx + (y - 2x + 1)dy = 0$.
4. $(5x - 7y + 1)dy + (x + y - 1)dx = 0$.

5. $y^3dx + 2(x^2 - xy^2)\,dy = 0.$
6. $3x^5ydx + (y^4 - x^6)\,dy = 0.$
7. $(x - 2y + 1)dy - (2x - y + 1)dx = 0.$
8. $(8x + 25y - 62)dx - (11x + 4y - 11)dy = 0.$
9. $(x + 2y + 1)dx - (2x + 4y + 3)dy = 0.$
10. $4xy^2dx + (3x^2y - 1)\,dy = 0.$
11. $(x + y - 2)dx + (x - y + 4)dy = 0.$
12. $2\left(\sqrt{x^4y^2 + 1} - x^2y\right)dx - x^3dy = 0.$
13. $(2x\sin y - y^2\sin x)\,dx + (x^2\cos y + 2y\cos x + 1)\,dy = 0.$
14. $(x\ln y - x^2 + \cos y)\,dy + (x^3 + y\ln y - y - 2xy)\,dx = 0.$
15. $(x^2 - 4xy - 2y^2)\,dx + (y^2 - 4xy - 2x^2)\,dy = 0.$
16. $(6xy + x^2 + 3)\,y' + 3y^2 + 2xy + 2x = 0.$
17. $(2x\cos y - y^2\sin x)\,dx + (2y\cos x - x^2\sin y)\,dy = 0.$
18. $\dfrac{xdx + ydy}{\sqrt{1 + x^2 + y^2}} + \dfrac{ydx - xdy}{x^2 + y^2} = 0.$
19. $(2y + xy^3)\,dx + (x + x^2y^2)\,dy = 0.$
20. $(1 + x^2y)\,dx + x^2(x + y)\,dy = 0.$
21. $(x^2 + y^2 + 2x)\,dx + 2ydy = 0.$
22. $(x^2 + x^2y + 2xy - y^2 - y^3)\,dx + (y^2 + xy^2 + 2xy - x^2 - x^3)\,dy = 0.$
23. $(2x^3 + 3x^2y + y^2 - y^3)\,dx + (2y^3 + 3xy^2 + x^2 - x^3)\,dy = 0.$
24. $y^2dx + (xy - 1)dy = 0.$
25. $(2x^2y + x)\,dy + (y + 2xy^2 - x^2y^3)\,dx = 0.$
26. $(x^3 + xy^2 - y)\,dx + (y^3 + x^2y + x)\,dy = 0.$
27. $(x^2 + y)\,dy + x(1 - y)dx = 0.$
28. $(x^2y^3 + y)\,dx + (x^3y^2 - x)\,dy = 0.$
29. $2y^3y' + xy^2 - x^3 = 0.$
30. $y'\sec^2 y + x\tan y = x.$
31. $e^{-x}y' - e^{-x} = e^y.$
32. $\dfrac{y'}{\sqrt{1 + y}} - \dfrac{2\sqrt{1 + y}}{x} = 2(x + 5).$
33. $\dfrac{yy'}{\sqrt{1 + y^2}} + \sqrt{1 + y^2} = x^2 + 1.$
34. $3dy + (1 + e^{x+3y})dx = 0.$
35. $3y' - y\sin x + 3y^4\sin x = 0$
36. $y' + \sin x\tan y = \dfrac{\sin x}{\cos y}.$
37. $y'\cos x - y\sin x = y^4.$
38. $xy' + y = xy^2\ln x.$
39. $y' + \dfrac{y}{x} = y^4(1 - x^2);\ y(1) = 1.$
40. $y^2(x + a)dx + x(x^2 - ay)\,dy = 0.$
41. $(2xy - x^2y - y^3)\,dx - (x^2 + y^2 + x^3 - xy^2)\,dy = 0.$
42. $(3x^4y^2 + y^5)\,dx - (xy^4 + 2x^5y)\,dy = 0.$

43. $(x^2 + 2xy + x^2y - y^2 - y^3)\,dx + (y^2 + xy^2 + 2xy - x^2 - x^3)\,dy = 0.$
44. $x^2y' - x^2y^2 + 5xy - 3 = 0.$
45. $x^3y' - y^2 - x^2y + x^2 = 0.$
46. $y' + xy^2 + \frac{y}{x} - x^3 - 2 = 0.$
47. $y' = y^2 - x^2 + 1.$
48. $(x - x^4)\,y' - x^2 - y + 2xy^2 = 0;\ y_1(x) = x^2.$
49. $y' = y^2 + \frac{y}{x} + \frac{1}{x^2};\ y_1(x) = -\frac{1}{x}.$
50. $(1 - x^3)\,y' = y^2 - x^2y - 2x.$
51. $y' + y^2 = x^{-2}.$
52. $y' + y^2 \sin x = 2\frac{\tan x}{\cos x}.$
53. $y' + \frac{3}{x}y + xy^2 = x^{-1}.$
54. $dx - dy + x(xdy - ydx) = 0.$
55. $(x^2 + 2y^2)\,dx - xydy - (xdy - ydx) = 0.$
56. $(x^3 - y)\,dx + (x^2y + x)\,dy = 0.$
57. $xy' - y = \phi\left(\frac{y}{x}\right).$
58. $xy' = y^2 - 3y + 4x^2 + 2.$
59. $y' = y^2 - xy - x$
Hint: Find a particular solution in the form $y_1 = ax + b$.
60. $xy' = y^2 - (2x + 1)y + x^2 + 2x$
Hint: Find a particular solution in the form $y_1 = ax + b$.
61. $y(x) = x + \int_0^x e^{x-t}y(t)dt.$
62. $\int_0^x (x - t)t^2\sqrt[3]{y(t)}dt + \frac{1}{2}\int_0^x (5x - 6t)y(t)dt + 2x = 0.$
63. $y = \int_0^x y(t)dt + x + 1.$

Chapter 2

N-th ORDER DIFFERENTIAL EQUATIONS

2.1 Reduction of order

The order of an n-th order differential equation can be reduced in the following cases:

1) if an equation does not contain the independence variable, that is, one has the form

$$F\,(y,\, y',\, \ldots,\, y^{(n)}) = 0; \tag{1}$$

2) if an equation does not contain the function y and its derivatives up to order $k-1$ inclusive, that is, one has the form

$$F\,(x,\, y^{(k)},\, y^{(k+1)},\, \ldots,\, y^{(n)}) = 0; \tag{2}$$

3) if an equation of the form

$$F\,(x,\, y,\, y',\, \ldots\, y^{(n)}) = 0 \tag{3}$$

is homogeneous or generalized homogeneous;

4) if the left-hand side of the equation $F(x, y, y', \ldots, y^{(n)}) = 0$ can be represented as the first derivative of some expression $\Phi(x, y, y', \ldots, y^{(n-1)})$, that is,

$$F(x, y, y', \ldots, y^{(n)}) = \frac{d}{dx}\Phi(x, y, y', \ldots, y^{(n-1)}) = 0, \tag{4a}$$

or, in the more general case,

$$F(x, y, y', \ldots, y^{(n)})\mu(x, y, y', \ldots, y^{(n-1)}) = \frac{d}{dx}\Phi(x, y, y', \ldots, y^{(n-1)}) = 0, \tag{4b}$$

where $\mu(x, y, y', \ldots, y^{(n-1)})$ is an integrating factor of the initial equation;

5) if an equation is linear, that is, one has the form

$$p_0(x)y^{(n)}(x) + p_1(x)y^{(n-1)}(x) + \cdots + p_n(x)y(x) = 0, \tag{5}$$

where $p_i(x)$ are some functions, and some of its particular solution $y_p(x)$ is known.

2.1.1 *Simple Cases*

1. Equation not containing the variable x. By the substitution $y'(x) = p(y)$, where $p(y)$ is a new function, the order of Eq.(1) is reduced by unity.

Example 1. Reduce the order of the differential equation

$$y'''y^2 - \cot y' + \exp(y'') = 0.$$

Substituting $y'(x) = p(y)$ leads to

$$y''(x) = \frac{dp(y)}{dx} = p'(y)y'(x) = p'(y)p(y), \qquad y'''(x) = p''p^2 + p'^2 p.$$

Hence we obtain

$$\left(p''p^2 + p'^2 p\right) y^2 - \cot p + \exp(p'p) = 0,$$

that is, the second order differential equation.

Example 2. Find the general solution of the differential equation

$$yy'' - y'^2 = 0.$$

Proceeding as in Ex.1 we obtain, after some minor calculations,

$$yp'p - p^2 = 0, \quad \text{or} \quad \frac{p'}{p} = \frac{1}{y}.$$

An integration of this equation yields

$$p(y) = C_1 y = y'(x),$$

where C_1 is an arbitrary constant. An integration of this equation leads to the general solution of the given equation

$$y(x) = C_2 e^{C_1 x},$$

where C_2 is an arbitrary constant.

2. Equation not containing the variables $y, y', \ldots, y^{(k-1)}$. By the substitution $y^{(k)}(x) = z(x)$, where $z(x)$ is a new function, the order of Eq.(1) is reduced up to $n - k$.

Example 3. Reduce the order of the differential equation

$$xy'' + xy'^2 - y'^2 - y' = 0.$$

By making change of variable $y'(x) = z(x)$ we obtain

$$xz' + z^2\left(x - 1 - \frac{1}{z}\right) = 0,$$

that is, the first order differential equation.

Example 4. Find the general solution of the differential equation

$$y''' + y'' = e^x.$$

By making change of variable $y''(x) = z(x)$ we obtain

$$z' + z = e^x$$

that is, the linear nonhomogeneous first order equation. His solution is

$$z(x) = y'' = (C_1 + x)e^{-x},$$

where C_1 is an arbitrary constant. A two-fold integration then yields

$$y = (C_1 + x + 2)e^{-x} + C_2 x + C_3,$$

where C_i are arbitrary constants.

2.1.2 Homogeneous Equations

1. Homogeneous equations. The Eq.(3) is called homogeneous equation of degree p, if it possesses the property

$$F(x, ky, ky', \ldots, ky^{(n)}) = k^p F(x, y, y', \ldots, y^{(n)}) = 0$$

for every nonvanish constant k. It is easy to verify whether an equation is homogeneous or not. Then by the substitution $y'(x) = y(x)z(x)$, where $z(x)$ is a new function, the order of Eq.(3) is reduced by unity.

Example 5. Reduce the order of the differential equation

$$x^2\left(y''y + y'^2\right) = y^2.$$

It is easy to verify that this equation is homogeneous of degree 2 in the arguments y, y', y''. Then by substituting $y'(x) = y(x)z(x)$, where $z(x)$ is a new function, we obtain, after some minor calculations,

$$z' + 2z^2 = \frac{1}{x^2}.$$

This is the Riccati's equation.

2. Generalized homogeneous equation. The Eq.(3) is called the generalized homogeneous of degree p, if it possesses the property

$$F(kx, k^\alpha y, k^{\alpha-1}y', \ldots, k^{\alpha-n}y^{(n)}) = k^p F(x, y, y', \ldots, y^{(n)}) = 0$$

for every nonvanish constant k and a certain constant $\alpha(p)$. First we should investigate whether an equation is generalized homogeneous or not. For this purpose we make the change of variables $x = kt$ and $y(x) = y(x)k^\alpha$ in Eq.(3). The new n-th order differential equation has new coefficients depending on the constants α and k. Then equating powers of k in all terms of this equation leads to a system of algebraic equation for α. If this system is compatible and has a solution for α then that equation is generalized homogeneous. By substitution $x = \exp(t)$, $y = z(t)\exp(\alpha t)$ into Eq. (3) we obtain a new n-th order differential equation for $z(t)$ which does not contain t. Then proceeding as has been shown above we reduce its order by unity.

Example 6. Reduce the order of the differential equation

$$xyy'' + yy' - x^3y'^3 = 0.$$

By substituting $x = kt$, $y = k^\alpha z$, $y' = k^{\alpha-1}z'$, $y'' = k^{\alpha-2}z''$ in the given equation we obtain, after some minor calculations,

$$k^{2\alpha-1}tzz'' + k^{2\alpha-1}z' - k^{3\alpha}t^3z'^3 = 0.$$

Equating powers of k in all coefficients yields $2\alpha - 1 = 3\alpha$. Hence $\alpha = -1$. Further by making change of variables $x = \exp(t)$ and $y(x) = z(t)\exp(-t)$ we find, after some minor calculations,

$$z(z'' - 2z' + z) - (z' - z)^3 = 0.$$

This equation has the form of Eq.(1). Hence by substituting $z'(t) = p(z)$ we obtain

$$z\left(p'p - 2p + z\right) - (p - z)^3 = 0.$$

This is the first order equation.

2.1.3 Exact Equations

1. Exact Equations. Differentiating the right-hand side of Eq.(4a) we obtain the general form of an exact equation of higher order

$$F(x, y, y', \ldots, y^{(n)}) = \Phi'_x + \Phi'_y y' + \Phi'_{y'} y'' + \ldots + \Phi'_{y^{(n-1)}} y^{(n)} = 0, \qquad (6)$$

where $\Phi = \Phi(x, y, y', \ldots, y^{(n-1)})$. This equation is always linear in the highest order derivative. It is the necessary condition of the exactness of an n-th order equation. The function $\Phi(x, y, y', \ldots, y^{(n-1)})$, if it exists, can be found by the method of *successive exclusions of highest derivatives* as shown in Ex.7.

Example 7. Reduce the order of the differential equation

$$y'''y + y''(xy + y') + y'\left(y + xy' + 2x^2y\right) + 2xy^2 + 3x^2 = 0.$$

This equation is linear in y'''. Hence $\Phi'_{y''} = y$ and an integration by y'' leads to

$$\Phi(x, y, y', y'') = yy'' + \Phi_1(x, y, y'),$$

where $\Phi_1(x,y,y')$ is a new unknown function. By substituting the initial equation and $\Phi(x,y,y',y'')$ in Eq.(6) we obtain, after some simplifications,

$$xyy'' + y'\left(y + xy' + 2x^2y\right) + 2xy^2 + 3x^2 = \Phi'_{1x} + \Phi'_{1y}y' + \Phi'_{1y'}y''. \tag{6.1}$$

This equation is linear in y''. Hence $\Phi'_{1y'} = xy$, and an integration by y' yields

$$\Phi_1(x,y,y') = xyy' + \Phi_2(x,y,),$$

where $\Phi_2(x,y)$ is a new unknown function. By substituting $\Phi_1(x,y,y')$ in Eq.(6.1) we obtain, after some simplifications,

$$2x^2yy' + 2xy^2 + 3x^2 = \Phi'_{2x} + \Phi'_{2y}y'. \tag{6.2}$$

This equation is linear in y'. Hence $\Phi'_{2y} = 2x^2y$, and an integration by y leads to

$$\Phi_2(x,y,) = x^2y^2 + \Phi_3(x),$$

where $\Phi_3(x)$ is a new unknown function. By substituting $\Phi_2(x,y)$ in Eq.(6.2) we obtain, after some simplifications,

$$3x^2 = \Phi'_{3x}. \tag{6.3}$$

Hence $\Phi_3(x) = x^3 + C$, where C is an arbitrary constant. Recalling all the substitutions we find the second order equation

$$\Phi(x,y,y',y'') = yy'' + xyy' + x^2y^2 + x^3 + C = 0.$$

This equation is linear in y'' too. Therefore, we make the attempt of order reduction. Assuming that

$$yy'' + xyy' + x^2y^2 + x^3 + c = \Psi'_x + \Psi'_y y' + \Psi'_{y'}(x,y,y')y'' = 0,$$

where $\Psi(x,y,y')$ is an unknown function, we obtain $\Psi'_{y'} = y$ and $\Psi = yy' + \Psi_1(x,y)$, where $\Psi_1(x,y)$ is also an unknown function. Further proceeding as above we find

$$-y'^2 + xyy' + x^2y^2 + x^3 + C = \Psi'_x + \Psi'_y y'.$$

This equation is nonlinear in y'. Therefore, it is not exact.

Thus the initial equation is reduced to the second order equation

$$yy'' + xyy' + x^2y^2 + x^3 + C = 0.$$

2. Integrating factor. Multiplying an equation by suitable integrating factor $\mu(x,y,y',\ldots,y^{(n-1)})$ we should reduce one to an exact equation. The question of the existence of μ and general techniques for finding it are beyond the scope of our text. However, choosing any function $\mu(x, y, y', \ldots, y^{(n-1)})$ we can verify whether it is an integrating factor or not by the method of successive exclusions of highest derivatives.

Example 8. Reduce the order of the differential equation

$$y'' - 2y^3 = 0.$$

It is easy to verify that this equation is not exact, however, multiplying it by $y'(x)$ and proceeding as in Ex.7 we obtain, after some minor simplifications,

$$y'^2 - y^4 = C_1,$$

where C_1 is an arbitrary constant. Then we find

$$\pm \int \frac{dy}{(C_1 + y^4)^{1/2}} = x + C_2,$$

where C_2 is an arbitrary constant. It is the solution of the given equation in the form of an inverse function.

2.1.4 Linear Equations

The order of Eq.(5) is reduced by unity by the substitution

$$y(x) = y_p(x) \int z(x)dx,$$

where $y_p(x)$ is a nontrivial particular solution of Eq.(5), and $z(x)$ is a new function. If Eq.(5) is a second order equation, then a second linearly independent solution of it can be found using the Abel's identity (the Norwegian mathematician N.H. Abel (1802-1829))

$$y_2(x) = y_1(x) \int \exp\left[-\int \frac{p_1(x)}{p_0(x)}\, dx\right] \frac{dx}{y_1^2(x)}. \tag{7}$$

Example 9. Reduce the order of the differential equation

$$y'''\left(x^2 - 2x + 2\right) - y''\left(x^2 - 2x\right) + y'\left(x^2 - 2\right) - y(2x - 2) = 0,$$

if its particular solution is $y_1(x) = x^2$.

By substitution $y = x^2 \int z(x)\, dx$ we obtain, after some minor calculations,

$$x^2\left(x^2 - 2x + 2\right) z'' - \left(x^4 + 2x^3 - 6x^2 + 12x - 12\right) z' + \left(4x^3 - 6x^2 + 12x\right) z = 0.$$

It is the second order differential equation.

Example 10. Solve the differential equation

$$x^3 y'' + (x - 1)y' - y = 0,$$

if its particular solution is $y_1(x) = x - 1$.

Using Eq.(7) we find, after some minor calculations,

$$y(x) = C_1(x-1) + C_2(x-1)\int \exp\left(\frac{1}{x} - \frac{1}{2x^2}\right)\frac{dx}{(x-1)^2}.$$

2.1.5 The Initial Value Problem

The general solution of an n-th order differential equation should depend on n arbitrary constants. These constants can be determined by substituting this solution into n prescribed initial conditions

$$\begin{aligned} &y(x_0,\, C_1,\, \ldots\, C_n) = y_0, \\ &y'(x_0,\, C_1,\, \ldots\, C_n) = y_1, \\ &\vdots \\ &y^{(n-1)}(x_0,\, C_1,\, \ldots\, C_n) = y_{n-1}, \end{aligned} \tag{8}$$

where y_i are prescribed numbers. In general, this system is nonlinear and its solution may not exist. However, if it is compatible and it has some solutions $C_i = C_i(y_0,\, y_1,\, \ldots,\, y_{n-1})$ then the initial value problem has also a solution. Notice that for a linear differential equation system (8) is always compatible and solvable, when x_0 is not singular point of one.

Example 11. Solve the initial value problem

$$\begin{aligned} &yy'' - 2xy'^2 = 0, \\ &y = 2, \quad y' = 0.5 \quad \text{at} \quad x = 2. \end{aligned}$$

Determine those initial conditions y_0 and y_1 when this problem does not have a solution.

The given equation is homogeneous of second degree in variables y, y' and y''. Then by the substitution $y' = yz$ we obtain

$$z' + (1-2x)z^2 = 0.$$

The solution of this equation is $z(x) = 1/(C + x - x^2)$, where C is an arbitrary constant. Further, the second integration yields

$$y(x) = C_2 \exp\left(\frac{1}{C_1}\ln\left|\frac{C_1 + 2x - 1}{C_1 - 2x + 1}\right|\right),$$

where C_2 is an arbitrary constant and $C_1 = \sqrt{4C+1}$ is also an arbitrary constant. By substituting this solution into the initial conditions we obtain

$$\begin{aligned} &C_2 \exp\left(\frac{1}{C_1}\ln\left|\frac{C_1+3}{C_1-3}\right|\right) = 2, \\ &C_2 \exp\left(\frac{1}{C_1}\ln\left|\frac{C_1+3}{C_1-3}\right|\right)\frac{4}{C_1^2 - 9} = 0.5. \end{aligned}$$

This system is compatible, and its solution is $C_1 = 5$ and $C_2 = 2^{3/5}$. Hence the solution of the given initial value problem is

$$y(x) = \left(8\frac{x+2}{3-x}\right)^{1/5}.$$

In general, the system (8) has the form

$$C_2 \exp\left(\frac{1}{C_1}\ln\left|\frac{C_1+2x_0-1}{C_1-2x_0+1}\right|\right) = y_0,$$
$$C_2 \exp\left(\frac{1}{C_1}\ln\left|\frac{C_1+2x_0-1}{C_1-2x_0+1}\right|\right)\frac{4}{C_1^2-(2x_0-1)^2} = y_1.$$

Its solution is

$$C_1 = \pm\sqrt{(1-2x_0)^2+4\frac{y_0}{y_1}},$$
$$C_2 = y_0 \exp\left(-\frac{1}{C_1}\ln\left|\frac{C_1+2x_0-1}{C_1-2x_0+1}\right|\right).$$

It is easy to verify that the constants C_1 and C_2 do not exist at $y_1 = 0$ and $y_0 \neq 0$, or at $y_0 = 0$, $x_0 = 0$ and $y_1 \neq 0$. Hence for these initial conditions the solution of the given initial value problem does not exist.

PROBLEMS

Reduce the order of the following differential equations. Solve obtained first order equations.

1. $yy''^2 = y'$.
2. $yy'' = y'^2$.
3. $yy'y''' + 2y'^2y'' = 3yy''^2$.
4. $y^2(y'y''' - 2y''^2) = 1$.
5. $yy'y''y''' = 1$.
6. $(yy''' - 3y''^2)\,y = y'^5$.
7. $2yy'' + y'^2 + y'''^4 = 0$.
8. $3yy''' + 4y'y^{iv} = 1$.
9. $y'' - 2yy' = 2y'$.
10. $8y'' + e^y - y'^2 = 0$.
11. $yy''' + y'^2y'' = y''^2$.
12. $yy'^2 + y''^3 + y'''^4 = 0$.
13. $y'y''' = yy''$.
14. $y + y' + y'' + y''' = 1$.
15. $y^2 + y''^2 = 1$.
16. $y^2 - y''^2 = 1$.
17. $y - 2(y'^2 + y''^2) = 0$.
18. $y''' = -\cos x + y''$.
19. $xy^{iv} = 1 + \ln|y''|$.

20. $y''' = \dfrac{2}{x^3} + y'^2$.
21. $y''' = 2xy''$.
22. $y'' = \sin x^2 + y'$.
23. $xy^{iv} + y'' = e^x$.
24. $x = e^{-y'} + y'''$.
25. $x^2 = y'y''$.
26. $xy'' + y' = 0$.
27. $xy'' + y'e^{y'/x} = 0$.
28. $xy'' - y'^2 = 0$.
29. $y' = xy''' + y''^2$.
30. $y''' = xy'^3$.
31. $y''' + 2xy' = 0$.
32. $3xy'' + e^x y''' = 0$.
33. $xy' - \dfrac{y''^2}{y'} = e^{y''}$.

Reduce the order of the following homogeneous and generalized homogeneous differential equations. Solve obtained first order equations.

34. $yy'' - y'^2 - \dfrac{yy'}{\sqrt{1+x^2}} = 0$.
35. $yy'' - y'^2 - \dfrac{xyy'}{1+x^2} = 0$.
36. $x^2yy'' - (y - xy')^2 = 0$.
37. $xyy'' - y'(y + xy') = 0$.
38. $x^2yy'' - (y + xy')^2 = 0$.
39. $xyy'' - xy'^2 + yy' = 0$.
40. $x^2(yy'' - y'^2) + xyy' - y\sqrt{y^2 + x^2y'^2} = 0$.
41. $x^2(yy' - y'^2) + xyy' + y\sqrt{y^2 - xy'^2} = 0$.
42. $yy''' - y'y'' - xy^2 = 0$.
43. $y(y - xy''') + xy'y'' = 0$.
44. $y^2y''' - 3yy'y'' + 2y'^2 - 3y^2y'' + 3yy'^2 + 2y^2y' - xy^3 = 0$.
45. $yy'' - y'^2 - f(x)y^2 = 0$, where $f(x)$ is a function of x alone.
46. $yy'' - y'^2 + g(x)yy' - f(x)y^2 = 0$, where $g(x)$ and $f(x)$ are some functions of x alone.
47. $y(xy'' + y') - xy'^2(1 - x) = 0$.
48. $x^2yy'' + y'^2 = 0$.
49. $x^2y'' + xy' + y = x^3$.
50. $xy'' + x^2yy' + \dfrac{y}{x} = \dfrac{1}{x^3}$.

Reduce the order of the following differential equations by the method of successive exclusions of highest derivatives. Choose a suitable integrating factors if it is necessary.

51. $y'' + \dfrac{3}{\cosh y} = 0$.

52. $x^2y'^2y''' + 2x^2y'y''^2 + xy'e^y - e^y = 0.$
53. $yy''' + 2y^2y' + y' = 0.$
54. $y'y''' - 2y''^2 = 0.$
55. $(1+y^2)\,y'' - 2yy'^2 = 0.$
56. $yy'' - y'^2 = 0.$
57. $y'' = f(y)$, where $f(y)$ is a function of y alone.
58. $yy''' - y'y'' = 0.$
59. $yy'' + y'^2 = 0.$
60. $(1+x^2)^2\,y'' + x\,(1+x^2)\,y' + (1-x^2)\,y = 0.$

Solve each of the following equations.

61. $y''^2 + y'^2 = 0.$
62. $y''^2 - y'^2 = 0.$
63. $y''^2 + y'^2 = 1.$
64. $y''^2 - y'^2 = 1.$
65. $xy'' + \dfrac{y'}{2} - y'^2 = 0.$
66. $2yy'' - y'^2 = 0.$
67. $xy''' + (3 - x^2y'')\,y'' - 2xy'y'' - 4y'^2 = 0.$
68. $x\,(1+x)\,y'' + (2+x)\,y' = 0.$
69. $y''\tan x - 2y'' = -2.$
70. $4yy'' - 3y'^2 - 4y^2 = 0.$
71. $xyy'' - yy' - 4xy'^2 = 0.$
72. $yy'' - y'^3 = 0.$
73. $x^2y'' - xy' + y = 1.$
74. $x^2y'' + x^2y'^2 + 2xy' = 2.$
75. $xyy'' + xy'^2 - yy' = 0.$
76. $xyy'' - 2xy'^2 - yy' - x^3y^2 = 0.$
77. $y'' - (1+y'^2)^{3/2} = 0.$
78. $y^2y'' - yy'^2 - y^3 = 0.$
79. $x^2\,(2yy'' - y'^2) + 2xyy' = 1.$
80. $y'' - xy'^3 = 0.$

Solve each of the following initial value problems.

81. $y''' - 3yy' = 0, \quad y = -2, \quad y' = 0, \quad y'' = \dfrac{9}{2} \quad \text{at} \quad x = 0.$
82. $y'y''' - 3y''^2 = 0, \quad y = 0, \quad y' = 1, \quad y'' = 0 \quad \text{at} \quad x = 0.$
83. $yy''' + 3y'y'' = 0, \quad y = 1, \quad y' = 0, \quad y'' = 0 \quad \text{at} \quad x = 1.$
84. $xy''' + (x-1)\,y'' = 0, \quad y = 0, \quad y' = 0, \quad y'' = 0 \quad \text{at} \quad x = 0.$
85. $y'' + y'^2 - 2e^{-y} = 0, \quad y = 1, \quad y' = \pi \quad \text{at} \quad x = 0.$
86. $y''' + 2xy'' = 0, \quad y = y_0, \quad y' = y_1, \quad y'' = y_2 \quad \text{at} \quad x = 0.$
87. $y\,(y'' - 2x) - 2y\,(y' - x^2) = 0, \quad y = 1, \quad y' = 1 \quad \text{at} \quad x = 1.$
88. $y'' = y^{3/2}, \quad y = 1, \quad y' = -1 \quad \text{at} \quad x = 0.$

Reduce the order of the following linear differential equations using the given particular solutions. Solve obtained first order equations.

89. $x^2(1+x)y'' - 2y = 0, \quad y_1 = \dfrac{1+x}{x}.$

90. $xy'' + 2y' - xy = 0, \quad y_1 = \dfrac{e^x}{x}.$

91. $(1+e^x)y'' - 2y' - e^x y = 0, \quad y_1 = e^x - 1.$

92. $xy''' - y'' - xy' + y = 0, \quad y_1 = e^x, \quad y_2 = x.$

93. $x^2(2x-1)y''' + (4x-3)xy'' - 2xy' + 2y = 0, \quad y_1 = x, \quad y_2 = \dfrac{1}{x}.$

94. $y'' - 2(1+\tan^2 x)y = 0, \quad y_1 = \tan x.$

2.2 Linear homogeneous equations

A linear homogeneous n-th order differential equation is an equation of the form

$$p_0(x)y^{(n)}(x) + p_1(x)y^{(n-1)}(x) + \dots + p_n(x)y(x) = 0, \tag{1}$$

where $p_i(x)$, $i = 1, \dots, n$ are continuous real-valued functions on some interval $a \le x \le b$ and $p_0(x)$ is nowhere zero in the interval. A point x^*, where $p_0(x^*) = 0$, is called the *singular point* of Eq.(1) (see Sec. 3.1).

The general solution of Eq.(1) is a sum

$$y(x) = \sum_{i=1}^{n} C_i y_i(x), \tag{2}$$

where C_i are arbitrary constants, $y_i(x)$ are the particular solutions of Eq.(1).

If $y_1(x)$, $y_2(x)$, ..., $y_n(x)$ are some set prescribed linearly independent functions then they are solutions of the differential equation

$$W(y_1, \dots, y_n, y) = \begin{vmatrix} y_1 & y_2 & \dots & y_n & y \\ y_1' & y_2' & \dots & y_n' & y' \\ \vdots & \vdots & \ddots & \vdots & \vdots \\ y_1^{(n)} & y_2^{(n)} & \dots & y_n^{(n)} & y^{(n)} \end{vmatrix} = 0, \tag{3}$$

where $W(y_1, \dots, y_n, y)$ is called the *Wronskian of functions* $y_1, \dots, y_n, y$, after the Polish mathematician Joseph Hoene Wronski (1778-1853). The Wronskian of the linearly independent solutions of Eq.(1) is nowhere zero on the interval $a \le x \le b$. So called *Abel's identity* is justified for these functions

$$W(y_1, y_2, \dots, y_n) = C \exp\left[-\int \frac{p_1(x)}{p_0(x)} dx\right],$$

where C is a constant.

Example 1. Find the differential equation whose solutions are $y_1 = 1$, $y_2 = \exp(x)$, $y_3 = \exp(-x)$.

According to Eq.(3) we obtain, after some minor algebraic calculations,

$$\begin{vmatrix} 1 & e^x & e^{-x} & y \\ 0 & e^x & -e^{-x} & y' \\ 0 & e^x & e^{-x} & y'' \\ 0 & e^x & -e^{-x} & y''' \end{vmatrix} = y''' - y' = 0.$$

Hence the given functions are solutions of this equation .

A set of the linearly independent solutions $y_1(x), \ldots, y_n(x)$, of Eq. (1) which satisfy the initial conditions

$$y_i^{(k)}(x_0) = 1, \quad \text{if} \quad k = i-1, \quad y_i^{(k)}(x_0) = 0, \quad \text{if} \quad k \neq i-1, \tag{4}$$

where $k = 0, 1, \ldots, n-1$ and $i = 1, 2, \ldots, n$ is called the *normalized fundamental set* of solutions of Eq.(1) at point x_0. The Wronskian of the normalized fundamental set of solutions

$$W(y_1, \ldots, y_n) \equiv W(x) = \begin{vmatrix} y_1 & y_2 & \ldots & y_n \\ y_1' & y_2' & \ldots & y_n' \\ \vdots & \vdots & \ddots & \vdots \\ y_1^{(n-1)} & y_2^{(n-1)} & \ldots & y_n^{(n-1)} \end{vmatrix} \tag{5}$$

is unity at point x_0. Notice that the matrix $\Psi(x)$

$$\Psi(x) = \begin{pmatrix} y_1 & y_2 & \ldots & y_n \\ y_1' & y_2' & \ldots & y_n' \\ \vdots & \vdots & \ddots & \vdots \\ y_1^{(n-1)} & y_2^{(n-1)} & \ldots & y_n^{(n-1)} \end{pmatrix}$$

is called the *fundamental matrix* of the differential equation. It is easy to see that $\det |\Psi(x)| = W(x)$.

Example 2. Determine the normalized fundamental set of solutions of a differential equation at the point $x_0 = 1$, if its particular linearly independent solutions are

$$y_1 = 1, \quad y_2 = e^x, \quad y_3 = e^{-x}.$$

Since functions $y_i(x)$ are linearly independent solutions of a linear differential equation hence $\varphi_1(x) = C_1y_1(x) + C_2y_2(x) + C_3y_3(x)$, where C_i are arbitrary constants, is also solution of that equation. By substituting this function into Eq-s (4) for $y_1(1)$ we termine the arbitrary constants C_i

$$\begin{aligned} &C_1 + C_2e^1 + C_3e^{-1} = 1, \\ &C_2e^1 - C_3e^{-1} = 0, \\ &C_2e^1 + C_3e^{-1} = 0. \end{aligned}$$

Solution of this system is $C_1 = 1$, $C_2 = C_3 = 0$. Thus $\varphi_1(x) = 1$. Similarly we find the second normalized solution as $\varphi_2(x) = C_1\varphi_1(x) + C_2y_2(x) + C_3y_3(x)$, where C_i

are arbitrary constants. Substituting $\varphi_2(x)$ into Eq-s (4) at $i = 2$, leads to $C_1 = 0$, $C_2 = e/2$, $C_3 = -e/2$. Hence $\varphi_2(x) = \sinh(x-1)$. Similarly we write $\varphi_3(x)$ in the form $\varphi_3(x) = C_1\varphi_1(x) + C_2\varphi_2(x) + C_3y_3(x)$. Using Eq-s (4) at $i = 3$ we find, after some minor algebraic calculations, $\varphi_3(x) = \cosh(x-1) - 1$. Hence the normalized fundamental set of solutions at the point $x = 1$ is

$$\varphi_1(x) = 1, \quad \varphi_2(x) = \sinh(x-1), \quad \varphi_3(x) = \cosh(x-1) - 1.$$

The mathematical theory and the techniques for solving Eq.(1) are highly developed. Now we consider some particular methods (see also Sec-s 2.5 and 3.1).

2.2.1 Exponential Solution

It is a solution of Eq.(1) of the form $y(x) = C\exp(kx)$, where C is an arbitrary constant and k is a certain constant. By substituting this solution in Eq.(1) we obtain, after dividing by $y(x)$,

$$k^n p_0(x) + k^{n-1}p_1(x) + \ldots + p_n(x) = 0.$$

Equating the coefficients of like terms of x to zero we obtain a system of equations whose coefficients depend on k. If this system is compatible and it has some solutions for k then here the exponential solutions exist. Notice that for a complex value of k we obtain two exponential solutions of Eq. (1); namely: $y_1 = \text{Re}\, e^{kx}$ and $y_2 = \text{Im}\, e^{kx}$. If obtained k is l-fold root of these system, where $l > 1$ is an integer number, then see Sec. 2.4.1.

Example 3. Assuming that a solution of the differential equation

$$x(2x+1)y'' + 2(1-2x^2)y' - 4(x+1)y = 0$$

has an exponential form, determine of it.

By substituting $y = \exp(kx)$ in the given equation we find, after dividing by y,

$$k^2 2x^2 + k^2 x - 2kx - 4kx^2 - 4x - 4 = 0.$$

Equating the coefficients of like powers of x to zero we obtain

$$2(k^2 - 2k) = 0,$$
$$k^2 - 4 = 0,$$
$$k - 2 = 0.$$

This system is compatible and has the solution $k = 2$. Therefore $y_1(x) = \exp(2x)$ is the exponential solution of the given equation. The second linearly independent solution $y_2(x) = 1/x$ can be obtained by using reduction of order (see Sec. 2.1). Hence the general solution of the given equation is

$$y = C_1 e^{2x} + C_2 \frac{1}{x},$$

where C_i are arbitrary constants.

Example 4. Assuming that a solution of the differential equation

$$2y'' + 4y' + 5y = 0$$

has an exponential form, determine of it.

By substituting $y = \exp(kx)$ in the given equation we find, after dividing by y,

$$2k^2 + 4k + 5 = 0.$$

The roots of this equation are

$$k_1 = -1 + i\sqrt{\frac{3}{2}}, \qquad k_2 = -1 - i\sqrt{\frac{3}{2}}.$$

Hence the given equation has two exponential solutions and its general solution is

$$y(x) = e^{-x}\left(C_1 \cos\sqrt{\frac{3}{2}}x + C_2 \sin\sqrt{\frac{3}{2}}x\right),$$

where C_i are arbitrary constants.

2.2.2 Power Solution

It is a solution of Eq.(1) on the form $y = x^k$, where k is some constant. An equation for the constant k we find by substituting $y = x^k$ in Eq.(1)

$$k(k-1)\ldots(k-n+1)p_0(x)x^{k-n} + \ldots + p_n(x) = 0.$$

and equating the coefficient of the highest power of x to zero. If obtained equation has some solutions for k, then here the power solution may exist. If the k obtained is not a positive integer number then the solution for $y(x)$ may exist as only one term x^k. This assumption we verify by substituting this term in Eq.(1). Notice that for a complex value of k we obtain two power solutions of Eq. (1); namely: $y_1 = \text{Re } x^k$ and $y_2 = \text{Im } x^k$. If the k obtained is a positive integer number then here may exist so-called *polynomial solution* of Eq.(1) in the form $y = x^k + a_1x^{k-1} + \ldots + a_k \equiv P_k(x)$. The coefficients a_i we determine by substituting $P_k(x)$ in Eq.(1) and equating the coefficients of like terms of x to zero. If obtained k is l-fold root of these equation, where $l > 1$ is an integer number, then see Sec. 2.5.1.

Example 5. Assuming that a solution of the differential equation

$$2x(2x\sin x + 3\cos x)y'' - (4x^2 + 3)\cos xy' + 3(2x\cos x - \sin x)y = 0.$$

has a power form, determine it.

By substituting $y = x^k$ in the given equation and equating the coefficient only of the highest power of x to zero we obtain

$$-4kx^{k+1}\cos x + 6x^{k+1}\cos x = 0.$$

This equation has one root $k = 3/2$. By substituting $y = x^{3/2}$ in the given equation we convince that this function is its particular solution.

Example 6. Assuming that a solution of the differential equation

$$2x^2y'' + 4xy' + 5y = 0$$

has a power form, determine it.

By substituting $y = x^k$ in the given equation we obtain, after dividing by x^k,

$$2k(k-1) + 4k + 5 = 0.$$

The roots of this equation are $k_1 = \bar{k}_2 = (-1+3i)/2$. Hence the given equation has two power solutions. It is easy to verify that both functions $y_1 = \mathrm{Re}x^{k_1}$ and $y_2 = \mathrm{Im}x^{k_1}$ are particular solutions of the given equation. Hence the general solution is

$$y(x) = x^{-1/2}\left(C_1\cos\left(\frac{3}{2}\ln|x|\right) + C_2\sin\left(\frac{3}{2}\ln|x|\right)\right),$$

where C_i are arbitrary constant.

Example 7. Assuming that a solution of the differential equation

$$x\left(4x^3 + 6x^2 - 1\right)y'' - 2\left(2x^3 + 1\right)y' - 12x(x+1)y = 0$$

has a power form, determine it.

By substituting $y = x^k$ in the given equation and equating coefficient only of the highest power of x to zero we obtain

$$4k^2 - 8k - 12 = 0.$$

The roots of this equation are $k_1 = 3$ and $k_2 = -1$. By substituting $y = 1/x$ in the given equation we convince that this function is its particular solution. By substituting the polynomial $y = x^3 + ax^2 + bx + c$ in the given equation we obtain, after some minor algebraic simplifications,

$$x^4(24 - 12a) - x^3 16b - x^2 12(1 + b + c) - 6x(a + 2c) - 2b = 0.$$

Equating the coefficients of like powers of x to zero leads to

$$\begin{aligned}
&2 - a = 0,\\
&b = 0,\\
&1 + b + c = 0,\\
&a + 2c = 0,\\
&b = 0.
\end{aligned}$$

This system is compatible and its solution is $a = 2$, $b = 0$, $c = -1$. Hence the polynomial solution is $y = x^3 + 2x^2 - 1$, and the general solution of the given equation is

$$y(x) = \frac{C_1}{x} + C_2\left(x^3 + 2x^2 - 1\right),$$

where C_i are arbitrary constants.

Example 8. Assuming that a solution of the differential equation

$$\left(x^3 + 3\right) y''' + x(x-1)y'' - 2(x-1)y' + 2y = 0$$

has a power form, determine it.

Proceeding as in Ex.7 we obtain

$$k(k-1)(k-2) + k(k-1) - 2k + 2 = 0.$$

The roots of this equation are $k_1 = 2$, $k_2 = k_3 = 1$. By substituting $y = P_1(x)$ and $y = P_2(x)$ in the given equation we find that only $y = x - 1$ is its particular solution .

2.2.3 Transformations of Equation

Some transformations of Eq. (1) are useful for solving of one.

1. By substituting

$$y(x) = z(x) \exp\left[-\frac{1}{n}\int \frac{p_1(x)}{p_0(x)} dx\right]$$

in Eq.(1) we obtain, after dividing by $p_0(x)$,

$$z^{(n)}(x) + g_2(x)z^{(n-2)}(x) + \cdots + g_n(x)z(x) = 0 \tag{6}$$

where $g_i(x)$, $i = 2, \ldots, n$ are some functions connected with the all functions $p_j(x)$ and their derivatives. This equation does not contain the $(n-1)$-th derivative.

2. If $n = 2$ then by making change of variable

$$t = \int \exp\left[-\int \frac{p_1(x)}{p_0(x)} dx\right] dx$$

we obtain

$$\frac{dy}{dx} = \frac{dy}{dt}\frac{dt}{dx} = \exp\left[-\int \frac{p_1(x)}{p_0(x)} dx\right] \frac{dy}{dt},$$

$$\frac{d^2y}{dx^2} = -\frac{p_1(x)}{p_0(x)} \exp\left[-\int \frac{p_1(x)}{p_0(x)} dx\right] \frac{dy}{dt} + \exp\left[-\int \frac{p_1(x)}{p_0(x)} dx\right] \frac{d^2y}{dt^2}.$$

Substituting these expressions in Eq.(1) then yields, after some minor algebraic calculations,

$$y''(t) + g(t)y(t) = 0, \tag{7}$$

where $g(t)$ is a function connected with all the $p_j(x)$. This equation does not contain the first derivative.

3. By making change of variable

$$t = \int \left(\frac{p_n(x)}{p_0(x)} \right)^{1/n} dx$$

in Eq.(1) we obtain, after dividing by $p_0(x)$ and some calculations,

$$y^{(n)}(t) + g_1(t)y^{(n-1)}(t) + \cdots + g_{n-1}(t)y'(t) + y(t) = 0, \tag{8}$$

where $g_i(t)$, $i = 1, \ldots, n-1$ are some functions connected with all the functions $p_j(x)$ and their derivatives. In Eq.(8) the coefficients at $y(t)$ and $y^{(n)}(t)$ are unity.

4. The adjoint equation of Eq.(1) is an equation of the form

$$(-1)^n(p_0 z)^{(n)} + (-1)^{n-1}(p_1 z)^{(n-1)} + \cdots - (p_{n-1}z)' + p_n z = 0. \tag{9}$$

This equation is used for solving of a linear boundary value problem (see Sec-s 3.2 and 3.3).

Example 9. Find the solution of the differential equation

$$y'' \sin^2 2x + y' \sin 4x - 4y = 0.$$

By making change of variable in the given equation

$$t = \int \exp\left[-\int \frac{\sin 4x}{\sin^2 2x} dx \right] dx = \frac{1}{2} \ln |\tan x|$$

we find

$$y''(t) - 4y(t) = 0.$$

The general solution of this equation is

$$y(t) = C_1 e^{2t} + C_2 e^{-2t},$$

where C_i are arbitrary constants. Hence the general solution of the given equation is

$$y(x) = C_1 \tan x + C_2 \cot x.$$

Example 10. Find the adjoint equation of

$$y'' - \left(2 + 3\cot^2 2x\right) y = 0.$$

According to Eq.(9) we obtain

$$z''(x) - \left(2 + 3\cot^2 2x\right) z(x) = 0.$$

Hence the given equation is self-adjoint.

2.2.4 The Initial Value Problem

The Eq.(1) with the initial conditions at the point x_0

$$y^{(k)}(x_0) = y_k, \quad k = 0, 1, \ldots, n-1, \tag{10}$$

where y_k are some prescribed numbers, has a solution given by Eq.(2), if and only if the arbitrary constants C_i satisfy the linear nonhomogeneous algebraic system

$$\sum_{i=1}^{n} C_i y_i^{(k)}(x_0) = y_k, \quad k = 0, 1, \ldots, n-1. \tag{11}$$

This system is always solvable for the constants C_i, since its determinant is Wronskian and does not vanish at the point $a \le x_0 \le b$. By using the *Cramer's rule* we find

$$C_i = \frac{W_i(x_0)}{W(x_0)}, \quad i = 1, 2, \ldots, n, \tag{12}$$

where $W_i(x_0)$ is the determinant obtained from the Wronskian $W(x_0)$ by replacing the i-th column by the column $(y_0, y_1, \ldots, y_{n-1})^T$. The Eq-s(12) we can write in the matrix form

$$\vec{C} = \Psi^{-1}(x_0)\vec{y}_0, \tag{13}$$

where $\vec{C}$ is a constant column vector with the components C_1, C_2,..., C_n, $\Psi^{-1}(x_0)$ is the inverse matrix of $\Psi(x_0)$ and $\vec{y}_0$ is a constant column vector with the components y_0, y_1, ..., y_{n-1}. Then the solution of the initial value problem is

$$y(x) = \vec{y}_h(x)\Psi^{-1}(x_0)\vec{y}_0, \tag{14}$$

where $\vec{y}_h(x)$ is a row vector with the components $y_1(x)$, $y_2(x)$, ..., $y_n(x)$.

Example 11. Solve the initial value problem

$$x(x-1)y'' - \left(x^2 - 2x - 1\right)y' - (x+1)y = 0,$$

$$y = 5, \quad y' = -1 \quad \text{at} \quad x = 2.$$

Proceeding as in Ex-s 3 and 4 we find, after some minor calculations, one solution $y_1 = \exp(x)$. Further, by using reduction of order we obtain the second solution $y = 1/(x-1)$. Hence the general solution of the given equation is

$$y = C_1 e^x + C_2 \frac{1}{x-1},$$

where C_i are arbitrary constants. Now according to Eq.(11) we obtain the algebraic system for these constants

$$C_1 e^2 + C_2 = 5,$$
$$C_1 e^2 - C_2 = -1.$$

Solution of this system is $C_1 = 2e^2$ and $C_2 = 3$. Hence the solution of the given initial value problem is

$$y(x) = 2e^{x-2} + \frac{3}{x-1}.$$

Note, that a solution of the given initial value problem does not exist when initial conditions are given in the singular point $x^* = 1$.

PROBLEMS

Find the linear homogeneous differential equations which have the following solutions.

1. $y_1 = \cos x, \quad y_2 = \sin x.$
2. $y_1 = \cosh x, \quad y_2 = \sinh x.$
3. $y_1 = \cos x, \quad y_2 = \sin x, \quad y_3 = \cosh x, \quad y_4 = \sinh x.$
4. $y_1 = e^x \cos x, \quad y_2 = e^x \sin x, \quad y_3 = e^{-x}.$
5. $y_1 = \cos x, \quad y_2 = e^x, \quad y_3 = e^{-x}.$
6. $y_1 = e^x, \quad y_2 = x, \quad y_3 = \ln|x|.$
7. $y_1 = \tanh x, \quad y_2 = \coth x.$
8. $y_1 = e^x - 1, \quad y_2 = e^x - 1 - x, \quad y_3 = e^x - 1 - x - \frac{x^2}{2}.$
9. $y_1 = x, \quad y_2 = x^2, \quad y_3 = e^x.$
10. $y_1 = x, \quad y_2 = e^x, \quad y_3 = e^{-x}.$

Verify whether the following functions are linearly independent in the prescribed point, or not.

11. $1, \quad x, \quad x^2, \quad x^3, \quad \text{at} \quad x_0 = 1.$
12. $1, \quad x, \quad x^2, \quad x^3, \quad \text{at} \quad x_0 = 0.$
13. $(1+x)^2, \quad (1-x)^2, \quad \text{at} \quad x_0 = 5.$
14. $e^x, \quad e^{2x}, \quad e^{3x}, \quad \text{at} \quad x_0 = 0.$
15. $1, \quad \ln x, \quad \ln x^2, \quad \ln x^3, \quad \text{at} \quad x_0 = 3.$
16. $0, \quad \cos x, \quad \sin x, \quad x, \quad \text{at} \quad x_0 = 2.$
17. $1, \quad x, \quad x\ln x, \quad \text{at} \quad x_0 = e.$
18. $x, \quad |x|, \quad x - \sqrt{x^2}, \quad \text{at} \quad x_0 = 2.$
19. $\sin x, \quad \sin(x+1), \quad \sin(x+2), \quad \text{at} \quad x_0 = 1.$
20. $\sin x, \quad \sin 2x, \quad \sin 3x, \quad \text{at} \quad x_0 = 1.$

Assuming that at least one solution of the following differential equations has a power or an exponential form, solve them.

21. $(x-1)\,y'' - xy' + y = 0.$
22. $x\,(3x^2 + 4x - 1)\,y'' + 2\,(2x-1)\,y' - 2\,(3x+2)\,y = 0.$
23. $(2x-1)\,y'' - 2\,(4x-1)\,y' + 8xy = 0.$
24. $(x-1)\,y'' - (x+1)\,y' + 2y = 0.$
25. $x\,(x-4)\,y'' - 2\,(x-2)\,y' + 2y = 0.$
26. $xy'' + (x-1)\,y' - y = 0.$
27. $x^3 y''' + x^2 y'' - 2xy' + 2y = 0.$
28. $(x^2+2)\,y''' - (x^2+2x+2)\,y'' + 2\,(x+1)\,y' - 2y = 0.$

29. $x^2 \ln(x)\, y'' - 2x(\ln x + 1)\, y' + 2y = 0.$
30. $(2x + (2 - x^2)\cot x)\, y''' - x^2 y'' + 2xy' - 2y = 0.$
31. $(2x^2 - x - 1)\, y'' + (4x^2 - 4x + 3)(xy' - y) = 0.$
32. $x(x - 2)\, y'' + (x - 4)\, y' - y = 0.$
33. $(3x^3 - 1)\, y'' - 3x(3x^3 + 2)(xy' - y) = 0.$
34. $x(x^2 + 6)\, y'' - 4(x^2 + 3)\, y' + 6xy = 0.$
35. $x(x - 1)\, y'' + (x + 1)\, y' - y = 0.$
36. $(x - 1)\, y'' - xy' - 2(x - 2)\, y = 0.$
37. $2x(x + 2)\, y'' + (2 - x)\, y' + y = 0.$
38. $x(x - 1)\, y'' - (x - 1)\, y' + y = 0.$
39. $(4x^3 - 6x^2 + 2x - 1)\, y''' - (8x^4 - 8x^3 + 10x^2 - 12x + 1)\, y''$
$+2x(4x^3 - 2x^2 + 6x - 1)\, y' - 2(4x^3 - 2x^2 + 6x - 1)\, y = 0.$
40. $(x^2 + 1)\, y'' - 2y - 0.$

By a suitable substitution transform each of the following differential equations to an equation which does not contain the $(n-1)$-th derivative.

41. $y''' + 2y'' - 3y' + 4y = 0.$
42. $y'' - \tan(x)\, y' + \cos(x)\, y = 0.$
43. $y'' + (x - 5)\, y' - (x^2 + 1)\, y = 0.$
44. $x^2 y'' - 2xy' + (x^2 + 2)\, y = 0.$
45. $y'' - 2y' + 7y = 0.$
46. $y''' - xy'' - 2xy' + x^3 y = 0.$
47. $y'' + \frac{y'}{x} - \frac{y}{x^2} = 0.$
48. $(1 + x^2)\, y'' + 4xy' + 2y = 0.$

By making a suitable change of variable transform each of the following differential equations to an equation which does not contain the first derivative.

49. $y'' + 2y' - y = 0.$
50. $y'' + \frac{y'}{x} - \frac{y}{x^2} = 0.$
51. $(x^2 + 1)\, y'' + xy' + y = 0.$
52. $y'' - \cot(x)\, y' + \cos(x)\, y = 0.$
53. $y'' + \frac{y'}{x} + x^2 y = 0.$
54. $x^2(1 - x^2)\, y'' + 2x(1 - x^2) - 2y = 0.$

By making a suitable change of variable transform each of the following differential equations to an equation with the unity coefficients at $y^{(n)}(t)$ and $y(t)$.

55. $(1 - x^2)\, y'' - xy' + y = 0.$
56. $y''' - 2xy'' + 4x^2 y' + 8x^3 y = 0.$
57. $y'' + x(1 - x)\, y' + e^x y = 0.$
58. $x^2 y'' + 2xy' + 4y = 0.$
59. $x^4 y^{iv} - x^2 y'' + y = 0.$
60. $(1 + x^2)\, y'' + xy' + y = 0.$

Find the adjoint equation for the following differential equations.

61. $y'' - y' - y = 0.$
62. $x^3 y''' + 2x^2 y'' - 3xy' + 4y = 0.$
63. $y'' + xy' + x^2 y = 0.$
64. $(1 - x^2)\, y'' - xy' + y = 0.$
Solve each of the following initial value problems.
65. $x\,(x+1)\, y'' + 2\,(x+1)\, y' - 2y = 0, \quad y\,(1) = 1, \quad y'\,(1) = 0.$
66. $x\,(x+1)\, y'' + 2\,(x+1)\, y' - 2y = 0, \quad y\,(0) = 1, \quad y'\,(0) = 0.$
67. $(1 + x^2)\, y'' + 2xy' - 2y = 0, \quad y\,(0) = 1, \quad y'(0) = 0.$
68. $(1 + x^2)\, y'' + 2xy' - 2y = 0, \quad y\,(0) = 0, \quad y'\,(0) = 1.$
69. $x^2 \ln(x)\, y'' - xy' + y = 0, \quad y\,(e) = 1, \quad y'\,(e) = 3.$
70. $(1 - x^2)\, y'' - 2xy' + 2y = 0, \quad y\,(0) = 1, \quad y'\,(0) = 3.$
71. $x^2 y'' - x\left(1 - \frac{1}{2}\ln|x|\right) y' + \left(1 - \frac{1}{2}\ln|x|\right) y = 0,$
$y\,(e^{-1}) = 0, \; y'\,(e^{-1}) = 1.$
72. $xy''' - y'' - xy' + y = 0, \quad y\,(0) = 0, \quad y'(0) = 1, \quad y''(0) = 0.$
73. $x^2\,(2x-1)\, y''' + (4x-3)\, xy'' - 2xy' + 2y = 0, \quad y\,(1) = 1,$
$y'\,(1) = 1, \quad y''\,(1) = 0.$
74. $(x^2 - 2x + 3)\, y''' - (x^2 + 1)\, y'' + 2xy' - 2y = 0,$
$y\,(0) = 0, \quad y'\,(0) = 0, \quad y''\,(0) = 1.$
Find the normalized fundamental set of solutions in the prescribed point for the following differential equations.
75. $(x+1)\, y'' + xy' - y = 0$ at point $x_0 = 0.$
76. $(x+1)\, y'' + xy' - y = 0$ at point $x_0 = 1.$
77. $xy'' - (x+1)\, y' - 2\,(x-1)\, y = 0$ at point $x_0 = 0.$
78. $(x-2)\, y'' - xy' + 2y = 0$ at point $x_0 = 0.$

2.3 Linear nonhomogeneous equations

The general solution on the interval $x \in [a, b]$ of the linear nonhomogeneous n-th order differential equation

$$p_0(x) y^{(n)}(x) + p_1(x) y^{(n-1)}(x) + \cdots + p_n(x) y(x) = f(x) \qquad (1)$$

with continuous (on the same interval) coefficients $p_i(x)$ and with right side $f(x)$ is equal to the sum of a general solution of the corresponding homogeneous equation and of some particular solution $y_p(x)$ of the nonhomogeneous equation

$$y(x) = \sum_{i=1}^{n} C_i y_i(x) + y_p(x). \qquad (2)$$

We assume that the fundamental set of solutions of the corresponding homogeneous equation is known (see Sec. 2.2).

The particular solution of Eq.(1) can be found by using any of the following methods:

1) method of variation of parameters, so called Lagrange's method;

2) method of undetermined coefficients;

3) method of influence function, so called Cauchi's method, after the French mathematician O. Cauchi (1789-1857).

2.3.1 Method of Variation of Parameters

A particular solution of Eq.(1) has ben founding in the form

$$y_p(x) = \sum_{i=1}^{n} C_i(x) y_i(x), \tag{3}$$

where $y_i(x)$ are solutions of the corresponding homogeneous equation and $C_i(x)$ are solutions of the system of linear nonhomogeneous first order differential equations

$$\begin{aligned} &\sum_{i=1}^{n} C_i'(x) y_i(x) = 0, \\ &\sum_{i=1}^{n} C_i'(x) y_i'(x) = 0, \\ &\vdots \\ &\sum_{i=1}^{n} C_i'(x) y_i^{(n-2)}(x) = 0, \\ &\sum_{i=1}^{n} C_i'(x) y_i^{(n-1)}(x) = \frac{f(x)}{p_0(x)}. \end{aligned} \tag{4}$$

The determinant of the coefficients in this system is Wronskian $W(x)$ and is nowhere zero on the interval $[a, b]$. Using the Cramer's rule we find the solution of Eq.(4)

$$C'(x) = \frac{W_i(x)}{W(x)},$$

where $W_i(x)$ is the determinant obtained from $W(x)$ by replacing the i-th column by the column $(0, 0, \ldots, f(x)/p_0(x))^T$

$$W_i(x) = \begin{vmatrix} y_1 & \cdots & y_{i-1} & 0 & y_{i+1} & \cdots & y_n \\ y_1' & \cdots & y_{i-1}' & 0 & y_{i+1}' & \cdots & y_n' \\ \vdots & \ddots & \vdots & \vdots & \vdots & \ddots & \vdots \\ y_1^{(n-1)} & \cdots & y_{i-1}^{(n-1)} & \frac{f}{p_0} & y_{i+1}^{(n-1)} & \cdots & y_n^{(n-1)} \end{vmatrix}. \tag{5}$$

Hence

$$C_i = \int^x \frac{W_i(s)}{W(s)} ds + \tilde{C}_i, \tag{6}$$

where $\tilde{C}_i$ are arbitrary constants. We put $\tilde{C}_i = 0$. Substituting Eq-s(5) and (6) in Eq.(3), expanding all determinants $W_i(s)$ along the i-th column and assuming that the obtained sum is the expansion of a determinant $W(x,s)$ along the n-th row we obtain, after some algebraic calculations,

$$y_p(x) = \int^x \frac{f(s)W(x,s)}{p_0(s)W(s)}ds, \tag{7}$$

where $W(x,s)$ is the determinant, obtained from $W(s)$ by replacing the n-th row by the row $(y_1(x), y_2(x), \ldots, y_n(x))$

$$W(x,s) = \begin{vmatrix} y_1(s) & y_2(s) & \ldots & y_n(s) \\ y_1'(s) & y_2'(s) & \ldots & y_n'(s) \\ \vdots & \vdots & \ddots & \vdots \\ y_1^{(n-2)}(s) & y_2^{(n-2)}(s) & \ldots & y_n^{(n-2)}(s) \\ y_1(x) & y_2(x) & \ldots & y_n(x) \end{vmatrix}. \tag{8}$$

Hence the general solution of Eq.(1) is

$$y(x) = \sum_{i=1}^{n} C_i y_i(x) + \int^x \frac{f(s)W(x,s)}{p_0(s)W(s)}ds. \tag{9}$$

Note that in this method we should calculate two determinants $W(s)$ and $W(x,s)$.

Example 1. Find the general solution of the differential equation

$$y'' + 4y' - 5y = 3\sin x.$$

It is easy to verify that the solutions of the corresponding homogeneous equation are $y_1 = \exp(x)$, $y_2 = \exp(-5x)$. Further, according to Eq-s(7) -(8) we find

$$W(s) = \begin{vmatrix} e^s & e^{-5s} \\ e^s & -5e^{-5s} \end{vmatrix} = -6e^{-4s},$$

$$W(x,s) = \begin{vmatrix} e^s & e^{-5s} \\ e^x & e^{-5x} \end{vmatrix} = e^{-5x+s} - e^{-5s+x},$$

and

$$y_p(x) = 3\int^x \frac{(e^{-5x+s} - e^{-5s+x})\sin s}{-6e^{-4s}}ds = -\frac{9}{26}\sin x - \frac{3}{13}\cos x.$$

Hence the general solution of the given equation is

$$y(x) = C_1e^x + C_2e^{-5x} - \frac{9}{26}\sin x - \frac{3}{13}\cos x,$$

where C_i are arbitrary constants.

Example 2. Find the particular solution of the differential equation

$$\left(x^2+2\right)y''' - \left(x^2+2x+2\right)y'' + 2(x+1)y' - 2y = \left(x^2+2\right)^2,$$

if the solutions of the corresponding homogeneous equation are $y_1 = x+1$, $y_2 = x(x+2)$, $y_3 = \exp(x)$.

According to Eq-s(7)-(9) we find

$$W(s) = e^s\left(s^2+2\right),$$

$$W(x,s) = e^x\left(s^2+2s+2\right) + e^s\left((s-2)(x+1) - sx(x+2)\right)$$

and

$$y_p(x) = \int^x \frac{(s^2+2)W(x,s)}{W(s)}\,ds = -\left(\frac{x^4}{6} + \frac{2x^3}{3} + 3x^2 + 6x + 6\right).$$

2.3.2 Method of Undetermined Coefficients

By this method we may find a particular solution of Eq.(1) when the fundamental set of solution of one is unknown. However, it is useful only when the function $f(x)$ and coefficients $p_i(x)$ have an appropriate simple form, namely:

1) all $p_i(x)$ are some constants and $f(x)$ is any of the following functions

a) $Q_m(x) = q_m x^m + q_{m-1}x^{m-1} + \cdots + q_0$;

b) $\exp(ax)$;

c) $Q_m(x)\exp(ax)$;

d) a linear combination of these functions.

Here all the q_i are some constants, m is an integer number and a is a complex constant.

2) all $p_i(x)$ and $f(x)$ are polynomials or other simple functions.

a) See Sec. 2.4

b) If each of the coefficients $p_i(x)$, $i = 0, 1, \ldots, n$ is a polynomial of degree k_i and the function $f(x)$ is a polynomial of degree r, then the particular solution of Eq.(1) may exist as a polynomial of the degree $l = r - \max(k_i + i - n) > 0$

$$y_p(x) = a_l x^l + a_{l-1}x^{l-1} + \cdots + a_0, \tag{10}$$

where a_i are undetermined constant coefficients. This coefficients can be determined by substituting the polynomial (10) for $y_p(x)$ into Eq. (1).

If $p_i(x)$ and $f(x)$ are more complicated functions we can choose $y_p(x)$ in the form of a suitable combination of these functions and their products with some undetermined coefficients. Then by substituting $y_p(x)$ in the given equation we can find these coefficients.

Examples 3. Find the particular polynomial solution of the nonhomogeneous differential equation

$$(x^9 + 2x^6 + 3)\, y''' + (x^3 - 2x^2 + 1)\, y'' - (x^8 + 3x - 1)\, y' \\ - (17x^2 + 3)\, y = x^4 - 8.$$

First we determine the degree of the particular polynomial solution $l = 4 - \max(k_i + i - 3) = 4 - (8 + 2 - 3) = -3 < 0$. Hence the particular polynomial solution of the given equation does not exist.

Examples 4. Find the particular polynomial solution of the nonhomogeneous differential equation

$$\left(x^3 - 1\right) y''' + \left(x^2 - 1\right) y'' - 2(x - 1)y' + 2y = x^4 - 1.$$

Proceeding as in Ex.3 we find $l = 4 - \max(k_i + i - 3) = 4 > 0$. Hence the particular polynomial solution of the given equation may exist in the form

$$y_p(x) = ax^4 + bx^3 + cx^2 + dx + e$$

where a, b, c, d, e are undetermined constant coefficients. A substitution $y_p(x)$ in the given equation then yields

$$\left(x^3 - 1\right)(24ax + 6b) + \left(x^2 - 1\right)\left(12ax^2 + 6bx + 2c\right) \\ +2(x - 1)\left(4ax^3 + 3bx^2 + 2cx + d\right) \\ +2\left(ax^4 + bx^3 + cx^2 + dx + e\right) = x^4 - 1.$$

Equating the coefficients of like powers of x we obtain

$$\begin{aligned} &30a = 1, \\ &b + a = 0, \\ &-2a + b = 0, \\ &-12a - 3b + 2c = 0, \\ &-6b - 2c + 2d + 2e = -1. \end{aligned}$$

This system is uncompatible. Hence the particular polynomial solution of the given equation does not exist.

Examples 5. Find the particular polynomial solution of the nonhomogeneous differential equation

$$\left(x^3 - 1\right) y''' + \left(x^2 - 1\right) y'' - 2(x - 1)y' + 2y = 30x^4 - 3x^2.$$

In a way analogous to Ex.4 we obtain the system

$$\begin{aligned} &30a = 30, \\ &b + a = 0, \\ &-2a + b = -3, \\ &-12a - 3b + 2c = 0, \\ &-6b - 2c + 2d + 2e = 0. \end{aligned}$$

This system is compatible and has the solution $a = 1, b = -1, c = 9/2, d + e = 3/2$. Since the corresponding homogeneous equation has the polynomial solution $y_1(x) = x - 1$ (see Sec. 2.2, Ex. 4) then the particular polynomial solution obtained should contain an arbitrary constant. Let $e = 0$, then $d = 3/2$ and particular polynomial solution of the given nonhomogeneous equation is

$$y_p(x) = x^4 - x^3 + \frac{9}{2}x^2 + \frac{3}{2}x.$$

Examples 6. Find the particular solution of the nonhomogeneous differential equation

$$xe^x y'' + e^{-x} y' + x^2 y = x\left(4\cosh x + x^3\right).$$

Since coefficients $p_i(x)$ and $f(x)$ in this equation are polynomial and exponential in character, we chose a particular solution in the form

$$y_p = ax^3 + bx^2 + cx + d + e^{kx}\left(Ax^3 + Bx^2 + Cx + D\right),$$

where A, B, $\ldots, d$, k are undetermined coefficients. By substituting this expression in the given equation and equating the coefficients of like terms of x in the both sides we find, after some minor calculations, $k = 0$, $a = c = d = 0$, $b = 1$, $A = B = C = D = 0$. Hence the particular solution of the given nonhomogeneous equation is

$$y_p(x) = x^2.$$

2.3.3 *The Influence Function*

Here it is assumed that we know the solution $K(x, s)$ (dependent on a single parameter s) of the corresponding homogeneous equation, which satisfies the conditions

$$K(x,s) = 0, \quad K'(x,s) = 0, \quad \ldots, \quad K^{(n-2)}(x,s) = 0, \quad K^{(n-1)}(x,s) = 1 \qquad (11)$$

at $x = s$. In this case the particular solution is

$$y_p(x) = \int_{x_0}^{x} K(x,s)\frac{f(s)}{p_0(s)}\,ds, \qquad (12)$$

where $x_0 \in [a, b]$. When the fundamental set of solutions of the corresponding homogeneous equation is known, $K(x, s)$ can be determined as

$$K(x,s) = \sum_{i=1}^{n} C_i(s) y_i(x), \qquad (13)$$

where coefficients $C_i(s)$ are solutions of the system of linear algebraic nonhomogeneous equations (compare with Eq.(4))

$$\begin{aligned}
&\sum_{i=1}^{n} C_i(s)y_i(s) = 0,\\
&\sum_{i=1}^{n} C_i(s)y_i'(s) = 0,\\
&\vdots\\
&\sum_{i=1}^{n} C_i(s)y_i^{(n-2)}(s) = 0,\\
&\sum_{i=1}^{n} C_i(s)y_i^{(n-1)}(s) = 1,
\end{aligned} \tag{14}$$

This system is compatible and it has an unique solution, since its determinant is Wronskian of the given equation. Further, proceeding as it has been shown above we find

$$K(x,s) = \frac{W(x,s)}{W(s)}.$$

Then all conditions (11) are satisfy.

Example 7. Find the particular solution of the nonhomogeneous equation

$$y'' - y = f(x).$$

The general solution of the corresponding homogeneous equation is $y(x) = C_1 \cosh x + C_2 \sinh x$. Hence $K(x,s)$ can be written as $K(x,s) = C_1(s)\cosh x + C_2(s)\sinh x$, where functions $C_i(s)$ are solution of the system

$$\begin{aligned}
C_1(s)\cosh s + C_2(s)\sinh s = 0,\\
C_1(s)\sinh s + C_2(s)\cosh s = 1.
\end{aligned}$$

Solution of this system is $C_1 = -\sinh s$, $C_2 = \cosh s$. Further using Eq.(13) we obtain $K(x,s) = \sinh(x-s)$. Now the particular solution of the given equation is

$$y_p(x) = \int_{x_0}^{x} f(s)\sinh(x-s)\,ds,$$

where x_0 is a constant.

2.3.4 The Initial Value Problem

The solution of the initial value problem

$$\begin{aligned}
&y^{(n)}(x) + p_1(x)y^{(n-1)}(x) + \ldots + p_n(x)y(x) = f(x)\\
&y(x_0) = y_0, \quad y'(x_0) = y_1, \ldots, y^{(n-1)}(x_0) = y_{n-1},
\end{aligned} \tag{15}$$

where y_i are prescribed numbers, can be written in the form of Eq.(2), where coefficients C_i are solutions of the system of linear nonhomogeneous algebraic equations

$$\sum_{i=1}^{n} C_i y_i^{(k)}(x_0) + y_p^{(k)}(x_0) = y_k, \qquad k = 0, \ldots, n-1. \tag{16}$$

This system is compatible and has an unique solution, since its determinant is Wronskian of the given equation. Further proceeding as in Sec. 2.2 we obtain again those results, but we should write $y_k - y_p^{(k)}(x_0)$ instead of y_k

$$y(x) = \vec{y}_h(x)\Psi^{-1}(x_0)\vec{z}_0 + y_p(x), \tag{17}$$

where $\vec{y}_h(x)$ is a row vector with the components $y_1(x)$, $y_2(x)$,..., $y_n(x)$, $\Psi^{-1}(x_0)$ is the inverse matrix of $\Psi(x_0)$, $\vec{z}_0$ is a column vector with the components $y_0 - y_p(x_0)$, $y_1 - y_p'(x_0)$,...,$y_{n-1} - y_p^{(n-1)}(x_0)$.

Notice that when a particular solution of the nonhomogeneous equation is given by Eq.(12) then according to Eq.(11) $y_p^{(k)}(x_0) = 0$ at all k, and we obtain again exactly those formulas as in Sec. 2.2

$$y(x) = \vec{y}_h(x)\Psi^{-1}(x_0)\vec{y}_0 + y_p(x), \tag{18}$$

where $\vec{y}_0$ is a column vector with the components y_0, y_1,...,y_{n-1}.

Example 8. Solve the initial value problem

$$y'' + y = \cos 2x,$$

$$y(0) = 1, \quad y'(0) = 1.$$

The general solution of the corresponding homogeneous equation is $y = C_1 \cos x + C_2 \sin x$. Hence the function of influence is $K(x,s) = \sin(x-s)$ (see Ex. 6). Now according to Eq.(16) we obtain $C_1 = 1$, $C_2 = 1$. Hence the solution of the given initial value problem is

$$y(x) = \cos x + \sin x + \int_0^x \cos 2s \sin(x-s)\, ds = \frac{4}{3}\cos x + \sin x - \frac{1}{3}\cos 2x.$$

PROBLEMS

Find the general solution of the following nonhomogeneous differential equations.

1. $y'' + xy' - y = 2xe^x - 1$.
2. $xy'' + xy' - y = x^2 + 2x$.
3. $x^2y'' + xy' - y = x^2 \sin x$.
4. $x^3y'' + xy' - y = \cos\left(\frac{1}{x}\right)$.
5. $x\,(x+1)\,y'' + (x+2)\,y' - y = x + \frac{1}{x}$.
6. $2xy'' + (x-2)\,y' - y = x^2 - 1$.

7. $x^2(x+1)y'' + x(4x+3)y' - y = x + \dfrac{1}{x}$.

8. $x^2(\ln x - 1)y'' - xy' + y = x(1-\ln x)^2$.

9. $xy'' + 2y' + xy = \sec x$.

10. $(1-x^2)y'' - xy' + \dfrac{1}{4}y = \dfrac{1}{2}(1-x^2)$.

11. $(\cos x + \sin x)y'' - 2\cos x\ y' + (\cos x - \sin x)y = (\cos x + \sin x)^2 e^{2x}$.

12. $(\sin x - \cos x)y'' - 2\sin x\ y' + (\cos x + \sin x)y = (\sin x - \cos x)^2$.

13. $(1-x)y'' + (1+x)y' - 2y = (1-x)^3$.

14. $x(x-3)y'' - (x^2-6)y' + 3(x-2)y = (x-3)^2$.

15. $(1-x^2)y'' - xy' + y = x^2$.

16. $(x^2-1)y'' + 4xy' + 2y = 6x$.

17. $x^3(x-1)y'' + x(2x^2-2x-1)y' - y = \dfrac{(x-1)^2}{x}$.

18. $xy'' - (1+2x^2)y' = 4x^3 e^{x^2}$.

19. $x(2-x)y'' + (x^2-2)y' + 2(1-x)y = x^2(2-x)^2 e^x$.

20. $x(x-1)y'' + (1+x)y' - y = x^2 - x + 1$.

21. $(3x^3+x)y'' + 2y' - 6xy = 4 - 12x^2$.

22. $x^2y'' + 4xy' + 2y = x^2 + 2x - 1$.

23. $(2x+1)y'' + (2x-1)y' - 2y = x^2 + x$.

24. $x^3y''' - 3xy'' + 6xy' - 6y = \dfrac{x^4}{\sqrt{x}+1}$.

25. $(x^2+2)y''' - (x^2+2x+2)y'' + 2(x+1)y' - 2y = x^4 + 4x^2 + 4$.

26. $(x^2-2x+2)y''' - x^2y'' + 2xy' - 2y = \left(x - 2 + \dfrac{2}{x}\right)^2$.

27. $(x^2+2)y^{iv} - 2xy''' - x^2y'' + 2xy' - 2y = x + \dfrac{4}{x}\left(1 + \dfrac{1}{x^2}\right)$.

28. $x^2y''' + 4xy'' + 2y' = 2\dfrac{1-3x^2}{(1+x^2)^3}$.

29. $(x^2+2x+2)y^{iv} - 2(x+1)y''' - x(x+2)y'' + 2(x+1)y' - 2y$
$= -6x(2x^4 + 5x^3 - 20x - 40)$.

30. $xy''' - y'' - xy' + y = \dfrac{1}{x}(x^2+3)$.

31. $x(x^2+x-1)y''' - (x^3-3x-3)y'' - x(x+3)y' + (x+3)y$
$= x(x^2-4)$.

32. $(4x^3 - 6x^2 + 2x - 1)y''' - (8x^4 - 8x^3 + 10x^2 - 12x + 1)y''$
$+2x(4x^3 - 2x^2 + 6x - 1)y' - 2(4x^3 - 2x^2 + 6x - 1)y$
$= 2(4x^5 - 10x^4 + 14x^3 - 11x^2 + 12x - 1)$.

Find the particular solution of the following differential equations by using the method of undetermined coefficients.

33. $x(x+2)y''' + (x^2-2x-2)y'' - 3(x+1)y' - 2y$
$= -\dfrac{1}{x} + \dfrac{1}{x^2} - \dfrac{8}{x^3}$.

34. $2xy''' - 4y'' + y' - \dfrac{y}{x} = 1 - \dfrac{6}{x}$.
35. $e^{-x}y''' - e^{-x}y'' - \sinh x\; y' + y = 1 - e^{x}(x + (x+1)\sinh x)$.
36. $x(x-2)y''' - 3xy'' + (2x^2-1)y' - 2x^2y = -x(x^3+4)$.
37. $(3x+2x^2)y'' - 6(1+x)y' + 6y = 6$.
38. $(1+x^2)y'' + xy' - y = -1$.
39. $x^2y'' - xy' - 3y = 5x^4$.
40. $(x-1)y'' - xy' + y = (x-1)^2e^x$.
41. $y'' + y' + e^{-2x}y = e^{-3x}$.
42. $y'' + y' + e^{2x}y = xe^{2x} - 1$.
43. $x(x-1)y'' - (2x-1)y' + 2y = x^2(2x-3)$.
44. $(x-3)y'' - xy' + 3y = -\frac{x^2}{2} - x^3$.
45. $(x^2-3x+1)y'' + (2x+5)y' - 12y = 3x(2-x)$.
46. $(2x + (2-x^2)\tan x)y''' - x^2y'' + 2xy' - 2y = (x-2)\sin x$.
47. $2x(x+2)y'' + (2-x)y' - 4y$
$= (x-2)(15x^2 + 24x(1+\ln(x-2)) + 4)$.
48. $(x^2+1)y'' + 2y = \dfrac{8x^2}{(x^2+1)^2}$.
49. $x(x-1)y'' + (x+1)y' + 2y = 11x^3 + 21x^2 - 6x + 4$.
50. $x^3y''' - 3xy'' + 6xy' - 6y = x^{7/2}$.
51. $(x^9+2x^6+3)y''' + (x^3-2x^2+1)y'' - (x^8+3x-1)y'$
$-(17x^2+3)y = -3x^{10} + 6x^9 + 12x^6 - 17x^5 + 6x^4$
$-24x^3 + 3x^2 + 6x + 18$.
52. $6y''' - xy'' - 2x^2y' - 3x^3y = 17xe^{x^2/2}$.

Find the influence function of the following differential equations.

53. $y'' + 4y' + 4y = 0$.
54. $y'' + \omega^2y = 0$, where ω is a constant.
55. $y'' - y' + e^{2x}y = 0$.
56. $x^8y'' + (6x^4 + 4x^7)y' + 9y = 0$.
57. $x^2y'' - 3xy' + 4y = 0$.
58. $x^2y'' + xy' + y = 0$.
59. $(1+x^2)^2y'' + 2x(1+x^2)y' + y = 0$.
60. $y'' + (3e^{-x}+1)y' + 2e^{-2x}y = 0$.
61. $y''' + y'' = 0$.
62. $x^3y''' + 5x^2y'' + 2xy' - 2y = 0$.

Solve each of the following initial value problems.

63. $(x+1)y'' + xy' - y = 1 + x + \dfrac{1}{x}$, $\quad y(1) = 1$, $\quad y'(1) = 0$.
64. $x^2y'' + xy' - y = x^2$, $\quad y(-1) = 0$, $\quad y'(-1) = 2$.
65. $x(x-1)y'' + (x+1)y' - y = \dfrac{1}{x^2}$, $\quad y(-1) = 0$, $\quad y'(-1) = 0$.
66. $xy'' - (10x^2+1)y' + 24x^3y = x^3(6x^2-5)$, $\quad y(1) = 1$, $\quad y'(1) = 0$.
67. $x^4y'' + x^2(2x-5)y' + 6y = e^{1/x}$, $\quad y(1) = 0$, $\quad y'(1) = 0$.

68. $y'' - 2\left(3e^{2x} + 1\right)y' - 8e^{4x}y = 4e^{6x}, \quad y(0) = \frac{5}{4}, \quad y'(0) = 1.$

2.4 Linear equation with constant coefficients

2.4.1 The Homogeneous Equation with Constant Coefficients

The n-th order linear homogeneous differential equation with *constant coefficients* is

$$L[y] = a_0 y^{(n)} + a_1 y^{(n-1)} + \cdots + a_{n-1}y' + a_n y = 0, \tag{1}$$

where $a_0,\ a_1,\ \cdots,\ a_n$ are real constants. Solution of (1) can be found by trying the *exponential substitution*

$$y = e^{\lambda x}, \tag{2}$$

where λ is a real or complex constant. Substituting into (1) and observing that

$$y^{(k)} = \lambda^k e^{\lambda x}$$

we have

$$L\left[e^{\lambda x}\right] = \left(a_0\lambda^n + a_1\lambda^{n-1} + \cdots + a_{n-1}\lambda + a_n\right)e^{\lambda x} = 0$$

after canceling $e^{\lambda x}$

$$f(\lambda) = a_0\lambda^n + a_1\lambda^{n-1} + \cdots + a_{n-1}\lambda + a_n = 0. \tag{3}$$

We have a solution, provided λ is a root of (3), which is called the *auxiliary equation* or *characteristic equation*. This equation has n roots $\lambda_1,\ \lambda_2,\ \ldots,\ \lambda_n$. Each gives a solution $e^{\lambda x}$.

Real Distinct Roots. If the roots of (3) are real and distinct, we have in this way n real and linearly independent solutions. The general solution then is

$$y = C_1 e^{\lambda_1 x} + C_2 e^{\lambda_2 x} + \cdots + C_n e^{\lambda_n x} \tag{4}$$

where $C's$ are arbitrary constants.

Example 1. Solve the differential equation

$$y'' + y' - 2y = 0.$$

Substituting $e^{\lambda x}$ for y, we find the auxiliary equation

$$\lambda^2 + \lambda - 2 = 0.$$

The roots are $\lambda_1 = 1,\ \lambda_2 = -2$; hence the general solution of the given equation is

$$y = C_1 e^{x} + C_2 e^{-2x}.$$

Repeated Real Roots. If one of the roots λ_k is a multiple root, say, an r-fold root, then

$$e^{\lambda_k x},\ xe^{\lambda_k x},\ x^2 e^{\lambda_k x}, \ldots,\ x^{r-1} e^{\lambda_k x}$$

are solutions of Eq.(1). The r terms in (4) due to the repeated root λ_k must then be replaced by

$$\left(B_1 + B_2 x + B_3 x^2 + \ldots + B_r x^{r-1}\right) e^{\lambda_k x},$$

where B's are arbitrary constants.

Example 2. Find the general solution of

$$y^v - 6y^{iv} + 9y''' = 0.$$

The auxiliary equation is

$$\lambda^5 - 6\lambda^4 + 9\lambda^3 = 0.$$

The roots are $\lambda = 0,\ 0,\ 0,\ 3,\ 3,$ and the general solution of the original equation is

$$y = B_1 + B_2 x + B_3 x^2 + (B_4 + B_5 x) e^{3x}.$$

Complex Roots. If some of the roots of (4) are complex, the preceding expressions still give us the general solution, but complex quantities are involved. If the coefficients in the equation are real, we can rewrite the particular solutions so that they are real.

If $\alpha + i\beta$ is a complex root so also is $\alpha - i\beta$, since the coefficients $a_0,\ a_1,\ \ldots,\ a_n$ are real numbers, and we have in (4) the two terms

$$C_+ e^{(\alpha+i\beta)x} + C_- e^{(\alpha-i\beta)x}.$$

Using the relations

$$e^{(\alpha+i\beta)x} = e^{\alpha x}(\cos\beta x + i \sin\alpha x),$$
$$e^{(\alpha-i\beta)x} = e^{\alpha x}(\cos\beta x - i \sin\alpha x)$$

and putting

$$A = C_+ + C_-, \quad B = i(C_+ - C_-)$$

the two terms may be written

$$e^{\alpha x}(A\cos\beta x + B\sin\beta x).$$

Example 3. Find the general solution of

$$y'' - 4y' + 5y = 0.$$

The auxiliary equation is

$$\lambda^2 - 4\lambda + 5 = 0.$$

The roots are $\lambda = 2+i,\ 2-i$, and the general solution of the given equation is

$$y = (C_1 \cos x + C_2 \sin x)e^{2x}.$$

If $\alpha + i\beta$ is a complex root of order r, so also is $\alpha - i\beta$. With the preceding reduction the $2r$ roots contribute the following to the general solution

$$e^{\alpha x}[\left(A_1 + A_2 x + A_3 x^2 + \cdots + A_r x^{r-1}\right)\cos\beta x$$

$$+\left(B_1 + B_2 x + \cdots + B_r x^{r-1}\right)\sin\beta x]$$

which contains $2r$ arbitrary constants.

Alternative forms, which are sometimes useful, are

$$A\cos\beta x + B\sin\beta x = C\cos(\beta x + \gamma) = D\sin(\beta x + \delta).$$

These involve more changes in the definition of the constants.

Example 4. Find the general solution of

$$y^{iv} + 8y'' + y = 0.$$

The auxiliary equation is

$$\lambda^4 + 8\lambda^2 + 1 = 0.$$

The roots are $\lambda = 2i,\ 2i,\ -2i,\ -2i$, and the general solution of the given equation is

$$y = (A_1 + A_2 x)\cos 2x + (A_3 + A_4 x)\sin 2x.$$

In determining the roots of the auxiliary equation it is often necessary to compute the cube roots, or fourth, or even higher roots of a (possibly complex) number. This can usually be done most conveniently by using *Euler's formula*

$$e^{ix} = \cos x + i\sin x,$$

and

$$\left[e^{(\alpha+i\beta)x}\right]^p = e^{(p\alpha+ip\beta)x}.$$

Observing that $e^{i(\theta+2m\pi)} = e^{i\theta}$ for m an integer, and that

$$\left[e^{i(\theta+2m\pi)}\right]^{1/n} = \exp\left[\frac{i(\theta+2m\pi)}{n}\right]$$

we can determine the n-th roots of a number. This is illustrated in the following example .

Example 5. Find the general solution of

$$y^{iv} + 16y = 0.$$

The auxiliary equation is

$$\lambda^4 + 16 = 0.$$

In this case the polynomial is not readily factored. We must compute the fourth roots of -16. Now

$$-1 = \cos\pi + i\sin\pi = e^{i\pi} = \cos(\pi + 2m\pi) + i\sin(\pi + 2m\pi) = e^{i(\pi+2m\pi)},$$

$$-16 = -1e^{\ln 16} = -1e^{4\ln 2} = e^{4\ln 2 + i(\pi+2m\pi)}$$

where m is zero or any positive or negative integer. Thus

$$(-16)^{1/4} = \exp\left[\frac{4\ln 2 + i(\pi + 2m\pi)}{4}\right] = \exp\left[\ln 2 + i\left(\frac{\pi}{4} + \frac{m\pi}{2}\right)\right]$$

$$= 2\exp\left[i\left(\frac{\pi}{4} + \frac{m\pi}{2}\right)\right].$$

The fourth roots of -16 are obtained by setting $m = 0, 1, 2$ and 3; they are

$$\sqrt{2}(1+i),\ \sqrt{2}(-1+i),\ -\sqrt{2}(1+i),\ \sqrt{2}(1-i).$$

It is easy to verify that for any other value of m we obtain one of these four roots. For example, corresponding to $m = 5$ we obtain $\sqrt{2}(-1+i)$. The general solution of the original equation is

$$y = e^{\sqrt{2}x}\left(C_1\cos\sqrt{2}x + C_2\sin\sqrt{2}x\right)$$

$$+e^{-\sqrt{2}x}\left(C_3\cos\sqrt{2}x + C_4\sin\sqrt{2}x\right).$$

If the constants a_0, a_1, $\ldots$, a_n in Eq.(1) are complex numbers, the solution of Eq.(1) is still of the form (4). In this case, however, the roots of the auxiliary equation will, in general, be complex numbers; and it is no longer true that the complex conjugate of a root will be a root. The corresponding solutions will be complex-valued.

In conclusion we mention that the problem of determining the roots of $f(\lambda) = 0$ for $n > 2$ may require the use of numerical methods if they cannot be found by inspection or by a simple process of trial and error. Although there are formulas similar to the quadratic formula for the roots of cubic (Cardano formula, after the Italian mathematician G. Cardano (1501-1576)) equations, there is no formula for $n > 4$ (it was established by N.H. Abel and the French mathematician E. Galois (1811-1832)); and even for $n = 3$ and $n = 4$ it is usually more efficient to use numerical methods than the exact formula for determining the roots.

PROBLEMS

In each of Problems 1 through 70 determine the general solution of the given differential equation. Where specified solve the initial value problem.

1. $y'' - 2y' = 0.$
2. $2y'' - 5y' + 2y = 0.$
3. $y'' + 2y' + 10y = 0.$
4. $y'' + 4y = 0.$
5. $y''' - 8y = 0.$
6. $y^{iv} - y = 0.$
7. $y^{iv} + 4y = 0.$
8. $y^{vi} + 64y = 0.$
9. $y'' - 2y' + y = 0.$
10. $4y'' + 4y' + y = 0.$
11. $y^{v} - 10y''' + 9y' = 0.$
12. $y^{iv} + y = 0.$
13. $y^{iv} + 2y'' + y = 0.$
14. $y''' - y'' - y' + y = 0.$
15. $y''' - 3y'' + 3y' - y = 0.$
16. $2y''' - 4y'' - 2y' + 4y = 0.$
17. $y^{iv} - 4y''' + 4y'' = 0.$
18. $y^{iv} + y = 0.$
19. $y^{iv} - 5y'' + 4y = 0.$
20. $y^{vi} - 3y^{iv} + 3y'' - y = 0.$
21. $y^{vi} - y'' = 0.$
22. $y^{v} - 3y^{iv} + 3y''' - 3y'' + 2y' = 0.$
23. $y^{iv} - 8y' = 0.$
24. $y^{viii} + 8y^{iv} + 16y = 0.$
25. $y^{iv} + 2y'' + y = 0.$
26. $y^{v} + 8y''' + 16y' = 0.$
27. $y''' - 3y' + 2y = 0.$
28. $y^{iv} + 4y'' + 3y = 0.$
29. $y'' + 5y' + 4y = 0.$
30. $y'' - 7y' + 10y = 0.$
31. $y'' - 2y' + 10y = 0.$
32. $y'' + 3y' = 0.$
33. $y'' - a^2y = 0.$
34. $y'' + 9y = 0.$
35. $y'' - 2y' = 0,\ y(0) = 0,\ y(0) = 2.$
36. $y'' - 6y' + 13y = 0.$
37. $y'' - 4y' + 4y = 0.$
38. $3y'' + 5y' + 2y = 0.$
39. $y'' + 4y' + 3y = 0.$
40. $y'' - 4y' + 29y = 0,\ y(0) = 1,\ y'(0) = 7.$
41. $4y'' - 4y' + y = 0,\ y(0) = 2,\ y'(0) = -1.$
42. $y'' + 6y' + 9y = 0,\ y(0) = 1,\ y'(0) = -1.$

43. $y'' + 2y' - 2y = 0,\ y(0) = 0,\ y'(0) = 2\sqrt{3}$.
44. $y'' + 2y' + 3y = 0,\ y(0) = 3,\ y'(0) = 0$.
45. $y''' + 2y'' - 5y' - 6y = 0$.
46. $y''' + 3y'' - 4y' - 12y = 0$.
47. $y''' - 5y'' + 3y' + 9y = 0$.
48. $y''' - 2y'' = 0$.
49. $y''' - 4y' = 0$.
50. $y''' + 27y = 0$.
51. $y^{iv} - 8y' = 0$.
52. $y^{iv} + 9y'' = 0$.
53. $y^{iv} - 3y''' = 0$.
54. $y^{iv} + 8y''' + 24y'' + 32y' + 16y = 0$.
55. $y^{iv} - 4y''' + 6y'' - 4y' + y = 0$.
56. $y^{iv} + 5y'' + 6y = 0$.
57. $y^{v} - y = 0$.
58. $y^{v} - 3y^{iv} + 2y''' = 0$.
59. $y^{v} + 8y'' = 0$.
60. $y^{vi} - 4y^{v} + 13y^{iv} = 0$.
61. $y''' + y' = 0,\ y(0) = 1,\ y'(0) = 1,\ y''(0) = 2$.
62. $y''' - 6y'' + 11y' - 6y = 0$.
63. $y''' - 5y'' + 17y' - 13y = 0$.
64. $y^{iv} + 5y'' + 4y = 0$.
65. $y''' - 6y'' + 12y' - 8y = 0$.
66. $y^{iv} + 2y''' + 3y'' + 2y' + y = 0$.
67. $y^{iv} + 4y''' + 6y'' + 4y' + y = 0$.
68. $y^{iv} - y''' + y'' - y' + 12y = 0$.
69. $y^{vii} + 3y^{vi} + 3y^{v} + y^{iv} = 0$.
70. $y^{iv} + 2y''' + 5y'' + 8y' + 4y = 0$.

2.4.2 *The Complete Equation with Constant Coefficients. Method of Undetermined Coefficients*

A particular solution of the *nonhomogeneous* n-th order linear equation with constant coefficients

$$L[y] = a_0 y^{(n)} + a_1 y^{(n-1)} + \cdots + a_{n-1} y' + a_n y = g(x) \tag{7}$$

can be obtained by the method of undetermined coefficients provided that $g(x)$ is of an appropriate form. While the method of undetermined coefficients is not as general as the method of variation of parameters that is described below, it is usually much easier to use when applicable.

a. First consider the case that $g(x)$ is polynomial of degree m

$$g(x) = b_0 x^m + b_1 x^{m-1} + \cdots + b_m, \tag{8}$$

where $b_0, b_1, \ldots, b_m$, are given constants. It is natural to look for a particular solution of the form

$$y_p = A_0 x^m + A_1 x^{m-1} + \cdots + A_m \tag{9}$$

Substituting for y in Eq.(1), and equating the coefficients of like powers of x, we find from the terms in x^m that $a_n A_0 = b_0$. Provided that $a_n \neq 0$ we have $A_0 = b_0/a_n$. The constants $A_1, A_2, \ldots, A_m$ are determined from the coefficients of the terms $x^{m-1}, x^{m-2}, \ldots, x^0$. More generally if zero is an s-fold root of the auxiliary polynomial ($a_n = a_{n-1} = \cdots = a_{n-s+1} = 0$), then a suitable form for $y_p(x)$ is

$$y_p(x) = x^s \left(A_0 x^m + A_1 x^{m-1} + \cdots + A_m \right) \tag{10}$$

Example 6. Solve

$$y'' - y = x^2 + 1.$$

The auxiliary equation is

$$\lambda^2 - 1 = 0.$$

The roots are $\lambda_{1,2} = \pm 1$ and the general solution of the homogeneous equation is

$$y = C_1 e^x + C_2 e^{-x}.$$

Since $g(x)$ is a polynomial of the second degree we look for a particular solution of the form

$$y_p = Ax^2 + Bx + C.$$

On substituting into the original equation we get

$$2A - Ax^2 - Bx - C = x^2 + 1.$$

Equating coefficients of like terms, we have

$$\begin{aligned} -A &= 1, \\ -B &= 0, \\ 2A - C &= 1. \end{aligned}$$

Solving, $A = -1,\ B = 0,\ C = -3$. Hence the complete solution is

$$y = C_1 e^x + C_2 e^{-x} - x^2 - 3.$$

Example 7. Find a particular solution of

$$y^{iv} - y'' = x + 1.$$

The auxiliary equation is

$$\lambda^2 \left(\lambda^2 - 1 \right) = 0$$

and the roots are $\lambda_{1,2} = 0,\ \lambda_{3,4} = \pm 1$. Since $\lambda = 0$ is 2-fold root, then we put

$$y_p = x^2(Ax + B).$$

Putting this into the given equation we have

$$-6Ax - 2B = x + 1.$$

Equating the coefficients of like powers of x, we find $A = -1/6,\ B = -1/2$. Hence a particular solution is

$$y_p = -\frac{x^3}{6} - \frac{x^2}{2}.$$

b. As a second case suppose that $g(x)$ is of the form

$$g(x) = e^{\alpha x}\left(b_0 x^m + b_1 x^{m-1} + \cdots + b_m\right) \tag{11}$$

Then we try $y_p(x)$ in the form

$$y_p(x) = e^{\alpha x}\left(A_0 x^m + A_1 x^{m-1} + \cdots + A_m\right)$$

provided that $e^{\alpha x}$ is not a solution of the homogeneous equation.

If α is an s-fold root of the auxiliary equation, a suitable form for $y_p(x)$ is

$$y_p(x) = x^s e^{\alpha x}\left(A_0 x^m + A_1 x^{m-1} + \cdots + A_m\right) \tag{12}$$

Example 8. Determine a suitable form for y_p. Do not evaluate the constants

$$y'' - y = e^{3x}.$$

The auxiliary equation is

$$\lambda^2 - 1 = 0.$$

Here e^x and e^{-x} are solutions of the homogeneous equation. Since e^{3x} is not a solution of the homogeneous equation we try a particular solution of the form

$$y_p = Ae^{3x}.$$

Example 9. The same problem for the equation

$$y'' - y = e^{-x}(x + 1).$$

Since e^{-x} is a solution of the homogeneous equation, then suitable form for y_p is

$$y_p = xe^{-x}(Ax + B).$$

Example 10. The same problem for the equation

$$y''' - 3y'' + y' - 3 = e^x x^2.$$

The auxiliary equation is

$$\lambda^3 - 3\lambda^2 + 3\lambda - 1 = (\lambda - 1)^3 = 0$$

and $\lambda = 1$ is 3-fold root. Then we try y_p of the form

$$y_p = x^3 e^x \left(Ax^2 + Bx + C\right).$$

Example 11. Find a particular solution of

$$y'' + 3y' - 4y = e^{-4x} + xe^{-x}.$$

First we solve the homogeneous equation. The auxiliary equation is

$$\lambda^2 + 3\lambda - 4 = 0$$

and the roots are $\lambda = 1, -4$; hence

$$y_c = C_1 e^{-4x} + C_2 e^x.$$

Using the superposition principle, a particular solution will be the sum of the particular solutions corresponding to the differential equations

$$y'' + 3y' - 4y = e^{-4x}$$

and

$$y'' + 3y' - 4y = xe^{-x}.$$

For the first equation, since e^{-4x} is a solution of the homogeneous equation, we assume that

$$y_{p_1}(x) = Axe^{-4x}.$$

For the second equation we guess

$$y_{p_2}(x) = e^{-x}(B + Cx)$$

and there is no need to modify this initial guess since e^{-x} is not a solution of the homogeneous equation. The constants are determined by substituting into the individual differential equations; they are

$$A = -\frac{1}{5}, \quad B = -\frac{1}{36}, \quad C = -\frac{1}{6}.$$

Hence a particular solution of the original equation is

$$y_p = -\frac{x}{5} e^{-4x} - \left(\frac{x}{6} + \frac{1}{36}\right) e^{-x}.$$

c. Similarly, if $g(x)$ is of the form

$$g(x) = e^{\alpha x}\left(b_0 x^m + b_1 x^{m-1} + \cdots + b_m\right)\begin{cases}\sin\beta x\\ \cos\beta x\end{cases} \tag{13}$$

then a suitable form for $y_p(x)$, provided that $\alpha + i\beta$ is not a root of the auxiliary equation, is

$$\begin{aligned} y_p(x) &= e^{\alpha x}\left(A_0 x^m + A_1 x^{m-1} + \cdots + A_m\right)\sin\beta x \\ &+ e^{\alpha x}\left(B_0 x^m + B_1 x^{m-1} + \cdots + B_m\right)\cos\beta x \end{aligned} \tag{14}$$

If $\alpha + i\beta$ is an s-fold root of the auxiliary equation, it is necessary to multiply the right-hand side of Eq.(14) by x^s.

Example 12. Determine a suitable form for y_p. Do not evaluate the constants

$$y'' + 4y = \cos x.$$

The auxiliary equation is

$$\lambda^2 + 4 = 0.$$

The roots are $\lambda_{1,2} = \pm 2i$.

Here $\alpha = 0,\ \beta = 1$. Since $\lambda = \pm i$ are not solutions of the auxiliary equation, we try a particular solution of the form

$$y_p = A\sin x + B\cos x.$$

Example 13. The same problem for the equation

$$y'' + 4y = \cos 2x.$$

Here $\alpha = 0,\ \beta = 2$. As $\lambda = \pm 2i$ are solutions of the auxiliary equation, the desired particular solution is of the form

$$y_p = x(A\sin 2x + B\cos 2x).$$

Example 14. The same problem for the equation

$$y^{iv} + 2y'' + y = \cos x.$$

The auxiliary equation is

$$\lambda^4 + 2\lambda^2 + 1 = \left(\lambda^2 + 1\right) = 0.$$

and the roots are $\lambda_{1,2} = i,\ \lambda_{3,4} = -i$. Here $\alpha = 0,\ \beta = 1$. As $\lambda \pm i\beta$ are 2-fold roots of the auxiliary equation then we try a particular solution of the form

$$y_p = x^2(A\sin x + B\cos x).$$

Example 15. The same problem for the equation

$$y'' - y = e^{2x} x \sin x$$

Since $2 \pm i$ are not solutions of the auxiliary equation, we put

$$y_p = e^{2x}\left[(Ax + B)\sin x + (Cx + D)\cos x\right].$$

Example 16. The same problem for the equation

$$y'' + 2y' + y = e^{x} x \sin x.$$

The auxiliary equation is

$$\lambda^2 + 2\lambda + 2 = 0.$$

and the roots are $\lambda_{1,2} = 1 \pm i$. Here $\alpha = 1$, $\beta = 1$, since $1 + i$ is the root of the auxiliary equation, then we put

$$y_p = xe^{x}\left[(Ax + B)\sin x + (Cx + D)\cos x\right].$$

Example 17. Find a particular solution of

$$y^{iv} + y'' = x^3 + x^2 + e^{x}(x + 1) + 2\cos x.$$

The corresponding homogeneous equation is

$$y^{iv} + y'' = 0.$$

The auxiliary equation is

$$\lambda^4 + \lambda^2 = 0.$$

and the roots are $\lambda = 0,\ 0,\ \pm i$. Here $1,\ x,\ \sin x,\ \cos x$ are solutions of the reduced equation. We put

$$y_p = x^2\left(Ax^3 + Bx^2 + Cx + D\right) + e^{x}(Ex + F) + x(H\sin x + G\cos x)$$

into the equation, getting for the first member

$$\begin{gathered} 20Ax^3 + 12Bx^2 + (120A + 6C)x + 24B + 2D \\ +e^{x}(2Ex + 6E + 2F) + 2G\sin x - 2H\cos x. \end{gathered}$$

Equating coefficients of like terms in the two members, we have

$$\begin{aligned} 20A &= 1, & 2E &= 1, \\ 12B &= 1, & 6E + 2F &= 1, \\ 120A + 6C &= 0, & 2H &= 1, \\ 24B + 2D &= 0, & 2G &= 0. \end{aligned}$$

Solving, $A = 1/20$, $B = 1/12$, $C = -1$, $D = -1$, $H = -1/2$, $E = 1/2$, $F = -1$, $G = 1$ and we have the particular solution

$$y = \frac{1}{20}x^5 + \frac{1}{12}x^4 - x^3 - x^2 + e^{x}\left(\frac{1}{2}x - 1\right) - \frac{1}{2}x\sin x.$$

These results are summarized in Table 2.1.

Table 2.1

$g(x)$	$y_p(x)$
$P_m(x) = b_0x^m + b_1x^{m-1} + \cdots + b_{m-1}x + b_m$	$x^s(A_0x^m + A_1x^{m-1} + \cdots + A_m)$
$P_m(x)e^{\alpha x}$	$x^s e^{\alpha x}(A_0x^m + A_1x^{m-1} + \cdots + A_m)$
$P_m(x)e^{\alpha x}\begin{cases}\sin\beta x\\ \cos\beta x\end{cases}$	$x^s e^{\alpha x}\{\cos\beta x(A_0x^m + \cdots + A_m) + \sin\beta x(B_0x^m + \cdots + B_m)\}$

Here s is the smallest nonnegative integer for which every term in $y_p(x)$ differs from every term in the complementary function $y_c(x)$.

PROBLEMS

In each of Problems 71 through 106 determine the general solution of the given differential equation. Where specified find the solution satisfying the given initial conditions.

71. $y'' - 2y' - 3y = e^{4x}$.
72. $y'' + y = 4xe^x$.
73. $y'' - y = 2e^x - x^2$.
74. $y'' + y' - 2y = 3xe^x$.
75. $y'' - 3y' + 2y = \sin x$.
76. $y'' + y = 4\sin x$.
77. $y'' - 5y' + 4y = 4x^2e^{2x}$.
78. $y'' - 3y' + 2y = x\cos x$.
79. $y'' + 2y' - 3y = x^2e^x$.
80. $y'' - 4y' + 8y = e^{2x} + \sin 2x$.
81. $y''' + 4y' = x$; $y(0) = y'(0) = 0$, $y(0) = 1$.
82. $y''' - y'' - y' + y = 2e^{-x} + 3$.
83. $y''' + y'' + y' + y = e^{-x} + 4x$.
84. $y''' - y = 2\sin x$.
85. $y^{iv} + 2y'' + y = 3x + 4$; $y(0) = y'(0) = 0$; $y''(0) = y'''(0) = 1$.
86. $y^{iv} - 4y'' = x^2 + e^x$.
87. $y^{iv} + 2y'' + y = 3 + \cos 2x$.
88. $y''' - 3y'' + 2y' = x + e^x$.
89. $y^{vi} + y'' = x$.
90. $y^{iv} + y''' = \sin 2x$.
91. $y^{iv} - y = 3x + \cos x$.
92. $y''' - 4y = x + 3\cos x + e^{-2x}$.
93. $y'' - 9y = e^{3x}\cos x$.
94. $y'' - 2y' + y = 6xe^x$.
95. $y'' + y = x\sin x$.
96. $y'' + 4y' + 4y = xe^{2x}$.
97. $y'' - 5y' = 3x^2 + \sin 5x$.

98. $y'' + a^2y = e^x$.
99. $y'' - 3y' + 2y = e^{3x}(x^2 + x)$.
100. $y'' - y = xe^x + e^{2x}$.
101. $y''' - 4y' = xe^{2x} + \sin x + x^2$.
102. $y'' + 2y' + y = e^{-x}\cos x + xe^{-x}$.
103. $y'' + 2y' + 2y = e^{-x}\cos x + xe^{-x}$.
104. $y'' - 2y' + y = \sin x + e^{-x} + e^x$.
105. $y'' + y = 2\sin x \sin 2x$.
106. $y'' + a^2y = \sin bx$.

In each of Problems 107 through 140 determine a suitable form for $y_p(x)$ if the method of undetermined coefficients is to be used. Do not evaluate the constants.

107. $y'' - 2y' + 2y = e^x + x\cos x$.
108. $y'' + 6y' + 10y = 3xe^{-3x} - 2e^{3x}\cos x$.
109. $y'' - 8y' + 20y = 5xe^{4x}\sin 2x$.
110. $y'' + 7y' + 10y = xe^{-x}\cos 5x$.
111. $y'' - 2y' + y = 2xe^x + e^x\sin 2x$.
112. $y'' - 8y' + 17y = e^{4x}(x^2 - 3x\sin x)$
113. $y''' + y' = \sin x + x\cos x$.
114. $y''' - 2y'' + 4y' - 8y = e^2x\sin 2x + 2x^2$.
115. $y'' - 6y' + 8y = 5xe^{2x} + 2e^{4x}\sin x$.
116. $y'' + 2y' + y = x(e^{-x} - \cos x)$.
117. $y''' - 2y'' + y' = x^3 + 2e^x$.
118. $y''' - y' = xe^{-x} + 2\cos x$.
119. $y^{iv} - 2y'' + y = e^x + \sin x$.
120. $y^{iv} - y''' - y'' + y' = x^2 + 4 + x\sin x$
121. $y^{iv} + 4y'' = \sin 2x + xe^x + 4$.
122. $y^{iv} + 2y''' + 2y'' = 3e^x + 2xe^{-x} + e^{-x}\sin x$.
123. $y''' - y'' - y' + y = 3e^x + 5x\sin x$.
124. $y'' - 6y' + 13y = x^2e^{3x} - 3\cos 2x$.
125. $y'' - 9y = e^{-3x}(x^2 + \sin 3x)$.
126. $y^{iv} + y'' = 7x - 3\cos x$.
127. $y'' + 4y = \cos x\cos 3x$.
128. $y''' - 4y'' + 3y' = x^2 + xe^{2x}$.
129. $y'' - 4y' + 5y = e^{2x}\sin^2 x$
130. $y'' + 3y' + 2y = e^{-x}\cos^2 x$.
131. $y'' - 2y' + 2y = (x + e^x)\sin x$.
132. $y^{iv} + 5y'' + 4y = \sin x\cos 2x$.
133. $y'' - 3y' + 2y = 2^x$.
134. $y'' - y = 4\sinh x$.
135. $y'' + 4y' + 3y = \cosh x$.
136. $y'' + 4y = \sinh x\sin 2x$.
137. $y'' + 2y' + 2y = \cosh x\sin x$.

138. $y'' + y' - 2y = xe^{2x} + \sin x + x^2$.
139. $y^{v} - y^{iv} - y' + y = x^2 e^x + x \sin x + e^{2x}$.
140. $y''' - 8y = e^{2x}(\sin x + x)$.

2.4.3 The Method of Variation of Parameters

A particular solution of the nonhomogeneous n-th order linear differential equation

$$L[y] = a_0 y^{(n)} + a_1 y^{(n-1)} + \cdots + a_{n-1} y' + a_n y = g(x) \tag{15}$$

can be obtained by the *method of variation of parameters*, which Euler devised in 1739 and Lagrange refined in 1774. This method is still more general than the method of undetermined coefficients in the following sense. The method of undetermined coefficients is usually applicable only for constant coefficients equations, and a limited class of function $g(x)$; for constant coefficient equations the homogeneous equation can be solved and hence a particular solution for any continuous function $g(x)$ can be determined by the method of variation of parameters. Suppose then that we know a fundamental set of solutions

$$y_1, \; y_2, \; \ldots \; , \; y_n$$

of the homogeneous equation. Then

$$y_c(x) = C_1 y_1(x) + C_2 y_2(x) + \cdots + C_n y_n(x) \tag{16}$$

As it was shown by Lagrange the method of variation of parameters for determining a particular solution of Eq.(15) rests on the possibility of determining n functions

$$u_1(x), \; u_2(x), \; \ldots, \; u_n(x)$$

such that $y_p(x)$ is of the form

$$y_p(x) = u_1(x) y_1(x) + u_2(x) y_2(x) + \cdots + u_n(x) y_n(x) \tag{17}$$

We have replaced the constants in the solution of (16) by variables

$$u_1, \; u_2, \; \ldots, \; u_n$$

which are to be found from the next n linear nonhomogeneous equations

$$\begin{aligned} u_1' y_1 + u_2' y_2 + \cdots + u_n' y_n &= 0, \\ u_1' y_1' + u_2' y_2' + \cdots + u_n' y_n' &= 0, \\ u_1' y_1'' + u_2' y_2'' + \cdots + u_n' y_n'' &= 0, \\ &\cdots \\ u_1' y_1^{(n-2)} + u_2' y_2^{(n-2)} + \cdots + u_n' y_n^{(n-2)} &= 0, \end{aligned} \tag{18}$$

$$a_0\left(u_1'y_1^{(n-1)} + u_2'y_2^{(n-1)} + \cdots + u_n'y_n^{(n-1)}\right) = g(x).$$

The determinant of Eq.-s(18) is *Wronskian*, which is not zero since

$$y_1,\ y_2,\ \ldots,\ y_n$$

are linear independent solutions of the homogeneous equation. Hence it is possible to determine $u_1',\ u_2',\ \ldots,\ u_n'$. Using *Cramer's rule*, the solution of the system of equations (18) is

$$u_m' = \frac{g(x)W_m(x)}{a_0W(x)},\ m = 1,\ 2,\ \ldots,\ n \tag{19}$$

where $W(x) = W(y_1, y_2, \ldots y_n)(x)$ and W_m is the determinant, obtained from $W(y_1, y_2, \ldots, y_n)$ by replacing the m-th column by the column

$$(0,\ 0,\ \cdots,\ 0,\ 1),$$

or

$$W_m = \begin{vmatrix} y_1 & \cdots & y_{m-1} & 0 & y_{m+1} & \cdots & y_n \\ \vdots & \vdots & \vdots & \vdots & \vdots & \ddots & \vdots \\ y_1^{(n-1)} & \cdots & y_{m-1}^{(n-1)} & 1 & y_{m+1}^{(n-1)} & \cdots & y_n^{(n-1)} \end{vmatrix}.$$

With this notation a particular solution of Eq.(15) is given by

$$y_p(x) = \sum_{m=1}^{n} y_m(x)\int^x \frac{g(t)W_m(t)}{a_0W(t)}dt.$$

The calculation may be simplified to some extent by using *Abel's identity*

$$W(x) = C\exp[-a_1x].$$

The constant C can be determined by evaluating $W(y_1,\ y_2,\ \cdots,\ y_n)$ at a conveniently chosen point.

Example 18. Find a particular solution by the method of the variation of parameters

$$y'' - 2y' + y = \frac{e^x}{x}.$$

First we solve the homogeneous equation

$$y'' - 2y' + y = 0.$$

The auxiliary equation is

$$\lambda^2 - 2\lambda + 1 = 0$$

and the roots are $\lambda_{1,2} = 1$; hence

$$y_c = (C_1 + C_2x)e^x.$$

Let us try a particular solution of the form

$$y_p = u_1(x)e^x + u_2(x)xe^x.$$

We have two functions $u_1,\ u_2$ to determine from the system (18), which takes the form

$$u_1'e^x + u_2'xe^x = 0$$

and

$$u_1'e^x + u_2'(e^x + xe^x) = \frac{e^x}{x}.$$

The solution of this system of equations is

$$u_1' = -1, \quad u_2' = \frac{1}{x}.$$

Integrating, we have

$$u_1 = -x, \quad u_2 = \ln x$$

and a particular solution of the given equation is

$$y_p = -xe^x + x\ln x.$$

Example 19. Solve the differential equation

$$y'' + y = \sin^{-1} x.$$

The auxiliary equation is $\lambda^2 + 1 = 0$, and the roots are $\lambda = \pm i$; hence the solution of the homogeneous equation is

$$y_c = C_1 \sin x + C_2 \cos x.$$

Then a particular solution of the original equation is of the form

$$y_p = u_1(x)\sin x + u_2(x)\cos x,$$

where functions u_1 , and u_2 are to be found from the system of equations

$$u_1'(x)\sin x + u_2'(x)\cos x = 0$$

and

$$u_1'(x)\cos x - u_2'(x)\sin x = \sin^{-1} x$$

The solution is

$$u_1' = \cot x, \quad u_2' = -1.$$

Integrating, we have

$$u_1 = C_1 + \ln\sin x$$

and

$$u_2 = -x + C_2$$

The solution of the given equation is

$$y = (C_1 + \ln \sin x) \sin x + (C_2 - x) \cos x.$$

PROBLEMS

In each of Problems 141 through 153 use the method of variation of parameters to determine a particular solution of the given differential equation.

141. $y''' + y' = \tan x, \quad 0 < x < \pi/2.$
142. $y''' - y' = x.$
143. $y''' - 2y'' - y' + 2y = e^{4x}.$
144. $y'' + 3y' + 2y = \dfrac{1}{e^x + 1}.$
145. $y'' + 4y = 2\tan x, \quad 0 < x < \pi/2.$
146. $y'' + 2y' + y = 3e^{-x}\sqrt{x+1}.$
147. $y'' + y = 2\sec^3 x, \quad 0 < x < \pi/2.$
148. $y'' - y = \dfrac{2e^x}{e^x - 1}.$
149. $y'' + y = \dfrac{1}{\cos 2x \sqrt{\cos 2x}}.$
150. $y'' - 6y' + 9y = \dfrac{9x^2 + 6x + 2}{x^3}.$
151. $y''' - y'' + y' - y = g(x).$
152. $y^{iv} - y = g(x).$
153. $x^3 y''' - 3x^2 y'' + 6xy' - 6y = g(x).$

2.4.4 Symbolic Methods

Representing derivatives by powers of D

$$Dg(x) = g'(x),\ D^2 g(x) = g''(x),\ \ldots,\ D^0 g(x) = g(x) \tag{20}$$

we can form operator polynomials with constant coefficients. Thus the equation

$$a_0 y^{(n)} + a_1 y^{(n-1)} + a_2 y^{(n-2)} + \cdots + a_n y = g(x)$$

can be written in the form

$$f(D)y = \left(a_0 D^n + a_1 D^{n-1} + \cdots + a_n\right) y = g(x)$$

and $f(D)$ may be broken into factors

$$f(D) = (D - \lambda_1)(D - \lambda_2)\cdots(D - \lambda_n),$$

where λ_k are the roots of the auxiliary equation.

It can be proved that

$$\frac{1}{D-\lambda}g(x) = e^{\lambda x}\int e^{-\lambda x}g(x)dx.$$

The particular solution of Eq.(20) may then be written in the symbolic form

$$y_p(x) = \frac{1}{f(D)}g(x) = \frac{1}{D-\lambda_1}\frac{1}{D-\lambda_2}\cdots\frac{1}{D-\lambda_n}g(x).$$

Example 20. Find a particular solution of

$$y'' - y' = \sin x.$$

We have

$$\left(D^2 - D\right) y = \sin x,$$

or

$$y_p = \frac{1}{D^2 - D}\sin x = \frac{1}{D-1}\frac{1}{D}\sin x = -\frac{1}{D-1}\cos x$$

$$= -e^x\int e^{-x}\cos x dx = -e^x\frac{1}{2}\left(-e^{-x}\cos x + e^{-x}\sin x\right) = \frac{\cos x - \sin x}{2}.$$

Example 21. The same problem for the equation

$$y'' + y' - 2y = x^2 - x.$$

We have

$$(D^2 + D - 2)y = x^2 - x$$

whence

$$y_p = \frac{1}{D^2 + D - 2}\left(x^2 - x\right) = \frac{1}{D+2}\frac{1}{D-1}\left(x^2 - x\right)$$

$$= \frac{1}{D+2}e^x\int e^{-x}\left(x^2 - x\right)dx = \frac{1}{D+2}(-x^2 - x - 1)$$

$$= -e^{-2x}\int e^{2x}\left(x^2 + x + 1\right)dx$$

$$= -e^{-2x}\left[\frac{e^{2x}}{8}\left(4x^2 - 4x + 2\right) + \frac{e^{2x}}{4}(2x-1) + \frac{e^{2x}}{2}\right]$$

$$= -\frac{x^2}{2} - \frac{1}{2}.$$

Example 22. Find a particular solution of

$$y'' + y' - 2y = e^x.$$

We have

$$(D^2 + D - 2)y = e^x,$$

$$y_p = \frac{1}{D-1}\frac{1}{D+2}e^x,$$

$$\frac{1}{D+2}e^x = e^{-2x}\int e^{2x}e^x dx = \frac{e^x}{3},$$

$$\frac{1}{D-1}\frac{e^x}{3} = e^x\int e^{-x}\frac{e^x}{3}dx = \frac{xe^x}{3}.$$

And a particular solution of the original equation is

$$y_p = \frac{xe^x}{3}.$$

The operator $1/f(D)$ may be broken into partial fractions. If the factors of $f(D)$ are distinct, we have the form

$$y_p = \frac{1}{f(D)}g(x) = \left[\frac{\alpha_1}{D-\lambda_1} + \frac{\alpha_2}{D-\lambda_2} + \cdots + \frac{\alpha_n}{D-\lambda_n}\right]g(x).$$

Here $f(D)$ applied after the operator in brackets gives 1, so that

$$f(D)y_p = g(x)$$

as required.

In Example 22

$$y_p = \frac{1}{D-1}\frac{1}{D+2}e^x = \frac{1}{3}\left(\frac{1}{D-1} - \frac{1}{D+2}\right)e^x$$

$$= \frac{1}{3}\left(e^x\int e^{-x}e^x dx - e^{-2x}\int e^{2x}e^x dx\right)$$

$$= \frac{1}{3}xe^x - \frac{1}{9}e^x.$$

As e^x is the solution of the homogeneous equation then we have the same $y_p = \frac{1}{3}xe^x$. If some of the factors are repeated, the form of the partial fractions is altered. If $D - \lambda_k$ is an r-fold factor, we have in the partial fractions expansion the terms

$$\frac{\alpha_k}{D-\lambda_k} + \frac{\alpha_{k+1}}{(D-\lambda_k)^2} + \cdots + \frac{\alpha_{k+r-1}}{(D-\lambda_k)^r}.$$

The application of these operators to $g(x)$ may be made in order, each involving an integration based on the preceding.

Example 23. Find a particular solution of

$$y''' - 2y'' + y' = 1.$$

We may write this equation as

$$\left(D^3 - 2D^2 + D\right) y = 1,$$

or

$$y_p = \frac{1}{D\left(D^2 - 2D + 1\right)} 1 = \frac{1}{D(D-1)^2} 1.$$

Using partial fractions we have

$$y_p = \left(\frac{A}{D} + \frac{B}{D-1} + \frac{C}{(D-1)^2}\right) 1.$$

Equating coefficients of like powers of D we find $A = 1$, $B = -1$, $C = 1$. Hence

$$y_p = \frac{1}{D} 1 - \frac{1}{D-1} 1 + \frac{1}{(D-1)^2} 1 = \int dx - e^x \int e^{-x} dx + \frac{1}{D-1} e^x \int e^{-x} dx$$

$$= x + 1 - \frac{1}{D-1} 1 = x + 1 - e^x \int e^{-x} dx = x + 2.$$

It can be proved that

(a) $\dfrac{1}{f(D)} e^{kx} = \dfrac{e^{kx}}{f(k)}$, if $f(k) \neq 0$;

(b) $\dfrac{1}{f(D^2)} \sin ax = \dfrac{\sin ax}{f(-a^2)}$, if $f(-a^2) \neq 0$;

$\dfrac{1}{f(D^2)} \cos ax = \dfrac{\cos ax}{f(-a^2)}$, if $f(-a^2) \neq 0$;

(c) $\dfrac{1}{f_1(D) f_2(D)} g(x) = \dfrac{1}{f_1(D)} \dfrac{1}{f_2(D)} g(x)$;

(d) $\dfrac{1}{f(D)} e^{kx} V(x) = e^{kx} \dfrac{1}{f(D+k)} V(x)$;

Example 24. Find a particular solution of

$$y'' + y = e^{2x}.$$

We may write this equation as

$$\left(D^2 + 1\right) y = e^{2x}$$

whence

$$y_p = \frac{1}{D^2 + 1} e^{2x},$$

or using the formula (a)

$$y_p = \frac{e^{2x}}{5}.$$

Example 25. Find a particular solution of

$$y''' + 3y'' + 3y' + y = e^{-x}.$$

We may write this equation as

$$\left(D^3 + 3D^2 + 3D + 1\right) y = (D+1)^3 y = e^{-x},$$

or

$$y_p = \frac{1}{(D+1)^3} e^{-x}$$

Here $f(-1) = 0$, then we can't use the relation (a). Then considering e^{-x} as $e^{-x}1$ and using the formula (d) we have

$$y_p = \frac{1}{(D+1)^3} e^{-x} 1 = e^{-x} \frac{1}{[(D-1)+1]^2} 1$$

$$= e^{-x} \frac{1}{D^3} 1 = e^{-x} \frac{1}{D} \frac{1}{D} \frac{1}{D} 1 = e^{-x} \frac{1}{D} \frac{1}{D} x$$

$$= e^{-x} \frac{1}{D} \frac{x^2}{2} = e^{-x} \frac{x^3}{6}.$$

Example 26. Find a particular solution of

$$y^{iv} + 16y = \cos 5x.$$

We have

$$\left(D^4 + 16\right) y = \cos 5x,$$

or using the formula (b)

$$y_p = \frac{1}{D^4 + 16} \cos 5x = \frac{1}{(D^2)^2 + 16} \cos 5x = \frac{\cos 5x}{[-(5^2)]^2 + 16} = \frac{\cos 5x}{641}.$$

Example 27. The same problem for the equation

$$y'' - y = \sin x.$$

Writing this equation as

$$\left(D^2 - 1\right) y = \sin x$$

and using the formula (b), we get

$$y_p = \frac{1}{D^2 - 1} \sin x = \frac{\sin x}{-1-1} = -\frac{\sin x}{2}.$$

Example 28. The same problem for the equation

$$y''' + y = \sin x.$$

We have

$$\left(D^3+1\right)y=\sin x,$$

or

$$y_p=\frac{1}{D^3+1}\sin x.$$

Since the operator contains odd-powers of D, we can't use the formula (b). Therefore, representing $\sin x$ in the exponential form, we get

$$\begin{aligned}
y_p &= \frac{1}{D^3+1}\frac{1}{2i}\left(e^{ix}-e^{-ix}\right)\\
&= \frac{1}{2i}\left(\frac{1}{D^3+1}e^{ix}-\frac{1}{D^3+1}e^{-ix}\right)\\
&= \frac{1}{2i}\left(\frac{1}{(i)^3+1}e^{ix}-\frac{1}{(-i)^3+1}e^{-ix}\right)\\
&= \frac{1}{2i}\left(\frac{1}{1-i}e^{ix}-\frac{1}{1+i}e^{-ix}\right)\\
&= \frac{1}{2i}\left(\frac{1+i}{2}e^{ix}-\frac{1-i}{2}e^{-ix}\right)\\
&= \frac{1}{2}\frac{e^{ix}-e^{-ix}}{2i}+\frac{1}{2}\frac{e^{ix}+e^{-ix}}{2}\\
&= \frac{1}{2}(\sin x+\cos x).
\end{aligned}$$

Example 29. The same problem for the equation

$$y''+4y=\cos 2x.$$

We have

$$\left(D^4+4\right)y=\cos 2x,$$

$$y_p=\frac{1}{D^2+4}\cos 2x$$

Here $f(-2^2)=0$ and we can't use the formula (b). Then we again substitute $\cos 2x$ in the exponential form

$$\cos 2x=\frac{1}{2}\left(e^{i2x}+e^{-i2x}\right)$$

and have for y_p

$$\begin{aligned}
y_p &= \frac{1}{D^2+4}\frac{1}{2}(e^{2ix}+e^{-2ix})=\frac{1}{2}\left(\frac{1}{D^2+4}e^{2ix}+\frac{1}{D^2+4}e^{-2ix}\right)\\
&= \frac{1}{2}\left(\frac{1}{(D-2i)(D+2i)}e^{2ix}+\frac{1}{(D+2i)(D-2i)}e^{-2ix}\right)
\end{aligned}$$

$$= \frac{1}{2}\left(\frac{1}{D-2i}\frac{e^{2ix}}{4i} + \frac{1}{D+2i}\frac{e^{-2ix}}{-4i}\right)$$

$$= \frac{1}{8i}\left(\frac{1}{D-2i}e^{2ix}1 - \frac{1}{D+2i}e^{-2ix}1\right)$$

$$= \frac{1}{8i}\left(e^{2ix}\frac{1}{D}1 - e^{-2ix}\frac{1}{D}1\right)$$

$$= \frac{1}{8i}\left(e^{2ix}x - e^{-2ix}x\right) = \frac{x}{4}\frac{e^{2ix} - e^{-2ix}}{2i} = \frac{x}{4}\sin 2x.$$

Example 30. Find a particular solution of

$$y'' + y' - 2y = e^x x^2.$$

We have

$$y_p = \frac{1}{D^2 + D - 2}e^x x^2 = e^x \frac{1}{(D+1)^2 + (D+1) - 2}x^2$$

$$= e^x \frac{1}{D^2 + 3D}x^2 = e^x \frac{1}{3}\left(\frac{1}{D} - \frac{1}{D+3}\right)x^2$$

$$= e^x \frac{1}{3}\left(\int x^2 dx - e^{-3x}\int x^2 e^{3x} dx\right) = \frac{x^3 e^x}{9} - e^x\left(\frac{1}{3}x^2 - \frac{2}{9}x + \frac{2}{27}\right).$$

As e^x is the solution of the homogeneous equation then we have a particular solution

$$y_p = xe^x\left(\frac{x^2}{9} - \frac{x}{3} - \frac{2}{9}\right).$$

If in a real equation, two roots, say λ_1, and λ_2, are conjugate imaginaries, then α_1 and α_2 are conjugates. The evaluation of

$$\frac{\alpha_2}{D - \lambda_2}g(x)$$

may be got from

$$\frac{\alpha_1}{D - \lambda_1}g(x)$$

by taking the conjugate function, and the sum of the two functions will be real.

A form which is particularly useful when $g(x)$ is a polynomial is the expansion of $1/f(D)$ in a power series in D:

$$y = \frac{1}{f(D)}g(x) = \left(1 + b_1 D + b_2 D^2 + \cdots\right)g(x).$$

Example 31. Find a particular solution of

$$y''' - y = x^7 + x^3 + x + 2.$$

We have

$$y_p = \frac{1}{D^3 - 1}\left(x^7 + x^3 + x + 2\right).$$

On dividing 1 by the denominator we find

$$\begin{aligned} y_p &= -\left(1 + D^3 + D^6 + D^9 + \cdots\right)\left(x^7 + x^3 + x + 2\right) \\ &= -x^7 - x^3 - x - 2 - 210x^4 - 6 - 5040x. \end{aligned}$$

The following example illustrates the possibilities of manipulation.

Example 32. Find a particular solution of

$$y^{v} + y^{iv} + y''' + y'' + y' + y = x^5 + 2x^2 + 3.$$

We have

$$\begin{aligned} y_p &= \frac{1}{D^5 + D^4 + D^3 + D^2 + D + 1}\left(x^5 + 2x^2 + 3\right) \\ &= \frac{1}{1 - D^6}(1 - D)\left(x^5 + 2x^2 + 3\right) \\ &= \frac{1}{1 - D^6}\left(x^5 + 2x^2 + 3 - 5x^4 - 4x\right) \\ &= \left(1 + D^6 + D^{12} + \cdots\right)\left(x^5 + 2x^2 + 3 - 5x^4 - 4x\right) \\ &= x^5 - 5x^4 + 2x^2 - 4x + 3. \end{aligned}$$

PROBLEMS

Find a particular solution by the use of symbolic methods.

154. $y'' - 3y + 2y = xe^x$.
155. $y''' + y'' + y = x^4 + 2x + 1$.
156. $y''' + y'' + y' + y = x^4 + 2x + 1$.
157. $y''' - 2y' + y = x^3 + (x+1)^2$.
158. $y''' + y'' = 6x^2 + 4$.
159. $y''' - 8y = x^2$.
160. $y''' + 2y'' + y' = x$.
161. $y'' - y = x^2 e^{2x}$.
162. $y'' + 4y = e^x$.
163. $y^{iv} + y = 2\cos 3x$.
164. $y'' + 9y = 5\sin x$.
165. $y'' - 4y' + 4y = x^2 e^{2x}$.
166. $y''' - 3y'' + 3y' - y = e^x$.
167. $y''' - y = \sin x$.
168. $y'' + y = \cos x$.
169. $y^{iv} - y = e^x$.
170. $y'' + y = x^2 - x + 2$.
171. $y'' + y = x\cos x$.
172. $y'' + 2y' + 2y = x^2 e^{-x}$.

2.4.5 Laplace Transform

The Laplace transformation (after the French mathematician Pierre Simon Laplace (1779 - 1827)) leads to a method of solving linear differential equations which yields particular solutions without the necessity of finding general solutions and evaluating constants.

Now suppose that $f(t)$ is defined for all $t \geq 0$. The *Laplace transform* of f, which we will denote by $L\{f(t)\}$ or $F(s)$, is defined by the equation

$$F(s) = \int_0^\infty e^{-st} f(t)dt \tag{21}$$

whenever this improper integral converges.

Example 33. Find $L\{e^{at}\}$.

We have

$$L\left\{e^{at}\right\} = \int_0^\infty e^{-st}e^{at}dt = \int_0^\infty e^{(a-s)t}dt$$

$$= \lim_{A\to\infty} \int_0^A e^{(a-s)t}dt = \lim_{A\to\infty} \frac{1}{a-s} e^{(a-s)t}|_0^A$$

$$= \lim_{A\to\infty} \frac{1}{a-s}\left(e^{(a-s)A} - 1\right) = \frac{1}{s-a} = F(s).$$

provided $s > a$.

Example 34. Find $L\{t^n\}$.

We have

$$L\{t^n\} = \int_0^\infty t^n e^{-st}dt = \frac{1}{s^{n+1}} \int_0^\infty (ts)^n e^{-st} d(st)$$

$$= \frac{1}{s^{n+1}} \int_0^\infty y^{(n+1)-1} e^{-y} dy$$

where $y = ts$.

The last improper integral is well-known *gamma-function*

$$\Gamma(x) = \int_0^\infty y^{x-1}e^{-y}dy.$$

Since $\Gamma(n+1) = n!$ we have

$$L\{t^n\} = \frac{n!}{s^{n+1}} = F(s).$$

PROBLEMS

Find Laplace transformations directly from the definition (21) (answers see Table 2.2).

1. $L\{1\}$.
2. $L\{\sin at\}$.
3. $L\{\cos at\}$.

4. $L\{\sinh at\}$.
5. $L\{\cosh at\}$.
6. $L\{e^{bt}\cos at\}$.
7. $L\{e^{bt}\sin at\}$.
8. $L\{t^n e^{at}\}$.
9. $L\{t\sin at\}$.
10. $L\{t\cos at\}$.
11. $L\{e^{at}\sinh bt\}$.
12. $L\{e^{at}\cosh bt\}$.

Properties of Laplace transformations

(a) $L\{af(t)+bg(t)\} = aF(s)+bG(s)$, where a, b are arbitrary constants.
(b) $L\{t^n f(t)\} = (-1)^n F^{(n)}(s)$, $\quad n = 1, 2, 3, \cdots$.
(c) $L\{e^{at}f(t)\} = F(s-a)$.
(d) $L\{f(at)\} = a^{-1}F\left(\frac{s}{a}\right)$.
(e) $L\{f(t-t_0)\} = e^{-st_0}F(s), \quad t_0 > 0$.
(f) $L\left\{\int_0^t f(t-\tau)g(\tau)d\tau\right\} = F(s)G(s)$, *convolution theorem.*

The integral in the left member is called the *convolution* of f and g. This formula states that the inverse Laplace transform of the product of two functions equals the convolution of two functions.

(g) $L\{t^{-1}f(t)\} = \int_s^\infty F(s)ds$.
(h) $L\left\{\int_0^t f(t)dt\right\} = s^{-1}F(s)$.
(i) $L\left\{\int_t^\infty t^{-1}f(t)dt\right\} = s^{-1}\int_0^s F(s)ds$.
(j) $L\{\delta(t)\} = 1$, where $\delta(t) = \begin{cases} \infty, & t = 0 \\ 0, & t > 0 \end{cases}$ is so called *Dirac delta-function.*

Example 35. Find $L\{t\sin t\}$.
Since from the Table

$$L\{\sin t\} = \frac{1}{s^2+1}$$

using (b), we have

$$L\{t\sin t\} = (-1)\frac{d}{ds}\left(\frac{1}{s^2+1}\right) = \frac{2s}{(s^2+1)^2}.$$

Table 2.2.

Formula	$F(s)$	$f(t)$
1.	$1/s$	1
2.	$\dfrac{1}{s-a}$	e^{at}
3.	$\dfrac{1}{s^n}$	$\dfrac{t^{n-1}}{(n-1)!}$
4.	$\dfrac{1}{(s-a)^{n+1}}$	$\dfrac{t^n e^{at}}{n!}$
5.	$\dfrac{a}{s^2+a^2}$	$\sin at$
6.	$\dfrac{s}{s^2+a^2}$	$\cos at$
7.	$\dfrac{a}{s^2-a^2}$	$\sinh at$
8.	$\dfrac{s}{s^2-a^2}$	$\cosh at$
9.	$\dfrac{a}{(s-b)^2+a^2}$	$e^{bt}\sin at$
10.	$\dfrac{s}{(s-b)^2+a^2}$	$e^{bt}\left(\cos at+\dfrac{b}{a}\sin at\right)$
11.	$\dfrac{1}{(s^2+a^2)^2}$	$\dfrac{\sin at - at\cos at}{2a^3}$
12.	$\dfrac{s}{(s^2+a^2)^2}$	$\dfrac{t\sin at}{2a}$
13.	$\dfrac{1}{a}F\left(\dfrac{s}{a}\right),\ a>0$	$f(at)$
14.	$F(s-a)$	$e^{at}f(t)$
15.	$F^{(n)}(s)$	$(-1)^n t^n f(t)$
16.	$s^n F^{(m)}(s)$	$(-1)^m \dfrac{d^n}{dt^n}\left[t^m f(t)\right],\ m\ge n$
17.	$e^{-st_0}F(s)$	$f(t-t_0)$
18.	$\dfrac{F(s)}{s^n}$	$\int_0^t \cdots \int_0^t f(t)(dt)^n$
19.	$F(s)G(s)$	$\int_0^t f(t-\tau)g(\tau)d\tau$
20.	$\int_s^\infty F(s)ds$	$\dfrac{f(t)}{t}$
21.	$s^{-1}\int_s^\infty F(s)ds$	$\int_0^t \dfrac{f(t)}{t}dt$
22.	$s^{-1}\int_0^s F(s)ds$	$\int_t^\infty \dfrac{f(t)}{t}dt$
23.	e^{-as}	$\delta(t-a)\ (t>0)$

Short Table of Laplace Transforms

Example 36. Find $L\{e^{at}\cos t\}$.
From the Table

$$L\{\cos t\} = \frac{s}{s^2+1}$$

and using (c)

$$L\{e^{at}\cos t\} = \frac{s-a}{(s-a)^2+1}.$$

Example 37. Find $L\left\{\frac{\sin t}{t}\right\}$.
Since

$$L\{\sin t\} = \frac{1}{s^2+1}$$

we have

$$L\left\{\frac{\sin t}{t}\right\} = \int_s^\infty \frac{1}{s^2+1}ds = \lim_{A\to\infty}\int_s^A \frac{1}{s^2+1}ds$$

$$= \lim_{A\to\infty} \arctan s|_s^A = \lim_{A\to\infty}(\arctan A - \arctan s) = \frac{\pi}{2} - \arctan s = \arctan\frac{1}{s}.$$

Example 38. Find the inverse Laplace transform of

$$\frac{1}{(s+1)(s-2)}.$$

From the Table we have

$$L^{-1}\left\{\frac{1}{s+1}\right\} = e^{-t}$$

and

$$L^{-1}\left\{\frac{1}{s-2}\right\} = e^{2t}.$$

Hence using convolution theorem we get

$$L^{-1}\left\{\frac{1}{(s+1)(s-2)}\right\} = L^{-1}\left\{\frac{1}{s+1}\frac{1}{s-2}\right\}$$

$$= \int_0^t e^{-(t-\tau)}e^{2\tau}d\tau = e^{-t}\int_0^t e^{3\tau}d\tau = e^{-t}\frac{1}{3}e^{3\tau}\Big|_0^t = \frac{1}{3}e^{-t}\left(e^{3t}-1\right).$$

THEOREM. *Suppose that the functions $f,\ f',\cdots,f^{(n-1)}$ are continuous, and that $f^{(n)}$ is piecewise continuous on any interval $0 \le t \le A$. Suppose further that there exist constants K, a, and M such that*

$$f(t) \le Ke^{at},\ |f'(t)| \le Ke^{at},\ \cdots,\ |f^{(n-1)}(t)| \le Ke^{at} \quad \text{for } t \ge M.$$

Then $L\left\{f^{(n)}(t)\right\}$ exists for $s > a$ and is given by

$$L\left\{f^{(n)}(t)\right\} = s^nL\{f(t)\} - s^{n-1}f(0) - \cdots - sf^{(n-2)}(0) - f^{(n-1)}(0). \tag{22}$$

Consider now the differential equation with constant coefficients

$$a_0y^{(n)} + a_1y^{(n-1)} + a_2y^{(n-2)} + \cdots + a_ny = g(t) \tag{23}$$

We wish to find the solution $y(t)$ with initial conditions

$$y(0) = y_0,\ y'(0) = y_1,\ \ldots,\ y^{(n-1)}(0) = y_{n-1}.$$

We take the Laplace transform of both members of (23). By application of (22), we have after transporting,

$$\left(a_0s^n + a_1s^{n-1} + \cdots + a_n\right)F(s) = y_0\left(a_0s^{n-1} + a_1s^{n-2} + \cdots + a_{n-1}\right)$$

$$+y_1\left(a_0s^{n-2} + a_1s^{n-3} + \cdots + a_{n-2}\right) + \cdots y_{n-2}\left(a_0s + a_1\right) + y_{n-1}a_0 + G(p) \tag{24}$$

where $G(p) = L\{g(t)\}$.

Formula (24) is known as the *subsidiary equation.* $G(p)$ must be found by integration or from a table.

The subsidiary equation is used in the following way: Solving for $F(s)$, let us suppose that the result can be written as the sum of terms - partial fractions, for instance - which are the Laplace transforms of a known functions, such as those appearing in the Table. We can then write down $y(t)$ itself, and the solution has been found.

Example 39. Find the solution of the differential equation

$$y''' - y' = 1$$

which satisfies the initial conditions

$$y(0) = y'(0) = y''(0) = 1.$$

We will assume that this initial value problem has a solution $y(t)$ which with its first two derivatives satisfies the conditions of the Theorem. Then, taking the Laplace transform of the differential equation gives

$$(s^3 - s)Y(s) = y(0)(s^2 - 1) + y'(0)s + y''(0) + \frac{1}{s}$$

where the transform $L\{1\} = 1/s$ has been obtained from Table. Substituting for $y(0)$, $y'(0)$ and $y''(0)$ from the initial conditions and solving for $Y(s)$ yields

$$Y(s) = \frac{s^3 + s^2 + 1}{s^2(s-1)(s+1)}.$$

Using partial fractions we can write $Y(s)$ in the form

$$Y(s) = \frac{A}{s} + \frac{B}{s^2} + \frac{C}{s-1} + \frac{D}{s+1}$$

$$= \frac{A(s^2-1)s + B(s^2-1) + C(s+1)s^2 + Ds^2(s-1)}{s^2(s-1)(s+1)}.$$

Equating coefficients of like powers of s, we have

$$A + C + D = 1,$$

$$B + C - D = 1,$$

$$-A = 0,$$

$$-B = 1.$$

Solving, we find $A = 0,\ B = -1,\ C = 3/2,\ D = -1/2$

Referring to Table, it follows that

$$y(t) = -t + \frac{3}{2}e^t - \frac{1}{2}e^{-t}.$$

Example 40. Find the solution of the differential equation

$$y'' - 2y' = e^t$$

given that $y(0) = y'(0) = 0$.

Since $L\{e^t\} = 1/(s-1)$, the subsidiary equation is

$$(s^2 - 2s)Y = \frac{1}{s-1}$$

whence

$$Y(s) = \frac{1}{s(s-2)(s-1)}$$

Breaking this into partial fractions, we find

$$Y(s) = \frac{A}{s} + \frac{B}{s-2} + \frac{C}{s-1}.$$

Equating coefficients of like powers of s, we find $A = B = 1/2,\ C = -1$. A reference to the Table produces the solution

$$y(t) = \frac{1}{2} + \frac{1}{2}e^{2t} - e^t.$$

Example 41. Solve the initial-value problem

$$y'' + y = 2,\ y(0) = 1,\ y'(0) = 0.$$

The subsidiary equation is

$$(s^2 + 1)Y(s) = s + \frac{2}{s}$$

whence

$$Y(s) = \frac{s}{s^2+1} + 2\frac{1}{s}\frac{1}{s^2+1},$$

$$y_p = L^{-1}\left\{\frac{s}{s^2+1}\right\} + 2L^{-1}\left\{\frac{1}{s}\frac{1}{s^2+1}\right\}.$$

From the Table

$$L^{-1}\left\{\frac{s}{s^2+1}\right\} = \cos t,$$

$$L^{-1}\left\{\frac{1}{s^2+1}\right\} = \sin t,$$

$$L^{-1}\left\{\frac{1}{s}\right\} = 1$$

using the formula (h), we have

$$y_p(t) = \cos t + 2\int_0^t \sin\tau d\tau = \cos t + 2\left(-\cos\tau|_0^t\right) = 2 - \cos t.$$

Example 42. Solve the initial-value problem

$$y'' + y = e^t,\ y(0) = 0,\ y'(0) = 0.$$

The subsidiary equation is

$$(s^2+1)Y(s) = \frac{1}{s-1}$$

whence

$$Y(s) = \frac{1}{s-1}\frac{1}{s^2+1},$$

or

$$y_p = L^{-1}\left\{\frac{1}{s-1}\frac{1}{s^2+1}\right\}$$

using the convolution theorem, we get

$$y_p = \int_0^t e^{t-\tau}\sin\tau d\tau = e^t\int e^{-\tau}\sin\tau d\tau$$

$$= -e^t\frac{1}{2}\left[e^{-\tau}\sin\tau + e^{-\tau}\cos\tau\right]_0^t = \frac{1}{2}\left(e^t - \sin t - \cos t\right). \quad (2.1)$$

We can write down a formula for the solution in a very common case. If $Y(s)$ is a rational function

$$Y(s) = \frac{P(s)}{Q(s)}$$

in which the polynomial $Q(s)$ has distinct roots $s_1, s_2, \ldots, s_k$, the numerator being of lower degree, we can use a well-known formula for the partial fractions,

$$Y(s) = \sum_{j=1}^{k}\frac{P(s_j)}{(s-s_j)Q'(s_j)}.$$

We can now write down the inverse function, getting an expansion due to the English physicist Oliver Heaviside (1850-1925)

$$y(t) = \frac{P(s_1)}{Q'(s_1)}e^{s_1 t} + \ldots + \frac{P(s_k)}{Q'(s_k)}e^{s_k t}.$$

In the case of repeated roots

$$Q(s) = (s-a_1)^{n_1}(s-a_2)^{n_2}\cdots(s-a_r)^{n_r}$$

we have

$$y(t) = \sum_{k=1}^{r}\sum_{j=1}^{n_k}\frac{A_{kj}}{(n_k-j)!}t^{n_k-j}e^{a_k t},$$

where

$$A_{kj} = \frac{1}{(j-1)!}\left[\frac{d^{j-1}}{ds^{j-1}}\frac{(s-a_k)^{n_k}P(s)}{Q(s)}\right]_{s=a_k}.$$

Example 43. Solve the initial value problem

$$y'' - y' - 2y = e^t; \quad y(0) = 1, \ y'(0) = 2.$$

The subsidiary equation is

$$(s^2 - s - 2)Y(s) = (s-1) + 2 + \frac{1}{s-1}.$$

Whence

$$Y(s) = \frac{s^2}{(s-1)(s+1)(s-2)}.$$

Here

$$\frac{P(s)}{Q'(s)} = \frac{s^2}{3s^2 - 4s - 1}$$

and this has at the roots $-1, \ 1, \ 2$ the values $1/6, \ -1/2, \ 4/3$ respectively. Hence

$$y(t) = \frac{1}{6}e^{-t} - \frac{1}{2}e^t + \frac{4}{3}e^{2t}.$$

Using the Laplace transform it is possible to solve some *integro - differential* equation.

Example 44. Solve

$$\int_0^x \left[(x-t)^2 y'(t) + y(t)\right] dt = x^2, \ y(0) = 0.$$

Using the convolution theorem, we take the Laplace transform of both members of the given equation and have

$$\frac{2}{s^3}sY(s) + \frac{Y(s)}{s} = \frac{2}{s^3},$$

or

$$(2+s)Y(s) = \frac{2}{s}$$

whence

$$Y(s) = \frac{2}{s\,(s+2)} = \frac{1}{s} - \frac{1}{s+2}.$$

Applying to the Table, we get the required solution

$$y = 1 - e^{-2x}.$$

Example 45. Solve

$$y' + y = e^x - \int_0^x e^{x-t} y(t)dt, \; y(0) = 0.$$

Taking the Laplace transform of both members, we obtain

$$(s+1)Y(s) = \frac{1}{s-1} - \frac{1}{s-1}Y(s)$$

whence

$$Y(s) = \frac{1}{s^2}.$$

Thus applying to the Table, we find the solution

$$y = x.$$

PROBLEMS

In each of Problems 173 through 198 use the Laplace transform to solve the given initial value problem.

173. $y'' - y' - 6y = 0;\; y(0) = 1,\; y'(0) = 1.$
174. $y'' + 3y' + 2y = 0;\; y(0) = 1,\; y'(0) = 0.$
175. $y'' - 2y' + 2y = 0;\; y(0) = 0,\; y'(0) = 1.$
176. $y'' - 4y' + 4y = 0;\; y(0) = 1,\; y'(0) = 1.$
177. $y^{iv} - 4y''' + 6y'' - 4y' + y = 0;\; y(0) = y''(0) = 0,\; y'(0) = y'''(0) = 1.$
178. $y^{iv} - y = 0;\; y(0) = y''(0) = 1,\; y'(0) = y'''(0) = 0.$
179. $y'' - 2y' - 2y = 0;\; y(0) = 2,\; y'(0) = 0.$
180. $y'' + \omega^2 y = \cos 2t;\; \omega^2 = 4;\; y(0) = 1,\; y'(0) = 0.$
181. $y'' - 2y' + 2y = \cos\theta;\; y(0) = 1,\; y'(0) = 1.$
182. $y'' - 2y' + 2y = e^{-t};\; y(0) = 0,\; y'(0) = 1.$
183. $y'' + y = \sin 2t;\; y(0) = 0,\; y'(0) = 1.$
184. $y'' - y' - 2y = 0;\; y(0) = 1,\; y'(0) = 1.$
185. $y''' - y' = 2\cos t;\; y(0) = 3,\; y'(0) = 2,\; y''(0) = 1.$
186. $y''' - y' = 2\cos t;\; y(0) = y'(0) = y''(0) = 0.$
187. $y'' - 3y' + 2y = 2e^{3t};\; y(0) = 2,\; y'(0) = 3.$
188. $y'' + 5y' + y = 2;\; y(0) = 2,\; y'(0) = 0.$

189. $y'' + 3y' + 2y = 2(t^2 + t + 1);\ y(0) = 2,\ y'(0) = 0.$
190. $y' + 2y = 2t - 1;\ y(0) = 3.$
191. $y'' + y' = 3t^2 - 6;\ y(0) = 0,\ y'(0) = 1.$
192. $y'' + 4y' + 5y = 0;\ y(0) = 2,\ y'(0) = 3.$
193. $y'' - 2y' + 5y = 10;\ y(0) = y'(0) = 0.$
194. $y'' - 2y' + y = 2e^t;\ y(0) = 3,\ y'(0) = 1.$
195. $y^{iv} - y''' = 0;\ y(0) = y'(0) = y''(0) = y'''(0) = 1.$
196. $y^{iv} - y''' = 0;\ y(0) = y'(0) = 2;\ y''(0) = y'''(0) = 1.$
197. $y''' + y = 3;\ y(0) = y'(0) = y''(0) = 0.$
198. $y'' + y = 2\sin t;\ y(0) = 0;\ y'(0) = 1.$

In each of Problems 199 through 207 use the Laplace transform and convolution theorem to solve the given integro - differential equations.

199. $y' + \int_0^x [\cos(x-t) - 2]y(t)dt = 0,\ y(0) = 4.$
200. $y' - 2\int_0^x e^{x-t}y(t)dt = 0,\ y(0) = 1.$
201. $y' - \int_0^x (x-t)y(t)dt = \cos x,\ y(0) = 1.$
202. $y' - y + \int_0^x \sin(x-t)y(t)dt = 1 - \sin x,\ y(0) = 0.$
203. $y'' + \int_0^x \sin(x-t)\,[y''(t) + y(t)]dt = 2\cos x,\ y(0) = 0,\ y'(0) = 0.$
204. $y'' - 4\int_0^x e^{-(x-t)}\,[y'(t) + y(t)]dt = 0,\ y(0) = 0,\ y'(0) = 12.$
205. $y'' - y - 4\int_0^x (x-t)\cos(x-t)y(t)dt = 0,\ y(0) = 4,\ y'(0) = 0.$
206. $y'' + y = \cos x + \int_0^x \sin(x-t)y(t)dt = 0,\ y(0) = 0,\ y'(0) = 2.$
207. $y''' + 4y'' - 3y' + 5y = \int_0^x \cos(x-t)y''(t)dt = 0,$
$y(0) = y'(0) = 0,\ y''(0) = 1.$

2.5 Equations with polynomial coefficients

The general form of these equations is

$$p_0(x)y^{(n)}(x) + p_1(x)y^{(n-1)}(x) + \cdots + p_n(x)y(x) = f(x), \tag{1}$$

where $p_i(x)$ $i = 0, 1, \ldots, n$ are some polynomials, $f(x)$ is a function. Exponentials and powers solution of these equations can be found as has been shown in Sec. 2.2.

Here the following are cases of reduction of Eq.(1) to an equation with constant coefficients:

1) by making change of variable $x = \varphi(t)$, where $\varphi(t)$ is a suitable function, we obtain a new equation for $y(t)$ with constant coefficients;

2) by substituting $y(x) = z(x)g(x)$, where $g(x)$ is a suitable function and $z(x)$ is a new unknown function, we obtain a new equation for $z(x)$ with constant coefficients;

3) by making change of variable and a substitution as above we obtain a new equation with constant coefficients;

4) the given equation is not reduced to an equation with constant coefficients.

2.5.1 Changes of Variable

Eq.(1) may be reduced to an equation with constant coefficients by change of variable $t = \varphi(x)$ if and only if

$$t = C \int \left(\frac{p_n(x)}{p_0(x)} \right)^{1/n} dx, \tag{2}$$

where C is an arbitrary constant.

Euler's Equation

Example 1. Find the general solution of the differential equation

$$x^2 y'' + 13xy' - 13y = 0.$$

According to Eq.(2) we find

$$t = C \int \sqrt{\frac{-13}{x^2}}\, dx = C\sqrt{-13} \ln |x| + C_1,$$

and setting $C_1 = 0$, $C\sqrt{-13} = 1$ we obtain $|x| = e^t$. Let $x > 0$, then we find, after some minor calculations,

$$y''(t) + 12y'(t) - 13y(t) = 0.$$

The coefficients in this equation are constants. Its solution is

$$y(t) = C_1 e^t + C_2 e^{-13t}.$$

Hence the general solution of the given equation is

$$y(x) = C_1 x + \frac{C_2}{x^{13}},$$

where C_i are arbitrary constants.

The equation with the coefficients $p_i(x) = p_i(x+b)^{n-i}$, $\quad i = 0, 1, \ldots, n$, where p_i and b are arbitrary constants, is called Euler-Lagrange's equation when $b \neq 0$, or Euler's equation when $b = 0$, after the Russian and Germany mathematician Leonard Euler (1707-1783). To solve the Euler's equation

$$p_0 x^n y^{(n)} + p_1 x^{n-1} y^{(n-1)} + \cdots + p_n y = 0 \tag{3}$$

in any interval not containing the origin, use the substitution $y = x^m$, and compute the roots m_1, m_2, ..., m_n of the indicial equation

$$F(m) = p_0 m(m-1)\ldots(m-n+1) + p_1 m(m-1)\ldots(m-n) + \cdots + p_n = 0. \quad (4)$$

If all roots m_k of Eq.(4) are real and distinct, then the general solution is

$$y = \sum_{k=1}^{n} C_k x^{m_k}. \quad (5)$$

If some roots are real and repeated, say $m_1 = m_2 = \ldots = m_l$, where $l < n$, and all the other roots are real and distinct, then

$$y = x^{m_1} \sum_{k=1}^{l} C_k (\ln|x|)^{k-1} + \sum_{k=l+1}^{n} C_k x^{m_k}. \quad (6)$$

If some pair of roots are complex, say $m_1 = \bar{m}_2 = \lambda + i\mu$, and all the other roots are real and distinct, then

$$y = x^{\lambda}\left(C_1 \cos(\mu \ln|x|) + C_2 \sin(\mu \ln|x|)\right) + \sum_{k=3}^{n} C_k x^{m_k}. \quad (7)$$

For more complicated cases of roots of Eq.(4) the solution of the Euler equation can be determined by using the solutions which has been obtained above.

Example 2. Solve each of the following Euler's equations:

$$\text{a)} \quad x^2 y'' - xy' + y = 0;$$

$$\text{b)} \quad x^2 y'' + 3xy' + 5y = 0.$$

By solving the indicial equation for each of the given equation we find, after some minor algebraic calculations, $m_1 = m_2 = 1$ and $m_1 = m_2 = -1 + 2i$, respectively. Hence the general solution of the given equations is

$$\text{a)} \quad y = x\left(C_1 + C_2 \ln|x|\right),$$

$$\text{b)} \quad y = \frac{1}{x}\left(C_1 \cos(2\ln|x|) + C_2 \sin(2\ln|x|)\right),$$

respectively, where C_i are arbitrary constants.

Tshebyshev's Equation

Example 3. Find the general solution of the differential equation

$$\left(1 - x^2\right) y'' - xy' - 4y = 0, \qquad |x| < 1.$$

In the same manner as in Ex. 1 we find

$$t = C\int\sqrt{\frac{-4}{1-x^2}}dx = C\sqrt{-4}\arcsin x + C_1.$$

Let $C\sqrt{-4} = 1$ and $C_1 = 0$, then we obtain $t = \arcsin x$. Further we find

$$y''(t) - 4y(t) = 0.$$

The general solution of this equation is

$$y(t) = C_1e^{2t} + C_2e^{-2t}.$$

Hence the general solution of the given equation is

$$y(x) = C_1\exp\left(2\arcsin x\right) + C_2\exp(-2\arcsin x).$$

where C_i are arbitrary constants.

Example 4. Find the general solution of the differential equation

$$\left(1+x^2\right)y'' + xy' + 4y = 0.$$

Likewise to Ex. 2 we find $t = \text{arcsinh}x$ and

$$y''(t) + 4y(t) = 0.$$

Solutions of this equation are $y_1(t) = \cos 2t$, $y_2(t) = \sin 2t$. Hence the general solution of given equation is

$$y(x) = C_1\cos(2\text{arcsinh}x) + C_2\sin(2\text{arcsinh}x),$$

where C_i are arbitrary constants.

The equations which are given in Ex. 3 and Ex. 4 are called Tshebyshev's equations, after the Russian mathematician P.L. Tshebyshev (1821-1894). Note that Tshebyshev's equations are always only even order. For a Tshebyshev's equation by substituting $y = \sin^m t$ or $y = \sinh^m t$ the indicial equation and all forms of its solutions can be found completely analogous to that of Euler's equation.

2.5.2 *Substitutions*

Eq.(1) may be reduced to an equation with constant coefficients by the substitution $y(x) = z(x)g(x)$ if and only if

$$g(x) = \exp\left(C\frac{1}{n}x - \frac{1}{n}\int\frac{p_1(x)}{p_0(x)}\,dx\right), \tag{8}$$

where C is an arbitrary constant.

Example 5. By substituting $y(x) = z(x)g(x)$ reduce the differential equation

$$x^3y''' + x^2(a_1x - 3\alpha)y'' + x\left(a_2x^2 - 2a_1\alpha x + 3\alpha(\alpha+1)\right)y'$$

$$+\left(a_3x^3 - a_2\alpha x^2 + a_1\alpha(\alpha+1)x - \alpha(\alpha+1)(\alpha+2)\right)y = 0$$

to an equation with constant coefficients. All a_i and α are constants.

According to Eq. (8) we find

$$g(x) = \exp\left(\frac{C}{3}x - \frac{1}{3}\int(a_1x^3 - 3\alpha x^2)\frac{dx}{x^3}\right) = |x|^\alpha,$$

at $C = a_1$. Let $x > 0$, then substituting $y = x^\alpha z$ in the given equation we find, after some minor calculations,

$$z''' + a_1z'' + a_2z' + a_3z = 0,$$

that is, an equation with constant coefficients.

Example 6. Find the general solution of the equation

$$y''' - 2(3x+1)y'' + \left(12x^2 + 8x - 7\right)y' - 2\left(4x^3 + 4x^2 - 7x - 3\right)y = 0.$$

According to Eq.(8) we find

$$g(x) = \exp\left(\frac{C}{3}x + \frac{2}{3}\int(3x+1)dx\right) = e^{x^2},$$

at $C = -2$. By substituting $y = \exp(x^2)z(x)$ in the given equation we obtain, after some minor calculations,

$$z''' - 2z'' - z' + 2z = 0.$$

Solutions of this equation are

$$z_1 = e^x, \quad z_2 = e^{-x}, \quad z_3 = e^{2x}.$$

Hence the general solution of the given equation is

$$y(x) = C_1e^{x^2+x} + C_2e^{x^2-x} + C_3e^{x^2+2x},$$

where C_i are arbitrary constants.

2.5.3 Substitutions and Changes of Variable

First by substituting $y(x) = z(x)g(x)$, where $g(x)$ is a suitable function and $z(x)$ is a new unknown function, we reduce the given equation to an Euler's equation, or an Euler-Lagrange's equation, or a Tshebyshev's equation. Then by making change of variable we reduce this new equation for $z(x)$ to an equation with constant coefficients.

1. Eq.(1) may be reduced to an Euler's equation by substituting $y(x) = z(x)g(x)$ if and only if

$$g(x) = \exp\left(\frac{C}{n}\ln|x| - \frac{1}{n}\int\frac{p_1(x)}{p_0(x)}\,dx\right), \tag{9}$$

where C is an arbitrary constant.

Example 7. Find the general solution of the differential equation

$$x^3y''' + x^2(3x+2)y'' + x\left(3x^2+4x-4\right)y' + \left(x^3+2x^2-4x+4\right)y = 0.$$

It is easy to verify that the change of variable according to Eq.(2) and the substitution according to Eq.(8) does not reduce the given equation to an equation with constant coefficients. However, according to Eq.(9) we find

$$g(x) = \exp\left(\frac{C}{3}\ln|x| - \frac{1}{3}\int(3x+3)\frac{dx}{x}\right) = e^{-x},$$

at $C = 2$. Then substituting $y(x) = \exp(-x)z(x)$ in the given equation yields, after some calculations,

$$x^3z''' + 2x^2z'' - 4xz' + 4z = 0,$$

that is, the Euler's equation. According to Eq.(4) we obtain the auxiliary equation

$$m(m-1)(m-2) + 2m(m-1) - 4m + 4 = 0$$

The roots are $m_1 = 1$, $m_2 = 2$, $m_3 = -2$. Hence the general solution of the given equation is

$$y(x) = e^{-x}\left(C_1x + C_2x^2 + C_3x^{-2}\right),$$

where C_i are arbitrary constants.

2. When Eq.(1) is a second order equation, then it may be reduced to a Tshebyshev's equation by substituting $y(x) = g(x)z(x)$ if and only if

$$g(x) = \exp\left(\frac{1}{4}\ln(1-x^2) - \frac{1}{2}\int\frac{p_1(x)}{p_0(x)}\,dx\right), \tag{10a}$$

when $p_0(x) \sim 1 - x^2$, or only if

$$g(x) = \exp\left(\frac{1}{4}\ln(1+x^2) - \frac{1}{2}\int\frac{p_1(x)}{p_0(x)}\,dx\right), \tag{10b}$$

when $p_0(x) \sim 1 + x^2$.

Example 8. Find the general solution of the differential equation

$$4\left(1-x^2\right)^2 y'' + 8\left(1-x^2\right)^2 y' + \left(4x^4 - 27x^2 + 26\right) y = 0, \quad |x| < 1.$$

According to Eq. (10a) we find

$$g(x) = \exp\left(\frac{1}{4}\ln\left(1-x^2\right) - x\right) = \left(1-x^2\right)^{1/4} e^{-x}.$$

The substitution $y(x) = (1-x^2)^{1/4} e^{-x} z(x)$ into the given equation then yields, after some minor calculations,

$$\left(1-x^2\right) z'' - xz' + 5z = 0.$$

Further, proceeding as in Ex.3 we obtain

$$z''(t) + 5z(t) = 0.$$

The linearly independent solutions of this equation are $z_1 = \cos(\sqrt{5}t)$ and $z_2 = \sin(\sqrt{5}t)$. Hence the general solution of the given equation is

$$y(x) = e^{-x}\left(1-x^2\right)^{1/4}\left(C_1 \cos\left(\sqrt{5}\arcsin x\right) + C_2 \sin\left(\sqrt{5}\arcsin x\right)\right),$$

where C_i are arbitrary constants.

2.5.4 *Series Solutions*

If Eq. (1) is not reduced to an equation with constant coefficients, then its solution can be found as a power series of $x - \zeta$, where ζ is a prescribed point. There is considered only that case when $p_0(x)$ is a constant. For the opposite case at $n = 2$ see Sec. 3.1, but, in general, it is beyond the scope of our text.

Homogeneous equation

The general solution of the homogeneous equation corresponding to Eq. (1) in a neighborhood of a point ζ has a power series

$$y(x) = \sum_{k=0}^{\infty} a_k (x-\zeta)^k, \tag{11}$$

where a_k are undetermined coefficients. This power series is convergence in the region $|x - \zeta| < \rho$, where ρ is the radius of convergence, which is determined as

$$\rho = \lim_{k\to\infty}\left|\frac{a_k}{a_{k+1}}\right|. \tag{12}$$

Note that radius of convergence of the series solution $y(x)$ is at least as large as minimum of the radii of convergence of the power series for each $p_i(x)$. Since all $p_i(x)$ are polynomials then they have the infinite radii of convergence. Therefore, the radius of convergence of the series (11) is also infinite. The coefficients a_k in the series solution are determined by substituting the series (11) for $y(x)$ in the corresponding homogeneous equation and equating the coefficients of like powers of $x - \zeta$ to zero.

Example 9. Determine the fundamental set of solutions of the differential equation

$$y'''(x) + \left(x^2 - 2x + 5\right) y''(x) + (x - 8)y'(x) - 4y(x) = 0$$

in a neighborhood of the point $\zeta = 0$. Determine the radius of convergence for each solution.

According to Eq.(11) we look for a solution of the form

$$y(x) = \sum_{k=0}^{\infty} a_k x^k.$$

Then

$$y' = \sum_{k=0}^{\infty} a_k k x^{k-1},$$

$$y'' = \sum_{k=0}^{\infty} a_k k(k-1)x^{k-2},$$

$$y''' = \sum_{k=0}^{\infty} a_k k(k-1)(k-2)x^{k-3}.$$

A substitution for y, y', y'' and y''' in the given equation yields, after some minor algebraic simplifications,

$$\sum_{k=0}^{\infty} \Big(a_k k(k-1)(k-2)x^{k-3} + a_k(k^2 - 4)x^k - 2a_k k(k+3)x^{k-1} + 5a_k k(k-1)x^{k-2} \Big) = 0.$$

Shifting the index of summation in the first, third and fourth series gives

$$\sum_{k=0}^{\infty} \Big(a_{k+3}(k+3)(k+2)(k+1) + a_k(k^k - 4) - 2a_{k+1}(k+1)(k+4) + 5a_{k+2}(k+2)(k+1) \Big) x^k = 0.$$

Equating the coefficients of like powers of x to zero we obtain

$$a_{k+3}(k+3)(k+2)(k+1) + a_k\left(k^2 - 4\right) - 2a_{k+1}(k+1)(k+4) + 5a_{k+2}(k+2)(k+1) = 0,$$

where $k = 0, 1, \ldots$. By making change of index of summation $k+3 = n+1$ we obtain the *recurrence relation*

$$a_{n+1} = -\frac{5a_n n(n-1) - 2a_{n-1}(n-1)(n+2) + a_{n-2}n(n-4)}{n\,(n^2-1)},$$

where $n \geq 2$. The coefficients a_0, a_1 and a_2 are arbitrary. Now we look for that solution $y_1(x)$, which satisfy the initial conditions

$$y_1(0) = 1, \quad y_1'(0) = 0, \quad y_1''(0) = 0$$

(see Sec. 2.2, Eq.(4)). By substituting the power series for $y(x)$ in these conditions we find $a_0 = 1$, $a_1 = 0$, and $a_2 = 0$. Further by the recurrence relation we determine term by term

$$a_3 = \frac{2}{3}, \quad a_4 = -\frac{5}{6}, \quad a_5 = \frac{37}{30}, \quad a_6 = -\frac{13}{9}$$

and so on. Hence

$$y_1(x) = 1 + \frac{2}{3}x^3 - \frac{5}{6}x^4 + \frac{37}{30}x^5 - \frac{13}{9}x^6 + \ldots$$

Similarly we look for the solutions $y_2(x)$ and $y_3(x)$ as the power series which satisfy the initial conditions

$$y_2(0) = 0, \quad y_2'(0) = 1, \quad y_2''(0) = 0,$$

$$y_3(0) = 0, \quad y_3'(0) = 0, \quad y_3''(0) = 1,$$

respectively. After some minor algebraic calculations, we obtain

$$y_2(x) = x + \frac{4}{3}x^3 - \frac{37}{24}x^4 + \frac{281}{120}x^5 + \ldots$$

$$y_3(x) = \frac{1}{2}x^2 - \frac{5}{6}x^3 + \frac{35}{24}x^4 + \frac{47}{24}x^5 + \ldots$$

For the ratio in the right-hand side of Eq.(12) we determine by dividing the obtained recurrence relation by a_n

$$\frac{a_{n+1}}{a_n} = -\left(\frac{5}{n+1} - 2\frac{a_{n-1}}{a_n}\frac{n+2}{n(n+1)} + \frac{a_{n-2}}{a_n}\frac{n-4}{n^2-1}\right).$$

Then we assume that

$$\frac{a_{n+1}}{a_n} = an^\alpha + o(n^\alpha)$$

at $n >> 1$, where a and α are unknown constants and we denote by $o(n^\alpha)$ the items of smaller order of n, i.e., $\lim\limits_{n\to\infty} \dfrac{o(n^\alpha)}{n^\alpha} = 0$. Further, we find

$$\frac{a_n}{a_{n-1}} = a(n-1)^\alpha + o\left((n-1)^\alpha\right) = an^\alpha + o(n^\alpha),$$

$$\frac{a_n}{a_{n-2}} = \frac{a_n}{a_{n-1}} \frac{a_{n-1}}{a_{n-2}} = a^2 n^{2\alpha} + o(n^{2\alpha}).$$

By substituting these expressions into obtained formula and collecting the equivalence terms of n we find

$$an^\alpha = -\frac{1}{n}\left(5 - \frac{2}{an^\alpha} + \frac{1}{a^2 n^{2\alpha}}\right) + o(n^\alpha).$$

It is easy to verify that this equation has a solution only if $\alpha < 0$. Equating the least powers of n and the coefficients at the least powers of n in both sides leads to

$$-\alpha = 1 + 2\alpha, \qquad a = -a^{-2}.$$

The roots of these equations are $a = -1$ and $\alpha = -1/3$. Now according to Eq. (12) we find the radius of convergence of the general solution

$$\rho = \lim_{k\to\infty} \left|\frac{a_k}{a_{k+1}}\right| = \lim_{k\to\infty} |-k^{1/3}| = \infty.$$

Hence the power series for $y_i(x)$ are convergent at all x. It is obvious since all $p_i(x)$ in the given equation are polynomials and its radii of convergence are infinite.

Nonhomogeneous equations

The series solution of Eq. (1) can be determined as has been shown above if the power series for $f(x)$ is known.

Example 10. Find the solution of the differential equation

$$y''(x) + x^3 y(x) = \sin x,$$

in a neighborhood of the point $\zeta = 0$ if $y(0) = 1$, $y'(0) = 0$.

We look for $y(x)$ as the power series (11). Taking into account that the power series for $\sin x$ in a neighborhood of the point $\zeta = 0$ is

$$\sin x = \sum_{k=0}^{\infty} (-1)^k \frac{x^{2k+1}}{(2k+1)!}.$$

By substituting the power series for $y(x)$ and $\sin x$ in the given equation we obtain

$$\sum_{k=0}^{\infty} \left(a_k k(k-1)x^{k-2} + a_k x^{k+3}\right) = \sum_{k=0}^{\infty} (-1)^k \frac{x^{2k+1}}{(2k+1)!}.$$

Let power of x be equal to $2l+1$, where l is a positive integer number. Equating the coefficients of this power of x then yields

$$a_{2l+3}(2l+3)(2l+2) + a_{2l-2} = \frac{(-1)^l}{(2l+1)!}.$$

Hence

$$a_{2l+3} = \frac{1}{(2l+3)(2l+2)}\left(\frac{(-1)^l}{(2l+1)!} - a_{2l-2}\right).$$

Likewise for power of x equal to $2l$ we obtain

$$a_{2l+2} = -\frac{a_{2l-3}}{(2l+2)(2l+1)}.$$

By using these recurrence relations and taking into account the initial conditions we find, after some calculations,

$$a_0 = 1, \quad a_1 = a_2 = 0, \quad a_3 = \frac{1}{3!}, \quad a_4 = 0,$$

$$a_5 = \frac{7}{6!}, \quad a_6 = 0, \quad a_7 = \frac{1}{7!}, \quad a_8 = -\frac{1}{336}, \ldots .$$

Hence the solution of the given initial value problem in the form of a power series is

$$y(x) = 1 + \frac{1}{3!}x^3 - \frac{7}{6!}x^5 + \frac{1}{7!}x^7 - \frac{1}{336}x^8 + \ldots .$$

PROBLEMS

Solve each of the following Euler's and Euler-Lagrange's equations.

1. $x^3y''' - 3x^2y'' + 6xy' - 6y = 0.$
2. $x^3y''' + x^2y'' = 0.$
3. $x^3y''' - 3x^2y'' + 7xy' - 8y = 0.$
4. $x^3y''' = 0.$
5. $x^3y''' + 7xy' + 13y = 0.$
6. $x^3y''' + 2x^2y'' + 4xy' - 4y = 0.$
7. $x^3y''' + 3x^2y'' = 0.$
8. $x^4y^{iv} + 6x^3y''' + 9x^2y'' + 3xy' + y = 0.$
9. $x^4y^{iv} + 6x^3y''' + 7x^2y'' + xy' + 4y = 0.$
10. $x^4y^{iv} + 6x^3y''' + 2x^2y'' - 4xy' + 4y = 0.$
11. $x^4y^{iv} + 8x^3y''' + 17x^2y'' + 5xy' - 5y = 0.$
12. $x^5y^{v} + 7x^4y^{iv} + 16x^3y''' + 32x^2y'' + 26xy' - 26y = 0.$
13. $2(6x-3)^3y''' + 10(6x-3)^2y'' + (6x-3)y' + 3y = 0.$
14. $9(x+8)^4y^{iv} + 21(x+8)^3y''' + 37(x+8)^2y''$
    $-74(x+8)y' + 74y = 0.$
15. $(4x+1)^3y''' - 6(4x+1)^2y'' + (4x+1)y' - 8y = 0.$

Solve each of the following Tshebyshev's equations.

16. $(1-x^2)y'' - xy' + 2y = 0, \qquad |x| \le 1.$
17. $2(1-x^2)y'' - 2xy' - 5y = 0, \qquad |x| \ge 1.$
18. $(1-x^2)^2y^{iv} - 6x(1-x^2)y''' - (4-7x^2)y'' + xy' - y = 0,$
    $|x| \le 1.$
19. $2(1-x^2)^2y^{iv} - 12x(1-x^2)y''' - (5-11x^2)y'' - xy' + y = 0,$

$|x| \le 1$.

20. $(1+x^2)\,y'' + xy' - 4y = 0$.
21. $(1+x^2)\,y'' + xy' + 5y = 0$.
22. $(1+x^2)^2\,y^{iv} + 6x\,(1+x^2)\,y''' + (4+7x^2)\,y'' + xy' - y = 0$.

Solve each of the following differential equations by substituting $y(x) = z(x)\,g(x)$.

23. $x^2y'' - x\,(3x+2)\,y' + (2x^2+3x+2)\,y = 0$.
24. $x^4y^{iv} - 4x^3y''' - 2x^2\,(x^2-6)\,y'' + 4x\,(x^2-6)\,y'$ $+ (x^4 - 4x^2 + 24)\,y = 0$.
25. $x^3y''' - 3x^2y'' - x\,(x^2-6)\,y' + (x^2-6)\,y = 0$.
26. $y''' + 3xy'' + 3\,(x^2+1)\,y' + x\,(x^2+3)\,y = 0$.
27. $x^3y''' - x^2\,(4x+3)\,y'' + x\,(x^2+8x+6)\,y'$ $+ (6x^3 - x^2 - 8x - 6)\,y = 0$.
28. $x^4y'' - 2x^2y' + (x^4 + 2x + 1)\,y = 0$.
29. $x^6y'' - x^3\,(3x^3-4)\,y' - 6\,(x^3+x^2-1)\,y = 0$.
30. $x^2y'' - 2x\,(x+3)\,y' + (x^2+6x+12)\,y = 0$.
31. $y''' - 6xy'' + 6\,(2x^2-1)\,y' - 4x\,(2x^2-3)\,y = 0$.
32. $x^2y'' - 2x\,(2x+1)\,y' + 2\,(2x^2+2x+1)\,y = 0$.
33. $xy''' + (3-4x)\,y'' + 8\,(x-1)\,y' + 8\,(1-x)\,y = 0$.
34. $x^6y''' + 3x^4\,(x^2-1)\,y'' + 3x^2\,(x^4-2x^2+2x+1)\,y'$ $-(3x^4 - 6x^3 + 3x^2 + 6x + 1)\,y = 0$.
35. $x^4y^{iv} + 16x^3y''' + x^2(72-5x^2)y'' + x(96-40x^2)y'$ $+4\,(x^4 - 15x^2 + 6)\,y = 0$.
36. $x^2y'' + 2x\,(x+2)\,y' + 2\,(x+1)^2\,y = 0$.

Using a substitution and change of variable solve each of the following equations.

37. $x^2y''' + 6xy'' + 6y' = 0$.
38. $xy''' + 3\,(x+1)\,y'' + 3\,(x+2)\,y' + (x+3)\,y = 0$.
39. $x^4y^{iv} + 2x^3\,(2x+3)\,y''' + 2x^2\,(3x^2+9x+1)\,y''$ $+2x\,(2x^3+9x^2+2x-2)\,y' + (x^4+6x^3+2x^2-4x+4)\,y = 0$.
40. $(1+x^2)\,y'' - (2x^2-x+2)\,y' + (x^2-x+8)\,y = 0$.
41. $2xy''' + (6x^2-1)\,y'' + 2x\,(3x^2+1) + x^2\,(2x^2+3)\,y = 0$.

Determine the power series solutions of the following equations.

42. $y''' + x^3y'' + x^2y' + xy = \cos x, \quad y(0) = 1, \quad y'(0) = 0, \quad y''(0) = 0$.
43. $2y''' + 4x^2y'' - 7xy' + 5y = x^2, \quad y(0) = 0, \quad y'(0) = 0, \quad y''(0) = 1$.
44. $y''' + (x^3 - 2x)\,y'' + xy = 0, \quad y(0) = 1, \quad y'(0) = 0, \quad y''(0) = 0$.
45. $y''' + (x^2-2x+3)\,y'' + (x-4)\,y' - y = 0, \quad y(1) = 0, \quad y'(1) = 0,$ $y''(1) = 2$.
46. $y'' - \left(\dfrac{x^2}{4} + 1\right) y = 0, \quad y(0) = 1, \quad y'(0) = 0$.
47. $y'' - \left(\dfrac{x^2}{4} + 1\right) y = 0, \quad y(0) = 0, \quad y'(0) = 1$.
48. $y^{iv} + x^3y = e^x, \quad y(0) = 0, \quad y'(0) = 0, \quad y''(0) = 0,$ $y'''(0) = 1$.

Solve each of the following initial value problems.

49. $y''' - 6(x-1)y'' + 6(2x^2 - 4x + 1)y'$
$-4(x-1)(2x^2 - 4x - 1)y = 0, \quad y(0) = 1, \quad y'(0) = 0,$
$y''(0) = 0.$

50. $x^3y''' + 5x^2y'' + 2xy' - 2y = 0, \quad y(1) = 3, \quad y'(1) = 2, \quad y''(1) = 1.$

51. $x^2y'' - (2x^2 - x)y' + (x^2 - x - 1)y = 0, \quad y(1) = 1, \quad y'(1) = -1.$

52. $(1 - x^2)y'' - (2x^2 + x - 2)y' - (x^2 + x - 4)y = 0, \quad y(0) = 0,$
$y'(0) = 1.$

53. $x^3y''' - 6x^2y'' + x(x^2 + 18)y' - 2(x^2 + 12)y = 0, \quad y(1) = 1,$
$y'(1) = 0, \quad y''(1) = 0.$

54. $(1 + x^2)y'' + xy' - 7y = 0, \quad y(0) = 0, \quad y'(0) = 1.$

Chapter 3

LINEAR SECOND ORDER EQUATIONS

3.1 Series solutions

The general form of a linear second order differential equation is the equation

$$a_2(x)y''(x) + a_1(x)y'(x) + a_0(x)y(x) = f(x), \tag{1}$$

where $a_i(x)$ and $f(x)$ are continuous functions of the independent variable x.

If $a_2(x) \neq \text{const}$ it is possible that

1) we will consider the solution of Eq. (1) in a neighborhood of the point ζ where $a_2(\zeta) \neq 0$ (an *ordinary point* of Eq. (1));

2) we will consider the solution of Eq. (1) in a neighborhood of the point ζ where $a_2(\zeta) = 0$ (a *singular point* of Eq. (1));

3) we will consider the solution of Eq.(1) at infinity. Note that this is done by making the change of variable $s = 1/x$ and studying the resulting equation at $s = 0$. For Eq.(1) at infinity we find

$$A_2(s)u''(s) + A_1(s)u'(s) + A_0(s)u(s) = F(s), \tag{1.i}$$

where $u(s) = y(1/s)$, $A_2(s) = s^4 a_2(1/s)$, $A_1(s) = 2s^3 a_2(1/s) - s^2 a_1(1/s)$, $A_0(s) = a_0(1/s)$, $F(s) = f(1/s)$. Hence the study of Eq.(1) at infinity is the same as at origin.

If $a_2(x)$ =const, we have only ordinary points of Eq. (1) at all x.

3.1.1 *Ordinary Point*

In a neighborhood of this point the corresponding homogeneous equation can be written in the form

$$y''(x) + p(x)y'(x) + q(x)y(x) = 0, \tag{2}$$

where $p(x) = a_1(x)/a_2(x)$, $q(x) = a_0(x)/a_2(x)$ are analytical functions at the point ζ. Hence they have the Taylor series expansions in a neighborhood of the point ζ:

$$p(x) = \sum_{k=0}^{\infty} p^{(k)}(\zeta)\frac{(x-\zeta)^k}{k!}, \qquad q(x) = \sum_{k=0}^{\infty} q^{(k)}(\zeta)\frac{(x-\zeta)^k}{k!}. \tag{3a}$$

If coefficients $p(x)$ and $q(x)$ are polynomials then see Sec.2.5.

The general solution of Eq.(2) is the power series

$$y(x) = \sum_{k=0}^{\infty} a_k (x-\zeta)^k = C_1 y_1(x) + C_2 y_2(x), \tag{3b}$$

where a_k are constant coefficients, C_1 and C_2 are arbitrary constants , $y_1(x)$ and $y_2(x)$ are the linearly independent series solutions of Eq.(2). By substituting the series (3a) and (3b) in Eq.(2) and equating the coefficients of like powers of $x-\zeta$ to zero the recurrence relation can be found

$$a_{k+2} = -\frac{1}{(k+1)(k+2)} \sum_{i=0}^{k} \frac{1}{(k-i)!} \left((i+1) p^{(k-i)}(\zeta) a_{i+1} + q^{(k-i)}(\zeta) a_i \right), \tag{4}$$

where $k \geq 0$ and the coefficients a_0 and a_1 are arbitrary. Hence we can put that $a_0 \equiv C_1$ and $a_1 \equiv C_2$.

The radius of convergence for each of the series solutions $y_1(x)$ and $y_2(x)$ given by Eq.(3b) is at least as large as the minimum of the radii of convergence for $p(x)$ and $q(x)$. One of the most useful test for the absolute convergence of a power series is the ratio test:

$$\frac{1}{\rho} = \lim_{k\to\infty} \left| \frac{a_{k+1}}{a_k} \right|, \tag{5}$$

where ρ is the radius of convergence of the power series (3b).

Example 1. Find the series solutions in powers of x of the differential equation

$$y'' + e^{-x} y' + e^{x} y = 0.$$

Since $a_2(x) = 1$ and $a_2(0) \neq 0$, then point $\zeta = 0$ is an ordinary point of this equation. Proceeding as has been stated above we obtain

$$p(x) = \sum_{k=0}^{\infty} \frac{(-1)^k}{k!} x^k, \qquad q(x) = \sum_{k=0}^{\infty} \frac{1}{k!} x^k, \qquad y(x) = \sum_{k=0}^{\infty} a_k x^k,$$

where a_k are unknown coefficients. Further, according to Eq.(4) we find, after some minor algebraic calculations,

$$\begin{aligned}
a_2 &= -\frac{1}{2}(a_0 + a_1), \\
a_3 &= -\frac{1}{6}(-a_1 + a_0 + 2a_2 + a_1) = \frac{a_1}{6}, \\
a_4 &= -\frac{1}{12}\left(a_0 + \frac{5}{2}a_1\right), \\
a_5 &= \frac{7}{120}a_0 + \frac{11}{120}a_1,
\end{aligned}$$

etc. Hence the series solution of the given equation is

$$\begin{aligned} y(x) &= a_0 + a_1 x - \frac{1}{2}(a_0 + a_1)x^2 + \frac{1}{6}a_1 x^3 - \frac{1}{12}\left(a_0 + \frac{5}{2}a_1\right)x^4 \\ &+ \left(\frac{7}{120}a_0 + \frac{11}{120}a_1\right)x^5 + \ldots \\ &= a_0\left(1 - \frac{1}{2}x^2 - \frac{1}{12}x^4 + \frac{7}{120}x^5 + \ldots\right) \\ &+ a_1\left(x - \frac{1}{2}x^2 + \frac{1}{6}x^3 - \frac{5}{24}x^4 + \frac{11}{120}x^5 + \ldots\right) \\ &\equiv a_0 y_1(x) + a_1 y_2(x), \end{aligned}$$

where

$$\begin{aligned} y_1(x) &= 1 - \frac{1}{2}x^2 - \frac{1}{12}x^4 + \frac{7}{120}x^5 + \ldots \\ y_2(x) &= x - \frac{1}{2}x^2 + \frac{1}{6}x^3 - \frac{5}{24}x^4 + \frac{11}{120}x^5 + \ldots \end{aligned}$$

are the linearly independent series solutions. Notice that $y_1(x)$ satisfies the initial conditions $y_1(0) = 1, \quad y_1'(0) = 0$ and $y_2(x)$ satisfies the initial conditions $y_2(0) = 0, \quad y_2'(0) = 1$.

Example 2. Find the series solutions in powers of x of the differential equation

$$\left(1 + x^2\right) y'' + 2xy' + 4x^2 y = 0.$$

Since $a_2(0) \neq 0$, then point $x = 0$ is an ordinary point of this equation. Proceeding as in Ex.1 we obtain

$$p(x) = 2x\sum_{k=0}^{\infty}(-1)^k x^{2k}, \qquad q(x) = 4x^2\sum_{k=0}^{\infty}(-1)^k x^{2k}, \quad \text{if } x^2 < 1.$$

The coefficients a_k in the series solution for $y(x)$ are determined by substituting the series for $p(x)$, $q(x)$ and $y(x)$ in the given equation:

$$\sum_{k=0}^{\infty} a_k\left(k(k-1)x^{k-2} + 2\sum_{n=0}^{\infty}(-1)^n k x^{2n+k} + 4\sum_{n=0}^{\infty}(-1)^n x^{2n+k+2}\right) = 0.$$

Equating the coefficients of like powers of x to zero we find

$$\begin{aligned} &a_2 2 = 0, \\ &a_3 6 + a_1 2 = 0, \\ &a_4 12 + a_2 4 + a_0 4 = 0, \\ &a_5 20 + 2(3a_3 - a_1) + 4a_1 = 0, \\ &a_6 30 + 2(4a_4 - 2a_2) + 4(a_2 - a_0) = 0, \end{aligned}$$

etc. Since $a_2 = 0$ we conclude that a_4, a_6, ..., a_{2k}, ... determines a_0, and a_3, a_5, ..., a_{2k+1}, ... determines a_1. Hence the series solution of the given equation is

$$\begin{aligned} y(x) &= a_0\left(1 - \frac{1}{3}x^4 + \frac{2}{9}x^6 + \ldots\right) + a_1\left(x - \frac{1}{3}x^3 + 0x^5 + \ldots\right) \\ &\equiv a_0 y_1(x) + a_1 y_2(x), \end{aligned}$$

where a_0 and a_1 are arbitrary constants and $y_i(x)$ are the linearly independent solutions of the given equation.

Example 3. Determine the radius of convergence each of the following power series:

a) $\sum_{n=0}^{\infty}(-1)^n\frac{x^n}{n!}$, b) $\sum_{n=0}^{\infty}\frac{x^n}{n!}$, c) $\sum_{n=0}^{\infty}(-1)^n x^{2n}$.

According to Eq.(5) we obtain

a) $\frac{1}{\rho} = \lim_{n\to\infty}\left|\frac{(-1)^{n+1}n!}{(-1)^n(n+1)!}\right| = 0$, hence $\rho \to \infty$.

b) $\frac{1}{\rho} = \lim_{n\to\infty}\left|\frac{1}{(n+1)}\right| = 0$, hence $\rho \to \infty$.

c) By making change of variable $x^2 = t$ we find a new power series $\sum(-1)^n t^n$, and its radius of convergence we find by Eq.(5)

$\frac{1}{\rho_t} = \lim_{n\to\infty}\left|\frac{(-1)^{n+1}}{(-1)^n}\right| = 1$, hence $\rho_t = 1$ and $\rho_x = 1$.

Example 4. Determine the radius of convergence by Taylor series expansion each of the following functions:

a) $(x^2 - 2x + 5)^{-1}$ about $\zeta = 0$; about $\zeta = -1$;

b) $(x^2 - 5x + 6)^{-1}$ about $\zeta = 0$; about $\zeta = 8$. a) First notice that equation $x^2 - 2x + 5 = 0$ has two roots $x_1 = 1 + 2i$ and $x_2 = 1 - 2i$. Hence if x is equal each of these roots the given function is nonanalytic and its Taylor series expansion does not exist. The distance from the point $\zeta = 0$ to either x_1 or x_2 in the complex plane is $\sqrt{5}$. Therefore the radius of convergence of Taylor series expansion $\sum a_n x^n$ about $\zeta = 0$ is $\sqrt{5}$. The distance from $\zeta = -1$ to either x_1 or x_2 is $2\sqrt{2}$. Hence the radius of convergence of the Taylor series expansion $\sum a_n(x+1)^n$ about $\zeta = -1$ is $2\sqrt{2}$.

b) The polynomial $x^2 - 5x + 6$ has zeros at $x_1 = 2$ and $x_2 = 3$. The distance from $\zeta = 0$ to x_1 is equal to 2 and the distance from $\zeta = 0$ to x_2 is equal to 3. Hence in the first case the series $\sum a_n x^n$ will converge at least for $|x| < 2$. Similarly, we find that in the second case the series $\sum a_n(x-8)^n$ will converge at least for $|x-8| < 5$.

Example 5. Estimate the radius of convergence of each solution found in Ex. 1 and Ex. 2.

a) According to Ex.3 the coefficients $p(x)$ and $q(x)$ in Ex.1 have the infinite radii of convergence. Hence the series solution $y(x) = \sum a_n x^n$ in this Example will converge for all x.

b) According to Ex.3 the coefficients $p(x)$ and $q(x)$ in Ex. 2 have the unity radii of convergence. Hence the series solution $y(x) = \sum a_n x^n$ in this Example will converge at least for $|x| < 1$. Note that application of Eq.(5) in Ex-s 1 and 2 is very difficult (see also Ex.9 in Sec. 2.5).

3.1.2 Regular Singular Point

First in the case of a singular point we should calculate

$$
\begin{aligned}
p_0 &= \lim_{x\to\zeta}(x-\zeta)\frac{a_1(x)}{a_2(x)},\\
q_0 &= \lim_{x\to\zeta}(x-\zeta)^2\frac{a_0(x)}{a_2(x)},\\
q_1 &= \lim_{x\to\zeta}\frac{d}{dx}\left((x-\zeta)^2\frac{a_0(x)}{a_2(x)}\right).
\end{aligned}
\tag{6}
$$

Then we should determine what sort of case from the following cases we have

1) if $p_0 = q_0 = q_1 = 0$, then point $x = \zeta$ is an ordinary point of Eq. (1);

2) if p_0, q_0 and q_1 are finite and $p_0^2 + q_0^2 + q_1^2 \neq 0$, then point $x = \zeta$ is so called *regular singular point* of Eq. (1);

3) if p_0 or q_0 does not exist, then point $x = \zeta$ is so called *irregular singular point* of Eq. (1).

Example 6. Determine the singular points of the differential equation

$$\left(x^2-2x-3\right)y'' + (x-3)y' + \frac{2x}{(x-3)^2}y = 0,$$

and classify them as regular or irregular.

Dividing this equation by $a_2(x) = (x^2-2x-3) = (x-3)(x+1)$ we have

$$y'' + \frac{1}{(x+1)}y' + \frac{2x}{(x-3)^3(x+1)}y = 0,$$

so $p(x) = 1/(1+x)$ and $q(x) = 2x/(x-3)^3(x+1)$. The singular points are $x = -1$ and $x = 3$. Consider $x = -1$. Then

$$p_0 = \lim_{x\to-1}\frac{x+1}{x+1} = 1,$$

and

$$q_0 = \lim_{x\to-1}\frac{2x(x+1)^2}{(x-3)^3(x+1)} = 0,$$

hence the point $x = -1$ is the regular singular point. For $x = 3$ we have

$$p_0 = \lim_{x\to3}\frac{x-3}{x+1} = 0,$$

and

$$q_0 = \lim_{x\to3}\frac{2x(x-3)^2}{(x-3)^3(x+1)}$$

does not exist. Hence the point $x = 3$ is the irregular singular point.

In a neighborhood of a regular singular point ζ it is justify that the functions $(x-\zeta)p(x)$ and $(x-\zeta)^2 q(x)$ are analytic, that is,

$$(x-\zeta)p(x) = \sum_{n=0}^{\infty} p_n (x-\zeta)^n, \qquad (x-\zeta)^2 q(x) = \sum_{n=0}^{\infty} q_n (x-\zeta)^n, \tag{7a}$$

where p_n and q_n are determined by

$$p_n = \lim_{x\to\zeta} \frac{1}{n!}\frac{d^n}{dx^n}\left((x-\zeta)p(x)\right), \quad q_n = \lim_{x\to\zeta} \frac{1}{n!}\frac{d^n}{dx^n}\left((x-\zeta)^2 q(x)\right), \tag{7b}$$

where $n = 0, 1, 2, \ldots$. These power series are convergence for $|x-\zeta| < \rho$ where $\rho > 0$.

Then the first solution of Eq.(2) takes always the following form

$$y_1(x) = |x-\zeta|^{m_1} \sum_{n=0}^{\infty} a_n(m_1)(x-\zeta)^n, \tag{8}$$

where m_1 is that root of the *indicial equation*

$$F(m) \equiv m(m-1) + mp_0 + q_0 = 0, \tag{9}$$

which has the largest real part; the coefficients $a_n(m_1)$ can be determined by substituting the series (7a) and (8) for $p(x)$, $q(x)$ and $y(x)$ in Eq.(2) and equating the coefficients of like powers of $(x-\zeta)$ to zero. The coefficient of the zeros power of $(x-\zeta)$ will be $a_0 F(m)$. Then putting $a_0 \neq 0$ gives Eq.(9). The coefficient of the first power of $(x-\zeta)$ will be

$$a_1 F(m+1) + a_0(mp_1 + q_1).$$

Hence

$$a_1 = -\frac{1}{F(m+1)} a_0(mp_1 + q_1).$$

The coefficient of the second power of $(x-\zeta)$ will be

$$a_2 F(m+2) + a_1\left(p_1(m+1) + q_1\right) + a_0(p_2 m + q_2).$$

Hence

$$a_2 = -\frac{1}{F(m+2)}\left(a_0(p_2 m + q_2) + a_1(p_1(m+1) + q_1)\right).$$

In general, we obtain the recurrence relation

$$a_n(m) = -\frac{1}{F(m+n)} \sum_{k=0}^{n-1} a_k(m)\left(p_{n-k}(m+k) + q_{n-k}\right), \tag{10}$$

where $m = m_1$ and $n \geq 1$. The form of the second solution of Eq.(2) depends on the second root of Eq.(9). There may be each of the following cases:

1) the roots of Eq.(9) are real and $m_1 - m_2$ is not positive integer;
2) the roots of Eq.(9) are real and $m_1 = m_2$;
3) the roots of Eq.(9) are real and $m_1 - m_2$ is a positive integer;
4) the roots of Eq.(9) are complex.

The roots m_1 and m_2 are real and $m_1 - m_2$ is not positive integer

Then the second solution of Eq.(2) is

$$y_2(x) = |x - \zeta|^{m_2} \sum_{n=0}^{\infty} a_n(m_2)(x - \zeta)^n, \tag{11}$$

where the coefficients $a_n(m_2)$ are given by the recurrence relation (10) with $a_0(m_2) = 1$.

Example 7. Find the regular singular points of the differential equation

$$4(1+x)^2 y'' - 3\left(1 - x^2\right) y' + 4y = 0,$$

and determine the linearly independent solutions in a neighborhood of these points.

It is easy to verify that

$$p(x) = \frac{3(x-1)}{4(x+1)}, \qquad q(x) = \frac{1}{(x+1)^2}.$$

Hence the given equation has one singular point $\zeta = -1$. This point is the regular singular point, since

$$p_0 = \lim_{x \to -1} \frac{3(x-1)(x+1)}{4(x+1)} = -\frac{3}{2},$$
$$q_0 = \lim_{x \to -1} \frac{(x+1)^2}{(x+1)^2} = 1.$$

Further according to Eq. (7b) we find $p_0 = -3/2$, $p_1 = 3/4$ and $p_k = 0$, if $k \geq 2$; and $q_0 = 1$, $q_k = 0$ if $k \geq 1$. Hence we have the indicial equation

$$F(m) \equiv m(m-1) - \frac{3}{2}m + 1 = m\left(m - \frac{5}{2}\right) + 1 = 0.$$

The roots are $m_1 = 2$ and $m_2 = 1/2$. Since these roots are not equal and do not differ by an positive integer hence there will be linearly independent solutions of the form

$$y_1(x) = (x+1)^2 \left(1 + \sum_{n=1}^{\infty} a_n(2)(x+1)^n\right),$$
$$y_2(x) = |x+1|^{1/2} \left(1 + \sum_{n=1}^{\infty} a_n(\tfrac{1}{2})(x+1)^n\right),$$

for $0 < |x+1| < \rho$. According to Eq. (10) we have the recurrence relation

$$a_n(m) = -a_{n-1}(m)\frac{3(m+n-1)}{4F(n+m)}$$
$$= -a_{n-1}(m)\frac{3(n+m-1)}{4(n+m)(n+m-\frac{5}{2})+4}.$$

Then for $m = m_1 = 2$ we find, after some minor algebraic calculations,

$$\begin{aligned} a_n(2) &= -a_{n-1}(2)\frac{3(n+1)}{2n(2n+3)} = a_{n-2}(2)\frac{3(n+1)3n}{2n(2n+3)2(n-1)(2n+1)} \\ &= \cdots = (-1)^n \frac{3(n+1)3n\ldots 3\cdot 2}{2n(2n+3)2(n-1)(2n+1)\ldots 2\cdot 1\cdot 5} \\ &= (-1)^n \left(\frac{3}{2}\right)^n \frac{3(n+1)}{(2n+3)!!}. \end{aligned}$$

Hence the first linearly independent solution is

$$y_1(x) = (x+1)^2 \left(1 + 3\sum_{n=1}^{\infty} (-1)^n \left(\frac{3}{2}\right)^n \frac{n+1}{(2n+3)!!}(x+1)^n\right).$$

Similarly for $m = m_2 = 1/2$ we have, take into account that $a_0(\frac{1}{2}) = 1$,

$$\begin{aligned} a_n(\tfrac{1}{2}) &= -a_{n-1}(\tfrac{1}{2})\frac{3(2n-1)}{4n(2n-3)} = a_{n-2}(\tfrac{1}{2})\frac{3(2n-1)3(2n-3)}{4n(2n-3)4(n-1)(2n-5)} \\ &= \cdots = (-1)^n \frac{3(2n-1)3(2n-3)\ldots 3\cdot 1}{4n(2n-3)4(n-1)(2n-5)\ldots 4\cdot 1\cdot(-1)} \\ &= (-1)^{n+1}\left(\frac{3}{4}\right)^n \frac{2n-1}{n!}. \end{aligned}$$

Hence the second linearly independent solution is

$$y_2(x) = |x+1|^{1/2}\left(1 - \sum_{n=1}^{\infty}(-1)^n \left(\frac{3}{4}\right)^n \frac{2n-1}{n!}(x+1)^n\right).$$

The roots m_1 and m_2 are equal

Then the solution $y_2(x)$ takes the form

$$y_2(x) = y_1(x)\ln|x-\zeta| + |x-\zeta|^{m_1}\sum_{n=1}^{\infty} b_n(m_1)(x-\zeta)^n, \tag{12}$$

where the coefficients $b_n(m_1)$ are given by the recurrence relation

$$\begin{aligned} b_n(m_1) = -\frac{1}{F(m_1+n)}\Big(a_n(m_1)2n &+ \sum_{k=0}^{n-1} a_k(m_1)p_{n-k} \\ &+ \sum_{k=1}^{n-1} b_k(m_1)\left(p_{n-k}(m_1+k) + q_{n-k}\right)\Big), \end{aligned} \tag{13}$$

where $n \geq 1$, and the coefficients $a_n(m_1)$ are given by Eq. (10). Note that $b_n(m_1)$ can be determined by the formal relation

$$b_n(m) = \frac{d}{dm}a_n(m), \tag{13a}$$

at $m = m_1$. The computation of $a'_n(m)$ can be carried out most conveniently by noting that when

$$f(x) = (x + a_1)^{b_1}(x + a_2)^{b_2} \cdots (x + a_n)^{b_n},$$

where a_i and b_i are some constants, then

$$f'(x) = f(x)\left(\frac{b_1}{x + a_1} + \frac{b_2}{x + a_2} + \cdots + \frac{b_n}{x + a_n}\right). \qquad (13b)$$

Example 8. Find the regular singular points of the differential equation

$$x^2y'' + x^3y' + \frac{1}{4}y = 0,$$

and determine its solutions in a neighborhood of these points.

It is easy to verify that $p(x) = x$, $q(x) = 1/4x^2$. Therefore, the point $\zeta = 0$ is the regular singular point of the given equation. The same way as in Ex. 7 we obtain $p_2 = 1$, $p_i = 0$ if $i \neq 2$, $q_0 = 1/4$, $q_i = 0$ if $i \neq 0$. The indicial equation

$$F(m) \equiv m(m-1) + \frac{1}{4} = \left(m - \frac{1}{2}\right)^2 = 0$$

has the two-fold roots $m_1 = m_2 = 1/2$. Hence the linearly independent solutions are given by Eq-s (8) and (12). According to Eq.(10) we find

$$a_1 = -\frac{1}{F(\frac{3}{2})}a_0\left(\frac{p_1}{2} + q_1\right) = 0,$$

$$a_2 = -\frac{1}{F(\frac{5}{2})}\left(a_0\left(\frac{p_2}{2} + q_2\right) + a_1\left(p_1\frac{3}{2} + q_1\right)\right) = -a_0\frac{1}{4\cdot 2},$$

and in general

$$a_n = -\frac{1}{F(n + \frac{1}{2})}a_{n-2}\left(n - \frac{3}{2}\right) = -a_{n-2}\frac{2n-3}{2n^2}.$$

Hence the coefficients $a_3, a_5, a_7, \ldots, a_{2k+1}, \ldots$ are equal to zero, and for the coefficients a_{2k} we obtain, after some simplifications,

$$\begin{aligned} a_{2k} &= -a_{2(k-1)}\frac{2k - \frac{3}{2}}{4k^2} = a_{2(k-2)}\frac{(2k - \frac{3}{2})(2k - \frac{7}{2})}{4k^2 4(k-1)^2} = \cdots \\ &= (-1)^k\frac{(2k - \frac{3}{2})(2k - \frac{7}{2})\cdots(\frac{1}{2})}{4k^2 4(k-1)^2 \cdots 4\cdot 1^2} \\ &= (-1)^k\frac{(k - \frac{3}{4})(k - \frac{7}{4})\cdots(\frac{1}{4})\Gamma(\frac{1}{4})}{2k^2 2(k-1)^2\cdots 2\cdot 1^2\Gamma(\frac{1}{4})} = (-1)^k\frac{\Gamma(k + \frac{1}{4})}{2^k(k!)^2\Gamma(\frac{1}{4})}, \end{aligned}$$

where $\Gamma(x)$ is the Gamma-function

$$\Gamma(x+1) = \int_0^\infty e^{-t} t^x \, dt = x\Gamma(x).$$

Hence the first solution of the given equation is

$$y_1(x) = |x|^{1/2} \frac{1}{\Gamma(\frac{1}{4})} \sum_{k=0}^{\infty} (-1)^k \frac{\Gamma(k+\frac{1}{4})}{(k!)^2} \left(\frac{x^2}{2}\right)^k.$$

Further, Eq.(13) takes the form

$$\begin{aligned} b_n &= -\frac{1}{n^2}\left(2na_n + a_{n-2} + \left(n - \frac{3}{2}\right) b_{n-2}\right) \\ &= -\frac{1}{n^2}\left(\left(\frac{3}{n} - 1\right) a_{n-2} + \left(n - \frac{3}{2}\right) b_{n-2}\right), \end{aligned}$$

where $n \geq 1$ and it is assumed that $a_i \equiv 0$ and $b_i \equiv 0$ at $i < 0$. Since $a_{2k+1} = 0$ then we obtain also that $b_{2k+1} = 0$ at $k \geq 0$. For the b_{2k} we find using the obtained recurrence relation

$$b_2 = -\frac{1}{8}, \quad b_4 = \frac{9}{512}, \quad b_6 = -\frac{71}{36864},$$

etc. Hence the second solution of the given equation is

$$y_2(x) = y_1(x) \ln|x| + |x|^{1/2} \left(-\frac{1}{8}x^2 + \frac{9}{512}x^4 - \frac{71}{36864}x^6 + \ldots\right).$$

The radii of convergence of Taylor's series for $xp(x) = 1$ and $x^2 q(x) = 1/4$ are infinite. Hence the radii of convergence of the obtained solutions are again infinite.

Example 9. Find the regular singular points of the differential equation

$$x^2 y'' + x^2 y' + \frac{y}{4} = 0$$

and determine the linearly independent solutions in a neighborhood of these points.

The same way as in Ex. 8 we find the regular singular point $\zeta = 0$ and the coefficients $p_1 = 1$, $p_i = 0$, if $i \neq 1$; $q_0 = 1/4$, $q_i = 0$ if $i \neq 0$. The indicial equation $F(m) \equiv m^2 - m + 1/4$ has the roots $m_1 = m_2 = 1/2$. The coefficients a_n of the first power solution are given by Eq.(10)

$$\begin{aligned} a_n(m) &= -a_{n-1}(m) \frac{m+n-1}{(m+n-\frac{1}{2})^2} = a_{n-2}(m) \frac{(m+n-1)(m+n-2)}{(m+n-\frac{1}{2})^2 (m+n-\frac{3}{2})^2} \\ &= \cdots = (-1)^n \frac{(m+n-1)(m+n-2)\ldots m}{(m+n-\frac{1}{2})^2 (m+n-\frac{3}{2})^2 \ldots (m+\frac{1}{2})^2} \\ &= (-1)^n \frac{(m+n-1)(m+n-2)\ldots m\Gamma(m)\Gamma^2(m+\frac{1}{2})}{(m+n-\frac{1}{2})^2 (m+n-\frac{3}{2})^2 \ldots (m+\frac{1}{2})^2 \Gamma^2(m+\frac{1}{2})\Gamma(m)} \\ &= (-1)^n \frac{\Gamma^2(m+\frac{1}{2})\Gamma(m+n)}{\Gamma(m)\Gamma^2(m+n+\frac{1}{2})}. \end{aligned}$$

Then substituting $m = 1/2$ and taking into account that $\Gamma(1) = 1$, $\Gamma(\frac{1}{2}) = \sqrt{\pi}$, we obtain

$$a_n(\tfrac{1}{2}) = (-1)^n \frac{\Gamma(n+\frac{1}{2})}{\sqrt{\pi}(n!)^2}.$$

Hence the first linearly independent solution is

$$y_1(x) = |x|^{1/2} \frac{1}{\sqrt{\pi}} \sum_{n=0}^{\infty} (-1)^n \frac{\Gamma(n+\frac{1}{2})}{(n!)^2} x^n.$$

The coefficients $b_n(m_1)$ we find by using Eq-s (13a) and (13b)

$$\begin{aligned} b_n(m) &= \frac{d}{dm}(-1)^n \frac{(n-1+m)(n-2+m)\dots m}{(n-\frac{1}{2}+m)^2(n-\frac{3}{2}+m)^2\dots(m+\frac{1}{2})^2} \\ &= a_n(m)\Big(\frac{1}{n-1+m} + \frac{1}{n-2+m} + \dots + \frac{1}{m} \\ &- \frac{2}{n-\frac{1}{2}+m} - \frac{2}{n-\frac{3}{2}+m} - \dots - \frac{2}{m+\frac{1}{2}}\Big) \\ &= a_n(\tfrac{1}{2})\left(\frac{1}{n-\frac{1}{2}} + \frac{1}{n-\frac{3}{2}} + \dots + 2 - \frac{2}{n} - \frac{2}{n-1} - \dots - 2\right). \end{aligned}$$

Hence the second linearly independent solution of the given equation is

$$y_2(x) = y_1(x)\ln|x| + \sqrt{\frac{|x|}{\pi}} \sum_{n=1}^{\infty} (-1)^n \frac{\Gamma(n+\frac{1}{2})}{(n!)^2} \left(S_{1/2}(n) - 2S_0(n)\right) x^n,$$

where by $S_p(n)$ is denoted the sum

$$S_p(n) = \frac{1}{n-p} + \frac{1}{n-p-1} + \dots + \frac{1}{1-p} = \sum_{k=1}^{n} \frac{1}{k-p}.$$

The radii of convergence of Taylor's series for $xp(x) = x$ and $x^2q(x) = 1/4$ are infinite. Hence the radii of convergence of the obtained solutions are again infinite.

The number $m_1 - m_2 = N$ is a positive integer

Then the solution $y_2(x)$ takes the form

$$y_2(x) = \gamma y_1(x)\ln|x-\zeta| + |x-\zeta|^{m_2}\left(1 + \sum_{n=1}^{\infty} C_n(m_2)(x-\zeta)^n\right), \qquad (14)$$

where coefficients $C_n(m_2)$ and constant γ can be determined by substituting the series solution (14) for y_2 in Eq. (2) and equating coefficients of like powers of $(x-\zeta)$ to zero. Note that γ and $C_n(m_2)$ can be found by using the formal relations

$$\gamma = \lim_{m \to m_2} (m - m_2) a_N(m), \qquad (15a)$$

$$C_n(m_2) = \lim_{m\to m_2} \frac{d}{dm}(m - m_2)a_n(m), \tag{15b}$$

where coefficients $a_n(m)$ are given by Eq. (10).

Example 10. Determine the linearly independent solutions of the differential equation

$$x^2y'' - x(1-x)y' + \frac{3}{4}y = 0$$

near the regular singular point $x = 0$.

Point $x = 0$ is the regular singular point, the coefficients p_i and q_i are $p_0 = -1$, $p_1 = 1$ and $p_i = 0$ if $k > 1$, $q_0 = 3/4$ and $q_i = 0$ if $k > 0$. The indicial equation

$$F(m) \equiv m(m-1) - m + \frac{3}{4} = 0$$

has the roots $m_1 = 3/2$, $m_2 = 1/2$. Hence the number $N = m_1 - m_2 = 1$ is positive integer. Proceeding as in Ex. 9 we find the coefficients $a_n(m)$

$$\begin{aligned} a_n(m) &= -a_{n-1}(m)\frac{n+m-1}{(m+n-\frac{3}{2})(m+n-\frac{1}{2})} \\ &= a_{n-2}(m)\frac{(n+m-1)(n+m-2)}{(m+n-\frac{3}{2})(m+n-\frac{1}{2})(m+n-\frac{5}{2})(m+n-\frac{3}{2})} = \cdots \\ &= \frac{(-1)^n(n+m-1)(n+m-2)...m}{(m+n-\frac{3}{2})(m+n-\frac{1}{2})(m+n-\frac{5}{2})(m+n-\frac{3}{2})...(m-\frac{1}{2})(m+\frac{1}{2})} \\ &= (-1)^n\frac{\Gamma(m+n)\Gamma(m-\frac{1}{2})\Gamma(m+\frac{1}{2})}{\Gamma(m+n-\frac{1}{2})\Gamma(m+n+\frac{1}{2})\Gamma(m)}. \end{aligned}$$

Substituting $m = m_1 = 3/2$ we obtain

$$a_n(\tfrac{3}{2}) = (-1)^n\frac{\Gamma(n+\frac{3}{2})}{(n!)^2(n+1)\Gamma(\frac{3}{2})}.$$

Then

$$y_1(x) = |x|^{3/2}\left(1 + \frac{2}{\sqrt{\pi}}\sum_{n=1}^{\infty}(-1)^n\frac{\Gamma(n+\frac{3}{2})}{(n+1)(n!)^2}x^n\right).$$

Further, using Eq-s (15) gives

$$\gamma = \lim_{m\to 1/2}(m - \frac{1}{2})a_1(m) = -\lim_{m\to 1/2}\frac{(m-\frac{1}{2})m}{(m+\frac{1}{2})(m-\frac{1}{2})} = -\frac{1}{2},$$

$$C_n(m) = \lim_{m\to 1/2}\frac{d}{dm}\left(m - \frac{1}{2}\right)a_n(m)$$

$$= \lim_{m\to 1/2}\frac{d}{dm}\frac{(-1)^n(m-\frac{1}{2})(n+m-1)(n+m-2)...}{(m+n-\frac{3}{2})(m+n-\frac{1}{2})(m+n-\frac{5}{2})(m+n-\frac{3}{2})...}$$

$$\times \frac{m}{(m-\frac{1}{2})(m+\frac{1}{2})}$$

Taking into account Eq.(13b), we obtain

$$\begin{aligned}
C_n(m) &= \lim_{m\to 1/2}\left(m-\frac{1}{2}\right)a_n(m)\Big(\frac{1}{m-\frac{1}{2}}+\frac{1}{n+m-1}+\frac{1}{n+m-2} \\
&+\cdots+\frac{1}{m}-\frac{1}{n+m-\frac{3}{2}}-\frac{1}{n+m-\frac{1}{2}}-\frac{1}{n+m-\frac{5}{2}}-\frac{1}{n+m-\frac{3}{2}} \\
&-\cdots-\frac{1}{m-\frac{1}{2}}-\frac{1}{m+\frac{1}{2}}\Big) \\
&= \frac{(-1)^n\Gamma(n+\frac{1}{2})}{\Gamma(\frac{1}{2})(n-1)n(n-2)(n-1)\ldots 1\cdot 1}\Big(\frac{1}{n-\frac{1}{2}}+\frac{1}{n-\frac{3}{2}} \\
&+\cdots+2-\frac{1}{n-1}-\frac{1}{n}-\frac{1}{n-2}-\frac{1}{n-1}-\cdots-1\Big) \\
&= (-1)^n\frac{\Gamma(n+\frac{1}{2})n}{(n!)^2\sqrt{\pi}}\left(S_{1/2}(n)-2S_0(n)+\frac{1}{n}\right),
\end{aligned}$$

where $S_p(n)$ is the same sum as in Ex. 9. Hence the second linearly independent solution of the given equation is

$$\begin{aligned}
y_2 &= -\frac{1}{2}y_1\ln|x| \\
&+\sqrt{|x|}\left(1+\sum_{n=1}^{\infty}(-x)^n\frac{\Gamma(n+\frac{1}{2})n}{(n!)^2\sqrt{\pi}}\left(S_{1/2}(n)-2S_0(n)+\frac{1}{n}\right)\right),
\end{aligned}$$

The complex roots

We assume that $m_1=\bar{m}_2=\lambda+i\mu$, where λ and μ are real. Then in the interval $0<|x-\zeta|<\rho$, where $\rho>0$, Eq.(2) has two linearly independent solutions

$$y_1(x)=\mathrm{Re}\left(|x-\zeta|^{\lambda}e^{i\mu\ln|x-\zeta|}\left[1+\sum_{n=1}^{\infty}a_n(\lambda+i\mu)(x-\zeta)^n\right]\right), \tag{16a}$$

$$y_2(x)=\mathrm{Im}\left(|x-\zeta|^{\lambda}e^{i\mu\ln|x-\zeta|}\left[1+\sum_{n=1}^{\infty}a_n(\lambda+i\mu)(x-\zeta)^n\right]\right), \tag{16b}$$

where complex coefficients $a_n(\lambda+i\mu)$ are given by Eq.(10).

Example 11. Determine the linearly independent solutions of the differential equation

$$x^2y''+x(x-1)y'+2y=0$$

near the regular singular point $x=0$.

Since $p_0=-1$, $p_1=1$, $q_0=2$ and all the other coefficients are vanished, we obtain the indicial equation $F(m)\equiv(m-1)^2+1=0$. Its roots are $m_1=1+i$, $m_2=1-i$.

Further by using Eq.(10) we find, after some minor algebraic calculations,

$$a_1(m) = -\frac{3}{5} + i\frac{1}{5},$$
$$a_2(m) = \frac{1}{5} - i\frac{3}{20},$$

and in general

$$a_n(m) = -a_{n-1}(m)\frac{n^2 - in + 2}{n\,(n^2+4)}.$$

Hence the linearly independent solutions of the given equation are

$$\begin{aligned} y_1(x) &= \mathrm{Re}\left[xe^{i\ln|x|}\left(1 + x\left(-\frac{3}{5} + i\frac{1}{5}\right) + x^2\left(\frac{1}{5} - i\frac{3}{20}\right) + \cdots\right)\right] \\ &= x\Big[\cos\ln|x|\left(1 - \frac{3}{5}x + \frac{1}{5}x^2 + \cdots\right) \\ &\quad - \sin\ln|x|\left(\frac{1}{5}x - \frac{3}{20}x^2 + \cdots\right)\Big], \\ y_2(x) &= \mathrm{Im}\left[xe^{i\ln|x|}\left(1 + x\left(-\frac{3}{5} + i\frac{1}{5}\right) + x^2\left(\frac{1}{5} - i\frac{3}{20}\right) + \cdots\right)\right] \\ &= x\Big[\cos\ln|x|\left(\frac{1}{5}x - \frac{3}{20}x^2 + \cdots\right) \\ &\quad + \sin\ln|x|\left(1 - \frac{3}{5}x + \frac{1}{5}x^2 + \cdots\right)\Big]. \end{aligned}$$

3.1.3 Irregular Singular Point

Let ζ be a finite irregular singular point. Then by making change of independent variable $s = (x-\zeta)^{-1}$, and function $u(s) = y(x)$ in Eq.(2) we find a new differential equation with irregular singular point at infinity

$$u''(s) + f(s)u'(s) + h(s)u(s) = 0, \tag{17}$$

where $f(s) = 2/s - p(\zeta + 1/s)/s^2$, $h(s) = q(\zeta + 1/s)/s^4$.

The infinity point is an irregular singular point of Eq.(17) with rank r, where $r \geq 1$ is an integer number, if and only if the functions $f(s)/s^{r-1}$ and $h(s)/s^{2r-2}$ are analytic at infinity, that is, they have the power series expansions of the forms

$$\frac{f(s)}{s^{r-1}} = \sum_{n=0}^{\infty}\frac{f_n}{s^n}, \qquad \frac{h(s)}{s^{2(r-1)}} = \sum_{n=0}^{\infty}\frac{h_n}{s^n}. \tag{18a}$$

Here the number r must be the least positive integer when f_0, h_0, h_1 exist. If $f_0 = h_0 = h_1 = 0$ at $r = 1$ then the infinity point is a regular singular point of Eq.(17).

The power series (18a) are convergent for $|1/s| < \rho$, where $\rho > 0$. The coefficients f_n and h_n can be determined as

$$f_n = \lim_{t\to 0}\frac{1}{n!}\frac{d^n}{dt^n}t^{r-1}f(t^{-1}), \qquad h_n = \lim_{t\to 0}\frac{1}{n!}\frac{d^n}{dt^n}t^{2r-2}h(t^{-1}). \tag{18b}$$

Example 12. Show that the differential equation

$$(1+x)y'' + 2y' - \left(1 + \frac{1}{(1+x)^5}\right) y = 0$$

has an irregular singular point and determine its rank.

Since q_0 does not exist at $x = -1$, then this point is the irregular singular point of the given equation. By making change of variable $x + 1 = 1/s$ and function $u(s) = y(x)$ we obtain, after some minor calculations,

$$u''(s) - \left(s^2 + \frac{1}{s^3}\right) u(s) = 0.$$

Hence $f(s) \equiv 0$ and $h(s) = -s^2 - s^{-3}$. Now according to Eq.(18b) and taking into account that h_0 should exist we obtain an equation for the rank r

$$h_0 = \lim_{t\to 0} t^{2r-2} h(t^{-1}) = -\lim_{t\to 0}\left(t^{2r-4} + t^{2r+1}\right) = const < \infty.$$

It is possible if $r \geq 2$. Hence the point $\zeta = -1$ is the irregular singular point of the second rank.

Solution of Eq.(17) in a neighborhood of the infinity is called *formally normal series*

$$u(s) = e^{P_r(s)} s^m v(s) \equiv e^{P_r(s)} s^m \sum_{n=0}^{\infty} \frac{a_n}{s^n}, \tag{19}$$

where $P_r(s) = p_r s^r + p_{r-1}s^{r-1} + \cdots + p_1 s$ is a polynomial of degree r, m is a certain constant and a_n are undetermined coefficients. The coefficients p_k of the polynomial $P_r(s)$, constant m and coefficients a_n can be determined by substitution of the solution (19) and series (18) in Eq.(17) and equating the coefficients of like powers of s to zero. The recurrence relation for coefficients p_k is

$$h_n + \sum_{j=1}^{r} \left[j\,(2r - j - n)\, p_j p_{2r-n-j} + j p_j f_{n-r+j} \right] = 0, \tag{20}$$

where $0 \leq n \leq r-1$ and $p_j \equiv 0$ for $j < 1$ and $j > r$. If $n = 0$ we obtain the quadratic equation for p_r, recalling that $f_i \equiv 0$ for $i < 0$,

$$h_0 + r^2 p_r^2 + r p_r f_0 = 0. \tag{21a}$$

It has two roots, corresponding to two solutions of the given equation. If $n = 1$ we find

$$p_{r-1} = -\frac{h_1 + r p_r f_1}{(r-1)(2rp_r + f_0)}, \tag{21b}$$

if $2rp_r + f_0 \neq 0$. Similarly we determine all the other coefficients p_k if $2rp_r + f_0 \neq 0$. The indicial equation for m is

$$m = -\frac{1}{2rp_r + f_0}\left(h_r + r\,(r-1)\, p_r + \sum_{j=1}^{r} j p_j \left(f_j + (r-j)\, p_{r-j} \right) \right), \tag{22}$$

if $2rp_r + f_0 \neq 0$. The recurrence relation for the coefficients a_n at $r = 1$ takes the form

$$a_n = \frac{1}{n(2p_1 + f_0)}\Big(a_{n-1}(n-m)(n-1-m) + \sum_{j=0}^{n-1} a_j\left(h_{n+1-j} + p_1 f_{n+1-j} - (j-m)f_{n-j}\right)\Big),$$

where $2p_1 + f_0 \neq 0$, $n \geq 1$. The similar relations for $r > 1$ will be fairly complicated. Therefore in these cases we find, at first, the polynomial $P_r(s)$ and constant m by using Eq-s (20)-(22); and then, by substituting the solution (19) in Eq.(17), we obtain a new equation for $v(s)$. Further we find its solution in the form of power series of s^{-1}. Notice that the power series in Eq.(19), if it is not polynomial, diverges, as a rule, for all s and define so called *asymptotic solution* of Eq.(17) at infinity.

If $2rp_r + f_0 = 0$, then by making changes of variable $s = t^2$ and function $u(s) = \exp(-f_0t^2/2)w(t)$ we obtain a new differential equation for $w(t)$ in the form of Eq.(17). Then we study this equation at infinity as has been stated above.

Example 13. Determine the linearly independent solutions of the differential equation

$$y'' + \frac{1}{x}y' - \left(x + \frac{1}{x^2}\right)y = 0$$

at infinity and determine the radii of convergence of each series.

The infinity point is the irregular singular point of this equation with rank $r = 2$, since the functions

$$\frac{f(x)}{x} = \frac{1}{x^2} \quad \text{and} \quad \frac{h(x)}{x^2} = -\frac{1}{x} - \frac{1}{x^4}$$

are analytic functions at infinity. According to Eq. (18a) we find $f_0 = 0$, $f_1 = 0$, $f_2 = 1$, and $f_i = 0$ if $k > 2$, $h_0 = 0$, $h_1 = -1$, $h_4 = -1$ and all the other $h_i = 0$. Using Eq.(21a) we obtain $4p_2^2 = 0$. Hence $p_2 = 0$ and $2rp_r + f_0 = 0$. Then by making changes of variable $x = s^2$ and function $y(x) = u(s)$ we find the new differential equation

$$u'' + \frac{1}{s}u' - 4\left(s^4 + \frac{1}{s^2}\right)u = 0$$

with irregular singular point at infinity. Rank of this point is 3, since the functions

$$\frac{f(s)}{s^2} = \frac{1}{s^3}, \quad \text{and} \quad \frac{h(s)}{s^4} = -4 - \frac{4}{s^6}$$

are analytic at the infinity. According to Eq.(18b) we obtain $f_3 = 1$, $h_0 = -4$, $h_6 = -4$ and all other coefficients f_i and h_i are equal to zero. Further by recurrence relations (21) and (22) we obtain, after some minor calculations, $p_3 = \pm 2/3$, $p_2 = p_1 = 0$, $m = -3/2$. Then the substitution

$$u(s) = \exp(\pm\frac{2}{3}s^3)\frac{v(s)}{s^{3/2}}$$

in the equation for $u(s)$ yields, after some calculations,

$$v''(s) + \left(\pm 4s^2 - \frac{2}{s}\right) v'(s) - \frac{7}{4s^2} v(s) = 0.$$

By substituting the power series $\sum a_n s^{-n}$ in this equation and equating the coefficients of like powers of s^{-1} to zero, we find

$$\begin{aligned} &\mp 0a_0 = 0, \\ &\mp 4a_1 = 0, \\ &\mp 8a_2 = 0, \\ &-\frac{7}{4}a_0 \mp 12a_3 = 0 \end{aligned}$$

and in general

$$\left(n^2 - 3n - \frac{7}{4}\right) a_{n-3} \mp 4na_n = 0.$$

Hence we obtain that a_0 is an arbitrary constant, $a_1 = a_2 = 0$, $a_3 = \mp 7a_0/48$, and, in general,

$$a_n = \pm a_{n-3}\frac{1}{4n}\left(n^2 - 3n - \frac{7}{4}\right),$$

where $n \geq 3$. According to this relation the coefficient a_0 determines a_3, which in turn determines a_6 etc. Similarly a_1 determines a_4, a_7 etc.; and a_2 determines a_5, a_8 etc. Since $a_1 = a_2 = 0$, we conclude that $a_4 = a_7 = a_{10} = \ldots = 0$ and $a_5 = a_8 = a_{11} = \ldots = 0$. Hence

$$\begin{aligned} a_{3k} &= \pm a_{3(k-1)}\frac{3}{4k}\left(k - \frac{7}{6}\right)\left(k + \frac{1}{6}\right) = \cdots \\ &= (\pm 1)^k \left(\frac{3}{4}\right)^k \frac{\Gamma(k - \frac{1}{6})\Gamma(k + \frac{7}{6})}{k!\Gamma(-\frac{1}{6})\Gamma(\frac{7}{6})}, \end{aligned}$$

when $a_0 = 1$. Taking into account that $\Gamma(x)\Gamma(1-x) = \pi/\sin(\pi x)$, we obtain for $u_i(s)$

$$\begin{aligned} u_1(s) &= -\exp(2s^3/3)s^{-3/2}\frac{1}{2\pi}\sum_{n=0}^{\infty}\left(\frac{3}{4}\right)^n \frac{\Gamma(n - \frac{1}{6})\Gamma(n + \frac{7}{6})}{s^{3n}n!}, \\ u_2(s) &= -\exp(-2s^3/3)s^{-3/2}\frac{1}{2\pi}\sum_{n=0}^{\infty}\left(-\frac{3}{4}\right)^n \frac{\Gamma(n - \frac{1}{6})\Gamma(n + \frac{7}{6})}{s^{3n}n!}. \end{aligned}$$

Recalling the change of variable, we find the solution of the given equation in the form

$$\begin{aligned} y_1(x) &= -\frac{\exp(2x^{3/2}/3)}{x^{3/4}2\pi}\sum_{n=0}^{\infty}\left(\frac{3}{4}\right)^n \frac{\Gamma(n - \frac{1}{6})\Gamma(n + \frac{7}{6})}{n!}x^{-3n/2}, \\ y_2(x) &= -\frac{\exp(-2x^{3/2}/3)}{x^{3/4}2\pi}\sum_{n=0}^{\infty}\left(-\frac{3}{4}\right)^n \frac{\Gamma(n - \frac{1}{6})\Gamma(n + \frac{7}{6})}{n!}x^{-3n/2} \end{aligned}$$

Now according to Eq.(5) we find the radii of convergence of these power series

$$\rho = \lim_{n\to\infty}\left|\frac{a_{3n}}{a_{3(n+1)}}\right| = \lim_{n\to\infty}\left|\frac{4n}{3(n+\frac{1}{6})(n-\frac{7}{6})}\right| = 0.$$

Hence obtained series solutions are asymptotic solutions of the given differential equation at infinity.

Example 14. Determine the linearly independent solutions of the differential equation

$$xy'' + 2y' - \left(\frac{1}{4} + \frac{5}{16x}\right)y = 0$$

at infinity.

The infinity point is the irregular singular point of this equation with rank $r = 1$, since the functions

$$f(x) = \frac{2}{x} \quad \text{and} \quad h(x) = -\frac{1}{4x} - \frac{5}{16x^2},$$

are analytic at the infinity. It is easy to verify that $f_1 = 2$, $h_1 = -1/4$, $h_2 = -5/16$ and all the other coefficients f_i, h_i are vanished. Then according to Eq. (21) we find $p_1 = 0$, hence $2p_1 + f_0 = 0$. By making change of variable $x = s^2$ and function $y(x) = u(s)$ we find a new differential equation

$$u'' + \frac{3}{s}u' - \left(1 + \frac{5}{4s^2}\right)u = 0,$$

with the irregular singular point at infinity. Rank of this point is equal to one, since $f(s) = 3/s$ and $h(s) = -1 - 5/4s^2$. Then by using Eq-s(21) and (22) we obtain $p_1 = \pm 1$ and $m = -3/2$. Proceeding as in Ex.13 we find the recurrence relation for the coefficients a_k

$$a_k = \pm a_{k-1}\frac{k^2 - k - 2}{2k},$$

where $k \geq 1$. Then we obtain $a_1 = \pm a_0$, $a_2 = 0$, $a_3 = 0$ etc. and solutions for $u(s)$ are

$$u_1(s) = e^s s^{-3/2}\left(1 - \frac{1}{s}\right),$$

$$u_2(s) = e^{-s} s^{-3/2}\left(1 + \frac{1}{s}\right).$$

Hence the given differential equation has the linearly independent solutions

$$y_1(x) = e^{\sqrt{x}}\left(x^{-3/4} - x^{-5/4}\right),$$

$$y_1(x) = e^{-\sqrt{x}}\left(x^{-3/4} + x^{-5/4}\right).$$

PROBLEMS

Find the radius of convergence of the following power series.

1. $\sum_{n=0}^{\infty}(-1)^n(x-2)^n$.
2. $\sum_{n=0}^{\infty}\frac{2^n}{n!}(x+1)^n$.
3. $\sum_{n=0}^{\infty}n(x-1)^n$.
4. $\sum_{n=1}^{\infty}\frac{1}{n^2}(x-\zeta)^n$.
5. $\sum_{n=1}^{\infty}\frac{1}{n}(x^2-2x+3)^n$.
6. $\sum_{n=1}^{\infty}\frac{1}{n^{3/2}}(2x+3)^n$.

Proceeding formally, find the linearly independent power series solutions in powers of $(x-\zeta)$ of the following differential equations.

7. $y''-xy'-y=0$, at $\zeta=0$.
8. $y''-xy'-y=0$, at $\zeta=1$.
9. $y''+x^2y=0$, at $\zeta=0$.
10. $\cosh(x)y''+y=0$, at $\zeta=0$.
11. $y''-\left(b+\frac{x^2}{4}\right)y=0$, at $\zeta=0$, where $b=$const.
12. $(1-x)y''+y=0$, at $\zeta=0$.
13. $y''+(x-1)^2y'+(x^2-1)\,y=0$, at $\zeta=1$.
14. $y''-xy=0$, at $\zeta=0$.
15. $y''-xy=0$, at $\zeta=1$.
16. $(1-x^2)\,y''-xy'-y=0$, at $\zeta=0$.
17. $y''+\sin(x)y=0$, at $\zeta=0$.
18. $e^xy''+xy=0$, at $\zeta=0$.

Show that each of the following differential equations has a regular singular point $\zeta=0$. Determine the linearly independent solutions for $x-\zeta>0$.

19. $2xy''+y'+xy=0$.
20. $x^2y''+xy'+\left(x^2-\frac{1}{9}\right)y=0$.
21. $3xy''+2y'+xy=0$.
22. $x^2y''+xy'+(x-2)y=0$.
23. $xy''+y'-y=0$.
24. $xy''+y'+2y=0$.
25. $xy''+2y'+y=0$.
26. $x^2y''+3xy'+(1+x)y=0$.
27. $xy''+y=0$.
28. $2xy''+(1-2x)y'-y=0$.
29. $x^2y''+4xy'+(2+x)y=0$.
30. $x^2y''+xy'+\left(x^2-\frac{9}{4}\right)y=0$.
31. $x^2y''+xy'+(x^2-1)\,y=0$.

32. $x(x-1)y'' + (2x-1)y' + \frac{1}{4}y = 0.$

33. $x^2y'' + xy' + x^2y = 0.$

34. $x^2y'' + xy' + (x^2 - \nu^2)\,y = 0$, where ν is a number. This equation is called Bessel's equation, after the Germany astronomer F.W. Bessel (1784-1846). See also problems 20, 22, 30, 31, 33.

35. $x^2y'' + \frac{1}{2}(x + \sin x)y' + y = 0.$

36. $x^2y'' + x(1+x)y' + y = 0.$

Determine the linearly independent solutions of the following differential equations at infinity.

37. $x^5y'' + 2x^4y' - y = 0.$

38. $2x^4y'' + 3x^3y' + y = 0.$

39. $x(1-x)y'' + (1-2x)y' - \frac{1}{4}y = 0.$

40. $x^4y'' + 2y = 0.$

Show that each of the following differential equations has an irregular singular point at infinity. Determine the linearly independent solutions. Find the radius of convergence of each series.

41. $y'' + x^4y = 0.$

42. $y'' - x^2y = 0.$

43. $y'' + e^{2x}y = 0.$

44. $y'' + x^2y = 0.$

45. $y'' + (x^4+1)\,y = 0.$

46. $x^2y'' + y\ln^2 x = 0.$

47. $y'' - xy = 0.$

48. $y'' + xy = 0.$

3.2 Linear boundary value problem

A linear nonhomogeneous second order boundary value problem consists of the differential equation

$$\hat{L}y(x) \equiv a_2(x)y''(x) + a_1(x)y'(x) + a_0(x)y(x) = f(x), \tag{1}$$

given on the open interval $0 < x < 1$, and the boundary conditions

$$\hat{B}_k(y) \equiv \alpha_k y'(0) + \beta_k y(0) + \gamma_k y'(1) + \delta_k y(1) = \epsilon_k, \qquad k = 1, 2, \tag{2}$$

where $\hat{L}$, $\hat{B}_k$ are linear differential operators, $a_i(x)$ and $f(x)$ are some functions and $\alpha_1, \alpha_2, \ldots, \epsilon_2$ are prescribed constants.

If function $f(x)$ and constants ϵ_k are equal to zero then problem is called homogeneous boundary value problem. If $\alpha_2 = \beta_2 = \gamma_1 = \delta_1 = 0$ or $\alpha_1 = \beta_1 = \gamma_2 = \delta_2 = 0$, then the linear boundary conditions are given at each end of the basic interval.

The choice of the interval $0 \leq x \leq 1$ involves no loss of generality. If the interval was originally $c \leq x \leq d$, where c and d are finite numbers, it could be reduced to $0 \leq x \leq 1$ by transformation $x = (s-c)/(d-c)$. If each of the following numbers $a_2(0)$, $a_2(1)$, c^{-1} and d^{-1} do not vanish then the linear boundary value problem is called *regular boundary value problem.* In the opposite case the linear boundary value problem is called *singular boundary value problem.*

We consider only regular linear value problems. Then each of the following cases is possible:

1) the function $f(x)$ is identically zero and constants ϵ_k are equal to zero (so called *homogeneous boundary value problem*);

2) the function $f(x)$ or one of the constants ϵ_k does not vanish and is prescribed (so called *nonhomogeneous boundary value problem*);

3) the function $f(x)$ or one of the constants ϵ_k is arbitrary nonvanish.

3.2.1 *Homogeneous Problem*

First we find the general solution of the corresponding homogeneous differential equation $\hat{L}y(x) = 0$

$$y(x) = C_1 y_1(x) + C_2 y_2(x),$$

where C_i are arbitrary constants. A substitution for y into the boundary conditions (2) then yields

$$\begin{aligned} C_1\hat{B}_1(y_1) + C_2\hat{B}_1(y_2) = 0, \\ C_1\hat{B}_2(y_1) + C_2\hat{B}_2(y_2) = 0. \end{aligned} \tag{3}$$

It is a system of the homogeneous linear algebraic equations concerning constants C_1 and C_2. The matrix of this system is

$$B(y) = \begin{pmatrix} \hat{B}_1(y_1) & \hat{B}_1(y_2) \\ \hat{B}_2(y_1) & \hat{B}_2(y_2) \end{pmatrix} \tag{4}$$

and its solution we can found by the Cramer's rule. Hence we can immediately conclude that the nontrivial solution of the homogeneous boundary value problem exists if and only if

$$\det |B(y)| = 0. \tag{5}$$

The following cases may arise:

1) the rank of $B(y)$ is equal to two, then $C_1 = C_2 = 0$ and the given initial value problem has the trivial solution (note that in this case $\det |B(y)| \neq 0$);

2) the rank of $B(y)$ is equal to one, then one of the C_i is arbitrary, other C_i is vanished; and the given initial value problem has a single infinite family of solutions;

3) the rank of $B(y)$ is equal to zero, then C_1 and C_2 are arbitrary constants and the given initial value problem has a twice infinite family of solutions.

Example 1. Find the solution satisfying

$$\begin{aligned} y'' + y &= 0, \\ \hat{B}_1(y) &\equiv y(0) = 0, \\ \hat{B}_2(y) &\equiv y(1) = 0. \end{aligned}$$

The general solution of the differential equation is

$$y(x) = C_1 \sin x + C_2 \cos x .$$

The matrix $B(y)$ takes the form

$$B(y) = \begin{pmatrix} 0 & 1 \\ \sin 1 & \cos 1 \end{pmatrix}.$$

Its rank is equal to two. Hence the given initial value problem has the trivial solution $y(x) = 0$.

Example 2. Find the solution satisfying

$$\begin{aligned} y'' + \pi^2 y &= 0, \\ \hat{B}_1(y) &\equiv y(0) = 0, \\ \hat{B}_2(y) &\equiv y(1) = 0. \end{aligned}$$

The general solution of the differential equation is

$$y(x) = C_1 \sin(\pi x) + C_2 \cos(\pi x).$$

Then

$$B(y) = \begin{pmatrix} 0 & 1 \\ 0 & -1 \end{pmatrix}.$$

The rank of $B(y)$ is equal to one. The solution of Eq.(3) is $C_2 = 0$ and an arbitrary constant C_1. Hence the single infinite family of solutions of the given initial value problem is

$$y(x) = C_1 \sin(\pi x).$$

Example 3. Find the solution satisfying

$$\begin{aligned} y'' + \pi^2 y &= 0, \\ \hat{B}_1(y) &\equiv y(0) + y(1) = 0, \\ \hat{B}_2(y) &\equiv y'(0) + y'(1) = 0. \end{aligned}$$

Similarly to Ex.2 we find the general solution of the given equation and the matrix $B(y)$

$$B(y) = \begin{pmatrix} 0 & 0 \\ 0 & 0 \end{pmatrix}.$$

Its rank is equal to zero. Hence C_i are arbitrary constants and the given initial value problem has a twice infinite family of solutions

$$y(x) = C_1 \sin(\pi x) + C_2 \cos(\pi x).$$

3.2.2 Nonhomogeneous Problem

The general solution of the nonhomogeneous differential equation $\hat{L}(y) = f$ is (see Sec.2.3)

$$y(x) = C_1 y_1(x) + C_2 y_2(x) + y_p(x). \tag{6}$$

A substitution for y into the boundary conditions (2) then yields

$$\begin{aligned} C_1\hat{B}_1(y_1) + C_2\hat{B}_1(y_2) &= \epsilon_1 - \hat{B}_1(y_p), \\ C_1\hat{B}_2(y_1) + C_2\hat{B}_2(y_2) &= \epsilon_2 - \hat{B}_2(y_p), \end{aligned} \tag{7}$$

that is, a system of the linear nonhomogeneous algebraic equations concerning the constants C_1 and C_2. There we have the following cases:

1) the rank of $B(y)$ given by Eq.(4) and rank of the *augmented matrix*

$$B_a \equiv \begin{pmatrix} \hat{B}_1(y_1) & \hat{B}_1(y_2) & \epsilon_1 - \hat{B}_1(y_p) \\ \hat{B}_2(y_1) & \hat{B}_2(y_2) & \epsilon_2 - \hat{B}_2(y_p) \end{pmatrix} \tag{5a}$$

are equal. Then the given nonhomogeneous problem has a unique solution.

a) If their ranks are equal to two, then the coefficients C_i are solutions of the system (7)

$$\vec{C} = B^{-1}(y)(\vec{\epsilon} - \vec{B}(y_p)), \tag{8}$$

and the initial value problem has the unique solution

$$y = \vec{y}_h(x)B^{-1}(y)(\vec{\epsilon} - \vec{B}(y_p)) + y_p, \tag{9}$$

where $\vec{C}$ is a column vector with the components C_1 and C_2, $B^{-1}(y)$ is the inverse matrix of $B(y)$, $(\vec{\epsilon} - \vec{B}(y_p))$ is a column vector with the components $(\epsilon_1 - \hat{B}_1(y_p))$ and $(\epsilon_2 - \hat{B}_2(y_p))$ and $\vec{y}_h(x)$ is a row vector with the components $y_1(x)$ and $y_2(x)$.

b) If their ranks are equal to one then we should find the coefficients C_i using only one equation of system (7), and that problem has a single infinite family of solutions.

c) If their ranks are equal to zero then the coefficients C_i are arbitrary constants and that problem has a twice infinite family of solutions.

2) the ranks of $B(y)$ and B_a are not equal. Then the given problem does not have a solution .

Example 4. Find the solution satisfying

$$\begin{aligned} &y'' + y = x, \\ &\hat{B}_1(y) \equiv y(0) = 1, \\ &\hat{B}_2(y) \equiv y(1) = 0. \end{aligned}$$

The general solution of the nonhomogeneous differential equation is

$$y(x) = C_1 \sin\, x + C_2 \cos\, x + x.$$

The matrices $B(y)$ and B_a are

$$B(y) = \begin{pmatrix} 0 & 1 \\ \sin 1 & \cos 1 \end{pmatrix}$$

and

$$B_a = \begin{pmatrix} 0 & 1 & 1 \\ \sin 1 & \cos 1 & -1 \end{pmatrix}.$$

Their ranks are equal to two. Further we find

$$B^{-1}(y) = \begin{pmatrix} -\cot 1 & \sin^{-1} 1 \\ 1 & 0 \end{pmatrix},$$

and

$$(\vec{e} - \vec{B}(y_p)) = \left(1 - \hat{B}_1(x), \, -\hat{B}_2(x)\right)^T = (1, -1)^T$$

Now using Eq.(8) we obtain

$$\begin{pmatrix} C_1 \\ C_2 \end{pmatrix} = \begin{pmatrix} -\cot 1 & \sin^{-1} 1 \\ 1 & 0 \end{pmatrix} \begin{pmatrix} 1 \\ -1 \end{pmatrix} = \begin{pmatrix} -\cot(1/2) \\ 1 \end{pmatrix}.$$

Hence the solution of the given problem is

$$y(x) = x + \cos x - \cot\frac{1}{2} \, \sin x.$$

Example 5. Find the solution satisfying

$$\begin{aligned} y'' + \pi^2 y &= x\pi^2, \\ \hat{B}_1(y) &\equiv y(0) = 0, \\ \hat{B}_2(y) &\equiv y(1) = 1 \end{aligned}$$

The general solution of the differential equation is

$$y = x + C_1 \sin(\pi x) + C_2 \cos(\pi x).$$

The matrices $B(y)$ and B_a are

$$B(y) = \begin{pmatrix} 0 & 1 \\ 0 & -1 \end{pmatrix}$$

and

$$B_a = \begin{pmatrix} 0 & 1 & 0 \\ 0 & -1 & 0 \end{pmatrix}.$$

Their ranks are equal to one. The boundary conditions require that $C_2 = 0$ and C_1 is an arbitrary constant. Hence the solution of given problem is

$$y = C_1 \sin(\pi x) + x.$$

Example 6. Find the solution satisfying

$$\begin{aligned} &y'' + \pi^2 y = x, \\ &\hat{B}_1(y) \equiv y(0) + y(1) = \pi^{-2}, \\ &\hat{B}_2(y) \equiv y'(0) + y'(1) = 2\pi^{-2} \end{aligned}$$

The general solution of the differential equation is

$$y = x\pi^{-2} + C_1 \sin(\pi x) + C_2 \cos(\pi x).$$

The matrices $B(y)$ and B_a are

$$B(y) = \begin{pmatrix} 0 & 0 \\ 0 & 0 \end{pmatrix}$$

and

$$B_a(y) = \begin{pmatrix} 0 & 0 & 0 \\ 0 & 0 & 0 \end{pmatrix}.$$

Their ranks are equal to zero. Using system (7) we conclude that coefficients C_i are arbitrary. Hence the solution of the given problem is

$$y = x\pi^{-2} + C_1 \sin(\pi x) + C_2 \cos(\pi x).$$

Example 7. Find the solution satisfying

$$\begin{aligned} &y'' + \pi^2 y = 0, \\ &\hat{B}_1(y) \equiv y(0) + y(1) = 2\pi^{-2}, \\ &\hat{B}_2(y) \equiv y'(0) + y'(1) = \pi^{-2}. \end{aligned}$$

Similar to Ex. 5 we find

$$y = C_1 \sin(\pi x) + C_2 \cos(\pi x),$$

$$B(y) = \begin{pmatrix} 0 & 0 \\ 0 & 0 \end{pmatrix}$$

and

$$B_a(y) = \begin{pmatrix} 0 & 0 & \pi^2 \\ 0 & 0 & -\pi^2 \end{pmatrix}.$$

The ranks of $B(y)$ and $B_a(y)$ are not equal. Hence the given problem does not have a solution.

3.2.3 Green's Function

We consider the linear nonhomogeneous boundary value problem which has some arbitrary function $f(x)$ and any arbitrary constants ϵ_k (compare with Sec. 2.2.4). The general solution of this problem is

$$y(x) = \int_0^1 G(x,s)f(s)ds + \vec{y}_h(x)B^{-1}(y)\vec{\epsilon}, \tag{10}$$

where $G(x,s)$ is called Green's function, after the English mathematician G. Green (1793-1841), $\vec{y}_h(x)$, $B^{-1}(y)$ have been defined above and $\vec{\epsilon}$ is a column vector with the components ϵ_1 and ϵ_2. The Green's function is a solution of the nonhomogeneous boundary value problem

$$\begin{aligned} &\hat{L}G(x,s) = \delta(x,s) \\ &\hat{B}_k G(x,s) = 0, \quad k = 1,2, \end{aligned} \tag{11}$$

where $0 \le x \le 1$, the operators $\hat{L}$ and $\hat{B}$ are given in Eq-s (1) and (2), s is a parameter and $\delta(x,s)$ is so called *Dirac delta-function*

$$\delta(x,s) = \begin{cases} 0 & \text{if} \quad x \neq s, \\ \infty & \text{if} \quad x = s, \\ \int_0^1 f(s)\delta(x,s)ds = f(x). \end{cases}$$

The solution of Eq.(11) is

$$G(x,s) = \begin{cases} \vec{y}_h(x)\vec{A}(s), & 0 \le s \le x, \\ \vec{y}_h(x)\vec{C}(s), & x \le s \le 1, \end{cases} \tag{12}$$

where the column vectors $\vec{C}(s)$ and $\vec{A}(s)$ are determined by substituting the Green's function in the following conditions:

a) continuous of $G(x,s)$ in the point $x = s$;

b) the jump of magnitude -1 for the derivative $G'_x(x,s)$ in the point $x = s$;

c) the boundary conditions (2).

Thus

$$\begin{aligned} &G(s-0,s) = G(s+0,s), \\ &G'_x(s-0,s) = G'_x(s+0,s) - 1, \\ &B(y(0))\vec{C}(s) + B(y(1))\vec{A}(s) = 0, \end{aligned} \tag{13}$$

where $B(y(0))$ is the matrix $B(y)$ at $\gamma_k = \delta_k = 0$, $k = 1,\ 2$ and $B(y(1))$ is the same matrix at $\alpha_k = \beta_k = 0$, $k = 1,\ 2$. Note that $B(y) = B(y(0)) + B(y(1))$.

If the corresponding linear homogeneous boundary value problem has one and only one solution, then there exist an unique solution of the nonhomogeneous boundary value problem given by Eq-s (10)-(12), where

$$\begin{aligned} &\vec{C}(s) = -B^{-1}(y)B(y(1))\Psi^{-1}(y(s))\vec{\eta}, \\ &\vec{A}(s) = \vec{C}(s) + \Psi^{-1}(y(s))\vec{\eta}, \end{aligned} \tag{14}$$

$\Psi(y(s))$ is the fundamental matrix of the corresponding homogeneous differential equation, $\Psi^{-1}(y(s))$ is the inverse matrix of Ψ, $\vec{\eta} = (0, 1)^T$.

Example 8. Find the solution satisfying

$$y'' - y = f(x),$$
$$\hat{B}_1(y) \equiv y(0) = \epsilon_1,$$
$$\hat{B}_2(y) \equiv y(1) = \epsilon_2,$$

where $f(x)$ is some function and ϵ_k are arbitrary constants.

It is easy to verify that solutions of the corresponding homogeneous differential equation are $y_1 = \cosh x$ and $y_2 = \sinh x$. Hence $\vec{y}_h = (\cosh x, \sinh x)$, the fundamental matrix takes the form

$$\Psi(y(s)) = \begin{pmatrix} \cosh s & \sinh s \\ \sinh s & \cosh s \end{pmatrix}$$

and the inverse matrix is

$$\Psi^{-1}(y(s)) = \begin{pmatrix} \cosh s & -\sinh s \\ -\sinh s & \cosh s \end{pmatrix}.$$

Further, according to Eq.(4) we obtain the matrix $B(y)$

$$B(y) = \begin{pmatrix} 1 & 0 \\ \cosh 1 & \sinh 1 \end{pmatrix}$$

and the inverse matrix

$$B^{-1}(y) = \begin{pmatrix} 1 & 0 \\ -\coth 1 & \sinh^{-1} 1 \end{pmatrix}.$$

In this problem $\alpha_1 = 0$, $\beta_1 = 1$, $\gamma_1 = \delta_1 = 0$ and $\alpha_2 = \beta_2 = \gamma_2 = 0$, $\delta_2 = 1$ are given. Putting $\gamma_1 = \gamma_2 = \delta_1 = \delta_2 = 0$ we find the matrix $B(y(0))$

$$B(y(0)) = \begin{pmatrix} 1 & 0 \\ 0 & 0 \end{pmatrix},$$

and putting $\alpha_1 = \alpha_2 = \beta_1 = \beta_2 = 0$ we find the matrix $B(y(1))$

$$B(y(1)) = \begin{pmatrix} 0 & 0 \\ \cosh 1 & \sinh 1 \end{pmatrix}.$$

A substitution of these expressions into Eq-s (14) then yields

$$\vec{C}(s) = -\begin{pmatrix} 1 & 0 \\ -\coth 1 & \sinh^{-1} 1 \end{pmatrix} \begin{pmatrix} 0 & 0 \\ \cosh 1 & \sinh 1 \end{pmatrix}$$
$$\times \begin{pmatrix} \cosh s & -\sinh s \\ -\sinh s & \cosh s \end{pmatrix} \begin{pmatrix} 0 \\ 1 \end{pmatrix} = \begin{pmatrix} 0 \\ \frac{\sinh(s-1)}{\sinh 1} \end{pmatrix},$$

$$\vec{A}(s) = \begin{pmatrix} 0 \\ \frac{\sinh(s-1)}{\sinh 1} \end{pmatrix} + \begin{pmatrix} -\sinh s \\ \cosh s \end{pmatrix} = \begin{pmatrix} -\sinh s \\ \cosh s + \frac{\sinh(s-1)}{\sinh 1} \end{pmatrix}.$$

Now the Green's function of the given initial value problem takes the form

$$G(x,s) = \begin{cases} \vec{y}_h(x)\vec{A}(s) = \sinh s \, \frac{\sinh(x-1)}{\sinh 1}, & 0 \le s \le x \\ \vec{y}_h(x)\vec{C}(s) = \sinh(s-1)\, \frac{\sinh x}{\sinh 1}, & x \le s \le 1 \end{cases},$$

Finally by using Eq.(10) we find the solution of the given problem

$$y(x) = \frac{\sinh(x-1)}{\sinh 1}\int_0^x f(s)\sinh s\, ds + \frac{\sinh x}{\sinh 1}\int_x^1 f(s)\sinh(s-1)\, ds$$

$$+\frac{1}{\sinh 1}\left(\epsilon_1 \sinh(1-x) + \epsilon_2 \sinh x\right).$$

PROBLEMS

Find the solutions of the following homogeneous second order boundary value problems.

1. $y'' = 0, \quad y(0) = 0, \quad y(1) = 0.$
2. $y'' = 0, \quad y(0) + y'(0) = 0, \quad y(1) + y'(1) = 0.$
3. $y'' = 0, \quad y(0) - y'(0) = 0, \quad y(0) - y(1) + y'(1) = 0.$
4. $y'' = 0, \quad y'(0) - y'(1) = 0, \quad y(0) + y'(0) - y(1) = 0.$
5. $y'' + y = 0, \quad y(0) + y'(1) = 0, \quad y'(0) + y(1) = 0.$
6. $y'' + y = 0, \quad y(0) - y(1) = 0, \quad y'(0) - y'(1) = 0.$
7. $(x-1)y'' - xy' + y = 0, \quad y(0) = 0, \quad y'(\ln 2) = 0.$
8. $(x-1)y'' - xy' + y = 0, \quad y'(0) = 0, \quad y(1) - y'(1) = 0.$
9. $x(x-1)y'' + (x+1)y' - y = 0, \quad y(0) = 0, \quad |y(x)| \to \infty$ at $x \to 1$.
10. $x(x-2)y'' + (x-4)y' - y = 0, \quad |y(0)| < \infty, \quad y(4) = 0.$

Find the solutions of the following nonhomogeneous second order boundary value problems.

11. $y'' = x, \quad y(0) = 0, \quad y(1) = 0.$
12. $y'' = x, \quad y(0) + y'(0) = 0, \quad y(1) + y'(1) = 0.$
13. $y'' = x, \quad y(0) - y'(0) = 0, \quad y(0) - y(1) + y'(1) = 0.$
14. $y'' = \sin \pi x, \quad y'(0) - y'(1) = 0, \quad y(0) + y'(0) - y(1) = 0.$
15. $y'' = 0, \quad y(0) = 1, \quad y(1) = 0.$
16. $y'' = x, \quad y(0) = 1, \quad y(1) = 0.$
17. $y'' = x, \quad y(0) - y'(0) = 1, \quad y(0) - y(1) + y'(1) = 0.$
18. $y'' = x, \quad y(0) - y'(0) = 1, \quad y(0) - y(1) + y'(1) = \frac{1}{3}.$
19. $y'' + y = 1, \quad y(0) = 0, \quad y(\pi/2) = 0.$
20. $y'' + y = 1, \quad y(0) = 0, \quad y(\pi) = 0.$
21. $y'' + y = 1, \quad y'(0) = 0, \quad y(\pi) = 1.$
22. $y'' - y = 2x, \quad y(0) = 0, \quad y(1) = -1.$
23. $y'' + y' = 1, \quad y(0) = 0, \quad y(1) = 1.$

24. $y'' - y' = 0, \quad y(0) = -1, \quad y'(1) - y(1) = 2.$
25. $y'' + y = 2x - \pi, \quad y(0) = 0, \quad y(\pi) = 0.$
26. $xy'' + (x-1)y' - y = 0, \quad y(0) = 1, \quad y(1) = 0.$
27. $(1+x^2)\,y'' - 2y = 0, \quad y(0) = 0, \quad y(1) = \pi.$
28. $2xy'' + (x-2)y' - y = x^2 - 1, \quad y(0) = 0, \quad y(2) = 0.$

Proceeding formally find the solutions of the following second order singular boundary value problems.

29. $x^2y'' - 6y = 0, \quad |y(x)| < \infty$ at $x \to 0, \quad y(1) = 2.$
30. $x^2y'' - 6y = 0, \quad y(x) = 0$ at $x \to -\infty, \quad y(x) = 0$ at $x \to +\infty.$
31. $y'' - y = 0, \quad y(0) = 0, \quad |y(x)| < \infty$ at $x \to +\infty.$
32. $y'' - 2iy = 0, \quad y(0) = -1, \quad y(x) = 0$ at $x \to +\infty.$
33. $y'' - y' + 2y = 0, \quad y'(0) = 2, \quad y(x) = 0$ at $x \to +\infty$.
34. $x^2y'' - 2xy' + 2y = 0, \quad y(x) = o(x)$ at $x \to 0,$ $y(1) = 3.$
35. $x^2y'' - 2xy' + 2y = 0, \quad y(x) = o(x^2)$ at $x \to 0,$ $y(1) = 3.$
36. $y'' + y = 0, \quad y(0) = 0, \quad |y(x)| < \infty$ at$x \to +\infty$.
37. $y'' + y = 0, \quad |y(x)| < \infty$ at $x \to \pm\infty.$
38. $xy'' + 2y' - xy = 4xe^x, \quad y'(-1) = 0, \quad y(x) = 0$ at $x \to -\infty.$
39. $(1+x^2)y'' + 2xy' = \dfrac{1}{1+x^2}, \quad y'(0) = 0, \quad y(x) = \dfrac{\pi^2}{8}$ at $x \to +\infty.$
40. $(1-x)y'' + xy' - y = (x-1)^2e^x, \quad y(0) = 1, \quad y(x) = 0$ at $x \to -\infty.$
41. $2x^2(2 - \ln x)y'' + x(4 - \ln x)y' - y = (2 - \ln x)x^{-1/2}, \quad y(1) = 1, \quad y(x) = 0$ at $x \to +\infty.$
42. Determine the parameter a, when the boundary value problem

$$y'' + a^2y = 0, \quad y(0) = 0, \quad y(1) = 0$$

has: 1) a unique solution, 2) a single infinite family of solutions, 3) a twice infinite family of solutions. Determine the solutions in each case.

Find the Green's functions of the following second order regular boundary value problems.

43. $y'' + y = f(x), \quad y'(0) = 0, \quad y(\pi) = 0.$
44. $y'' + y = f(x), \quad y(0) - y(\pi) = 0, \quad y'(0) - y'(\pi) = 0.$
45. $y'' - y = f(x), \quad y'(0) = 0, \quad y'(2) + y(2) = 0.$
46. $y'' - y = f(x), \quad y(0) - y'(0) = 0, \quad y(2) - y'(2) = 0.$
47. $x^2y'' - 2y = f(x), \quad y(1) = 0, \quad y(2) + 2y'(2) = 0.$
48. $x^2y'' + 2xy' = f(x), \quad y(1) = 0, \quad y'(3) = 0.$
49. $y'' + y = f(x), \quad y(0) = 0, \quad y(1) = 0.$

50. $y'' = f(x), \quad y(0) = 0, \quad y(1) = 0.$
51. $y'' = f(x), \quad y'(0) = 0, \quad y(1) = 0.$
52. $y'' = f(x), \quad y(0) = 0, \quad y(1) + y'(1) = 0.$
53. $y'' = f(x), \quad y(0) = 0, \quad y'(1) = 0.$

Proceeding formally by using Eq-s(11), (12) find the Green's function of the following second order singular boundary value problems.

54. $y'' = f(x), \quad y(0) = 0, \quad |y(x)| < \infty, \quad \text{at} \quad x \to +\infty.$
55. $y'' + y' = f(x), \quad y'(0) = 0, \quad y(x) = 0 \quad \text{at} \quad x \to +\infty$.
56. $y'' - y = f(x), \quad |y(x)| < \infty, \quad \text{at} \quad x \to \pm\infty.$
57. $x^2y'' - 2y = f(x), \quad |y(x)| < \infty, \quad \text{at} \quad x \to 0, \quad x \to +\infty.$

3.3 Eigenvalues and eigenfunctions

A linear homogeneous second order boundary value problem with one parameter consists of the differential equation

$$\hat{L}(\lambda)y \equiv y'' + p(x,\lambda)y' + q(x,\lambda)y = 0, \tag{1}$$

given on the open interval $0 < x < 1$, and the boundary conditions

$$\hat{B}_k(\lambda)y \equiv \alpha_k(\lambda)y'(0) + \beta_k(\lambda)y(0) + \gamma_k(\lambda)y'(1) + \delta_k(\lambda)y(1) = 0, \tag{2}$$

where $k = 1, 2$ and whose functions $p(x,\lambda)$, $q(x,\lambda)$ and coefficients $\alpha_1(\lambda)$, $\alpha_2(\lambda)$, ..., $\delta_2(\lambda)$ depend on a parameter λ. According to Eq.(5), Sec. 3.2, there exist a nontrivial solution if and only if

$$D(\lambda) = \det|By| \equiv \begin{vmatrix} \hat{B}_1(\lambda)y_1 & \hat{B}_1(\lambda)y_2 \\ \hat{B}_2(\lambda)y_1 & \hat{B}_2(\lambda)y_2 \end{vmatrix} = 0, \tag{3}$$

where $y_1(x,\lambda)$, $y_2(x,\lambda)$ are the linearly independent solutions of Eq.(1). The complex values λ, if any, satisfying this determinantal equation are called *eigenvalues* of the homogeneous boundary value problem. Corresponding to each eigenvalues there may exist nontrivial solution of Eq. (1), so called *eigenfunction*, which is determined at most to within an arbitrary multiplicative constant.

Example 1. Find the eigenvalues and the corresponding eigenfunctions of the boundary value problem

$$y'' + \lambda y = 0,$$
$$\lambda y(0) + y'(1) = 0,$$
$$y(1) - \lambda y'(0) = 0.$$

First, if $\lambda = 0$, the general solution of the differential equation is $y(x,0) = C_1x + C_2$, where C_i are arbitrary constants. By using Eq. (3) we find $D(0) = 1 \neq 0$, hence $\lambda = 0$ is not eigenvalue of the given equation. On the other hand, if $\lambda \neq 0$, then the general solution of the differential equation is

$$y(x,\lambda) = C_1e^{\sqrt{-\lambda}x} + C_2e^{-\sqrt{-\lambda}x},$$

where λ is a complex number and C_i are arbitrary constants. In this case the determinantal equation takes the form

$$\sqrt{-\lambda}(1+\lambda^2) - \lambda(1+\lambda)\sinh\sqrt{-\lambda} = 0,$$

or

$$\frac{1+\lambda^2}{(1+\lambda)\sqrt{-\lambda}} = -\sinh\sqrt{-\lambda}.$$

In the complex plane we have $\lambda = \rho\exp(i\varphi)$, where $\rho \geq 0$ and $0 \leq \varphi \leq 2\pi$ are real numbers. Then the determinantal equation becomes complex and its real and imaginary parts are

$$\begin{aligned}(1+\rho)\sin\frac{\varphi}{2} - \rho^2\sin\frac{3\varphi}{2} + \rho^3\sin\frac{5\varphi}{2}\\ = -\sqrt{\rho}(1-\rho^2)\sinh\left(\sqrt{\rho}\sin\frac{\varphi}{2}\right)\cos\left(\sqrt{\rho}\cos\frac{\varphi}{2}\right),\\ (-1+\rho)\cos\frac{\varphi}{2} - \rho^2\cos\frac{3\varphi}{2} + \rho^3\cos\frac{5\varphi}{2}\\ = -\sqrt{\rho}(1-\rho^2)\cosh\left(\sqrt{\rho}\sin\frac{\varphi}{2}\right)\sin\left(\sqrt{\rho}\cos\frac{\varphi}{2}\right),\end{aligned}$$

respectively. It is assumed that $\sqrt{-\lambda} = \sqrt{\rho}\exp(i\varphi/2 + i\pi/2)$. This system is transcendental and very complicated for solving. Therefore, we consider only the real values of λ, that is, $\varphi = 0$, π and only the positive imaginative values of λ, that is, $\varphi = \pi/2$.

At $\varphi = 0$ we obtain the determinantal equation

$$\frac{1+\rho^2}{(1+\rho)\sqrt{\rho}} = \sin\sqrt{\rho}.$$

It is easy to verify that the left hand side of this equation is large as one at all $\rho \neq 1$. Since $\sin\sqrt{\rho} \leq 1$ at all $\rho > 0$, we conclude that here a solution does not exist. Hence the given initial value problem does not have the positive real eigenvalues.

At $\varphi = \pi/2$ we obtain, after dividing the real part of the determinantal equation by the imaginary part of one, and some minor algebraic simplifications,

$$\frac{1+\rho}{-1+\rho} = \tanh\sqrt{\frac{\rho}{2}}\,\cot\sqrt{\frac{\rho}{2}},$$

or

$$\tan\sqrt{\frac{\rho}{2}} = \frac{\rho-1}{\rho+1}\tanh\sqrt{\frac{\rho}{2}}.$$

The roots of this equation can be determined approximately by numerical methods. Since right hand side of this equation is close by one at $\rho \gg 1$, then roots of this equation are given with reasonable accuracy by

$$\sqrt{\frac{\rho_n}{2}} \approx \pi\frac{4n+1}{4}, \qquad n = 1, 2, \ldots$$

and the precision of this estimate increase with n. Hence the eigenvalues are

$$\lambda_n \approx i\pi^2 \frac{(4n+1)^2}{8}, \qquad n = 1, 2, \dots$$

Finally, by one of the boundary conditions, we find

$$C_2 = -C_1 \frac{\lambda_n + \sqrt{-\lambda_n} e^{\sqrt{-\lambda_n}}}{\lambda_n - \sqrt{-\lambda_n} e^{-\sqrt{-\lambda_n}}} \equiv \nu_n C_1$$

Hence the eigenfunction corresponding to λ_n is

$$\psi_n(x) = y(x, \lambda_n) = C_1 \left(e^{\sqrt{-\lambda_n x}} + \nu_n e^{-\sqrt{-\lambda_n x}} \right).$$

At $\varphi = \pi$, proceeding as in previous case we obtain the determinantal equation

$$\frac{1+\rho^2}{(1-\rho)\sqrt{\rho}} = -\sinh \sqrt{\rho}.$$

The unique root of this equation is large as one and can be determined approximately by numerical methods. Hence the eigenfunction corresponding to this eigenvalue also is unique.

It is obvious that the given problem may have other eigenvalues ρ_k depending on values of φ. Their number and values are determined by Eq. (3).

3.3.1 Self-adjoint Problems

The linear homogeneous differential operator of order n

$$\hat{L}^* y \equiv (-1)^n \frac{d^n}{dx^n}(a_0 y) + (-1)^{n-1} \frac{d^{n-1}}{dx^{n-1}}(a_1 y) + \cdots + a_n y, \tag{4}$$

is called the *adjoint operator* of $\hat{L}$, where $\hat{L}$ is the linear homogeneous differential operator of order n

$$\hat{L} y \equiv a_0(x) y^{(n)} + a_1(x) y^{(n-1)} + \cdots + a_n(x) y = 0.$$

If $\hat{L}^* y = \hat{L} y$ for every function $y(x)$, then the operator $\hat{L}$ is called *self-adjoint operator.*

Example 2. Find the adjoint operator for each of the following linear differential operators:

$$\begin{aligned} &\text{a)} \quad \hat{L} y \equiv a_0(x) y'' + a_1(x) y' + a_2(x) y; \\ &\text{b)} \quad \hat{L} y \equiv y''' + 2xy' + y; \\ &\text{c)} \quad \hat{L} y \equiv y^{iv} + 2xy'' + 2y' + y. \end{aligned}$$

According to Eq. (4) we find

$$\begin{aligned} \text{a)} \quad \hat{L}^* y &\equiv a_0 y'' + 2a_0' y' + a_0'' y - a_1 y' - a_1' y + a_2 y \\ &= a_0 y'' + (2a_0' - a_1) y' + (a_0'' - a_1' + a_2) y. \end{aligned}$$

Hence the second order differential operator $\hat{L}$ is self-adjoint if and only if $a_0' = a_1$.

$$\text{b)} \quad \hat{L}^*y \equiv -y''' + 2xy'' + 4y' + y \neq \hat{L}_3 y.$$

$$\text{c)} \quad \hat{L}^*y \equiv y^{iv} + 2xy'' + 4y' - 2y' + y = y^{iv} + 2xy'' + 2y' + y = \hat{L}_4 y.$$

Hence the given operator $\hat{L}$ is self-adjoint.

The linear homogeneous second order boundary value problem

$$\hat{L}y \equiv a_0(x)y'' + a_1(x)y' + a_2(x)y = -\lambda r(x)y, \qquad 0 < x < 1, \tag{6}$$

$$\hat{B}_k y \equiv \alpha_k y'(0) + \beta_k y(0) + \gamma_k y'(1) + \delta_k y(1) = 0, \qquad k = 1, 2, \tag{7}$$

where $r(x) > 0$ on the interval $0 \leq x \leq 1$ and $\alpha_1, \ldots, \delta_2$ are prescribed constants, is called *self-adjoint eigenvalue problem* if operator $\hat{L}$ is self-adjoint, and the relation

$$\begin{aligned} &\int_0^1 \left(v(x)\hat{L}u(x) - u(x)\hat{L}v(x)\right) dx \\ &\qquad = \left(a_0(x)\left(u'(x)v(x) - u(x)v'(x)\right)\right)\Big|_0^1 = 0 \end{aligned} \tag{8}$$

is valid for every pair of functions $u(x)$ and $v(x)$, which are two-times continuously differentiable on [0, 1], and which satisfy the boundary conditions given by Eq. (7). The constants $\alpha_1, \alpha_2, \ldots, \delta_2$, which are necessary and sufficient for validity of Eq. (8) can be determined by substituting $u(x)$ and $v(x)$ in the boundary conditions (7) and relation (8).

Example 3. Determine whether the following boundary value problems are self-adjoint or not

a) $y'' + y = -\lambda y, \quad y(0) - y'(1) = 0, \quad y'(0) - y(1) = 0.$
b) $(1 + x^2)y'' + 2xy' + y = -\lambda y, \quad y(0) - y'(1) = 0, \quad y'(0) + 2y(1) = 0.$

It is easy to verify that the given differential operators are self-adjoint. Further, by substituting the boundary conditions of problem a) into Eq.(8) we find, after some minor algebraic calculations,

$$\begin{aligned} &u'(1)v(1) - u(1)v'(1) - u'(0)v(0) + u(0)v'(0) \\ &\qquad = 2\left(u(0)v'(0) - u'(0)v(0)\right) \neq 0. \end{aligned}$$

Hence this problem is not self-adjoint. For the problem b) we obtain

$$2\left(u'(1)v(1) - u(1)v'(1)\right) - \left(u'(0)v(0) - u(0)v'(0)\right) \equiv 0.$$

Hence this problem is self-adjoint.

The eigenvalues and eigenfunctions of the self-adjoint boundary value problem given by Eq-s (6) and (7) possess the following properties:

1) the eigenvalues are real;

2) the eigenvalues can be ordered according to increasing magnitude so that $\lambda_1 < \lambda_2 < \ldots < \lambda_n < \ldots$ and $\lambda_n \to \infty$ as $n \to \infty$;

3) every pair of eigenfunctions ψ_n and ψ_k corresponding to eigenvalues λ_n and λ_k respectively, are orthogonals with respect to the weight function $r(x)$, that is

$$\int_0^1 r(x)\psi_n(x)\psi_k(x)\,dx = 0$$

if $n \neq k$.

Example 4. Find the eigenvalues and the corresponding eigenfunctions of the self-adjoint boundary value problem

$$\begin{aligned} y'' &= -\lambda y, \\ y(0) - y'(1) &= 0, \\ y'(0) + y(1) &= 0. \end{aligned}$$

It is easy to verify that this problem is self-adjoint and solution of its equation is

$$y = C_1 \cos\left(\sqrt{\lambda}x\right) + C_2 \sin\left(\sqrt{\lambda}x\right).$$

Further the determinantal equation takes the form

$$2\sqrt{\lambda} + (1+\lambda)\sin\sqrt{\lambda} = 0.$$

The positive real roots of this equation can be determined approximately by numerical methods and are given with reasonable accuracy by

$$\lambda_n \approx \pi^2 n^2 \left(1 - (-1)^n \frac{4}{n^2\pi^2}\right), \quad n = 1, 2, \ldots.$$

The precision of this estimate increase with n. Then proceeding as in Ex. 1 we find the eigenfunctions of the given problem

$$y_n(x) = C_n \left(\sin\left(\sqrt{\lambda_n}x\right) + \nu_n \cos\left(\sqrt{\lambda_n}x\right)\right),$$

where $\nu_n = (1 + \sqrt{\lambda_n}\sin\sqrt{\lambda_n})/\sqrt{\lambda_n}\cos\sqrt{\lambda_n}$.

3.3.2 The Sturm-Liouville Problem

A second order Sturm-Liouville problem, named after the French mathematicians C.F. Sturm (1803-1855) and J. Liouville (1809-1882), consist of the differential equation

$$-\left(p(x)y'(x)\right)' + q(x)y(x) = \lambda r(x)y(x), \quad 0 < x < 1, \tag{9}$$

together with the boundary conditions at the end points

$$\begin{aligned} a_1 y(0) + a_2 y'(0) &= 0, & |a_1| + |a_2| &\neq 0, \\ b_1 y(1) + b_2 y'(1) &= 0, & |b_1| + |b_2| &\neq 0, \end{aligned} \tag{10}$$

or together with the *periodic boundary conditions*

$$y(0) = y(1), \qquad y'(0) = y'(1). \tag{11}$$

We will assume that functions $p(x), p'(x), q(x)$ and $r(x)$ are continuous on the interval $0 \le x \le 1$ and further, that $p(x) > 0$ and $r(x) > 0$ at all points of the interval. It is easy to verify that the Sturm-Liouville boundary value problem is self-adjoint problem.

The eigenvalues and eigenfunctions of the Sturm-Liouville boundary value problem have the following properties:

1) the eigenvalues are real;

2) the eigenvalues are simples, that is, to each eigenvalue there corresponds only one linearly independent eigenfunction (exception conditions (11));

3) the eigenvalues can be ordered according to increasing magnitude so that $\lambda_1 < \lambda_2 < \ldots < \lambda_n < \ldots$ and $\lambda_n \to \infty$ as $n \to \infty$;

4) every pair of eigenfunctions ψ_n and ψ_k corresponding to eigenvalues λ_n and λ_k respectively, are orthogonals with respect to the weight function $r(x)$, that is

$$\int_0^1 r(x)\psi_n(x)\psi_k(x)\, dx = 0$$

if $n \neq k$.

Example 5. Find the eigenvalues and the corresponding eigenfunctions of the Sturm-Liouville boundary value problem

$$y'' = -\lambda y, \quad y(0) = 0, \quad y(1) = 0.$$

Proceeding as in Ex. 4 we find the general solution of the given equation

$$y(x, \lambda) = C_1 e^{\sqrt{-\lambda}x} + C_2 e^{-\sqrt{-\lambda}x},$$

and the determinantal equation

$$\sin\sqrt{\lambda} = 0,$$

where λ is a real number. Its roots are

$$\lambda_n = \pi^2 n^2.$$

By the first boundary condition we find $C_2 = -C_1$, hence the eigenfunction corresponding to the λ_n is

$$\psi_n(x) \equiv y(x, \lambda_n) = C\sin(\pi n x),$$

where $C = 2iC_1$ is an arbitrary multiplicative constant. Note that when $n = 0$ we have $\lambda_0 = 0$ and $\psi_0 = 0$, but this function is not an eigenfunction of the given problem in accordance with the definition of one.

3.3.3 Nonhomogeneous Problem

We consider only nonhomogeneous Sturm-Liouville problem

$$-(p(x)y'(x))' + q(x)y(x) = \mu r(x)y(x) + f(x), \tag{12}$$

where μ is a prescribed number and $f(x)$ is a given function on the interval $0 < x < 1$, together with the boundary conditions (10). The choice of the homogeneous boundary conditions (10) involves no loss of generality. If the boundary conditions had originally the form

$$\begin{aligned} a_1 y(0) + a_2 y'(0) &= d_1, \\ b_1 y(1) + b_2 y'(1) &= d_2, \end{aligned} \tag{10a}$$

where a_i, b_i, d_i are some prescribed numbers, they could be reduced to the homogeneous conditions by the transformation $y(x) = z(x) + u(x)$, where $u(x)$ is any twice differentiable function satisfying only the boundary conditions (but not necessarily the differential equation). Then $z(x)$ is a solution of Eq. (12) with function $\tilde{f}(x) = f(x) + (pu')' - qu + \mu ru$, and it satisfies the homogeneous boundary conditions (10).

Example 6. Reduce the nonhomogeneous problem

$$\begin{aligned} &(1+x^2)y'' + 2xy' + 7y = -\mu y + x^2, \\ &y(0) + y'(0) = 1, \\ &y(1) + y'(1) = 3, \end{aligned}$$

to a problem with homogeneous boundary conditions.

Choosing $u(x) = ax + b$, where a and b are undetermined coefficients, and substituting its into the boundary conditions, we obtain the system of linear nonhomogeneous algebraic equations for a and b

$$\begin{aligned} a + b &= 1, \\ 2a + b &= 3. \end{aligned}$$

This system is compatible and it has the solution $a = 2$, $b = -1$. Hence $u = 2x - 1$. Then we obtain

$$\begin{aligned} &(1+x^2)z'' + 2xz' + 7z = -\mu z + (x^2 - 2x(\mu + 9) + 7 + \mu), \\ &z(0) + z'(0) = 0, \\ &z(1) + z'(1) = 0, \end{aligned}$$

that is, the problem with homogeneous boundary conditions.

The formal solution of the problem given by Eq-s (12) and (10) has the form

$$y(x) = \sum_{i=1}^{\infty} \frac{C_i}{\lambda_i - \mu} \psi_i(x), \tag{13}$$

where λ_i are the eigenvalues of the corresponding homogeneous problem given by Eq-s (9) and (10), $\psi_i(x)$ are the corresponding eigenfunctions and C_i are given by

$$C_i = \int_0^1 f(x)\psi_i(x)dx. \tag{14}$$

Notice that the corresponding homogeneous problem when $f(x) \equiv 0$ has always the trivial solution at $\mu \neq \lambda_n$. In the case, when μ is equal to one of the eigenvalues, say $\mu = \lambda_k$, and $C_k = 0$, we should replace the term $C_k/(\lambda_k - \mu)$ in Eq. (13) by an arbitrary constant C, that is,

$$y(x) = \sum_{i=1, i\neq k}^{\infty} \frac{C_i}{\lambda_i - \mu}\psi_i(x) + C\psi_k(x). \tag{13a}$$

If $\mu = \lambda_k$ and $C_k \neq 0$, then a solution of that problem does not exist.

Example 7. Find the solution satisfying

$$y'' + y = x,$$
$$y(0) = 0, \quad y(1) = 0.$$

First the given equation we rewrite as $y'' = -y + x$, concluding that $\mu = 1$. Then we find eigenvalues and eigenfunctions of the corresponding homogeneous problem (see Ex. 5)

$$\lambda_n = \pi^2 n^2, \quad \psi_n = \sin(\pi n x).$$

Further we calculate

$$C_i = \int_0^1 x\sin(\pi n x)dx = (-1)^{n+1}\frac{1}{\pi n}.$$

Since $\mu \neq \lambda_n$ at all n, then the solution of the given problem is

$$y(x) = -\sum_{i=1}^{\infty} \frac{(-1)^n}{\pi n(\pi^2 n^2 - 1)}\sin(\pi n x).$$

Here the right hand side is the sum of Fourier series of a function. Notice that proceeding as has been shown in Ex. 4 of Sec. 3.2 we find the solution of this problem in the form

$$y(x) = x - \frac{\sin x}{\sin 1}.$$

Hence the sum of obtained Fourier series is equal to this function.

PROBLEMS

Find the adjoint operator of the following linear differential operators.

1. $\hat{L}y \equiv x^2y' - y$.
2. $\hat{L}y \equiv x^2y'' + 2xy' + (n^2 - x^2)\,y = 0$.
3. $\hat{L}y \equiv x^2y'' - 2xy' - (n^2 + x^2)\,y$.

4. $\hat{L}y \equiv (1-x^2)\,y'' - 2xy' + 4y.$
5. $\hat{L}y \equiv (1+x^2)\,y'' + 2xy' - y.$
6. $\hat{L}y \equiv x^3y'' - 2xy' + 4y.$

Determine whether the following boundary value problems are self-adjoint or not.

7. $x^2y'' - 2xy' + y = \lambda\,(1+x^2)\,y, \quad y(0) = 0, \quad y(1) = 0.$
8. $x^2y'' + 2xy' + y = \lambda\,(1-2x^2)\,y, \quad y(0) = 0, \quad y(1) = 0.$
9. $x^2y'' + 2xy' + y = \lambda y, \quad y(0) = 0, \quad y(1) + y'(1) = 0.$
10. $x^2y'' - 2xy' + y = \lambda y, \quad y(0) = 0, \quad y'(1) - y(0) = 0.$

Find the eigenvalues and corresponding eigenfunctions of the following boundary values problems.

11. $y'' = -\lambda y, \quad y(0) = 0, \quad y(1) = 0.$
12. $y'' = -\lambda y, \quad y'(0) = 0, \quad y'(1) = 0.$
13. $y'' = -\lambda y, \quad y(0) = 0, \quad y'(1) = 0.$
14. $y'' = -\lambda y, \quad y(0) - y(1) = 0, \quad y'(0) - y'(1) = 0.$
15. $y'' + \lambda y = 0, \quad y(0) - y'(1) = 0, \quad y'(0) - y(1) = 0.$
16. $y'' + \lambda y = 0, \quad y(0) + y'(0) = 0, \quad y(1) = 0.$
17. $y'' + \lambda y = 0, \quad y(0) = 0, \quad y'(\pi) + y'(\pi) = 0.$
18. $y'' + \lambda y = 0, \quad y'(0) = 0, \quad y'(1) + y(1) = 0.$
19. $y'' + \lambda y = 0, \quad y(0) - y'(0) = 0, \quad y'(1) + y(1) = 0.$
20. $y'' + y' + \lambda y = 0, \quad y(0) = 0, \quad y(1) = 0.$
21. $(1+x^2)y'' + xy' + \lambda y = 0, \quad y'(0) = 0, \quad y(1) + 2y'(1) = 0.$
22. $x^2y'' = \lambda y, \quad y(1) = 0, \quad y(a) = 0, \quad a > 1.$
23. $x^2y'' - \lambda xy' + \lambda y = 0, \quad y(1) = 0, \quad y(2) = 0.$
24. $y'' - 2y' + (1+\lambda)y = 0, \quad y(0) = 0, \quad y(1) = 0.$
25. $y'' + 4y' + (4+9\lambda)y = 0, \quad y(0) = 0, \quad y'(1) = 0.$
26. $y'' + (1+\lambda)y' + \lambda y = 0, \quad y'(0) = 0, \quad y(1) = 0.$

Solve each of the following nonhomogeneous problems.

27. $y'' + y = 1, \quad y(0) = 0, \quad y(1) = 0.$
28. $y'' + y = 2\pi - x, \quad y(0) = 0, \quad y(1) = 0.$
29. $y'' - y = 2x, \quad y(0) = 0, \quad y(1) = -1.$
30. $y'' + y = 1, \quad y(0) = 0, \quad y(1) = 1.$

Chapter 4

SYSTEMS OF DIFFERENTIAL EQUATIONS

4.1 Linear systems with constant coefficients

A linear system of n first order differential equations with constant coefficients in the form

$$\vec{y}'(x) = A\vec{y}(x) + \vec{f}(x), \tag{1}$$

where $\vec{y}(x)$ is a column vector with the components $y_1(x)$, $y_2(x), \ldots,$ $y_n(x)$, A is a constant matrix with the elements $a_{11}, a_{12}, \ldots,$ a_{nn}, and $\vec{f}(x)$ is a column vector with the components $f_1(x)$, $f_2(x), \ldots,$ $f_n(x)$ is called *system of the normal form.* The solution of Eq.(1) is a sum of the general solution of the corresponding homogeneous system

$$\vec{y}'(x) = A\vec{y}(x), \tag{2}$$

and a particular solution of Eq.(1).

If the linear system is given in so called *general form*

$$A_0\vec{y}'(x) = A_1\vec{y}(x) + \vec{\varphi}(x), \tag{3}$$

where A_i are constant matrices and $\det A_0 \neq 0$, then by left multiply by A_0^{-1} we obtain the system of the normal form with $A = A_0^{-1}A_1$ and $\vec{f}(x) = A_0^{-1}\vec{\varphi}(x)$.

4.1.1 *Homogeneous Systems*

A general solution of Eq.(2) has the form

$$\vec{y} = \Psi(x)\vec{C}, \tag{4}$$

where $\Psi(x)$ is called *fundamental matrix* of system (2) (see also Sec. 2.2) and $\vec{C}$ is an arbitrary constant column vector. Substituting Eq.(4) in Eq.(2) leads to the equation for $\Psi(x)$

$$\Psi'(x) = A\Psi(x). \tag{2a}$$

Note that we can rewrite Eq.(4) as

$$\vec{y}(x) = \Psi(x)\vec{C} = \Psi(x)E\vec{C} = \Psi(x)SS^{-1}\vec{C} = (\Psi(x)S)(S^{-1}\vec{C}) \equiv \Psi_n(x)\vec{C}_n,$$

where E is the identity matrix, S is a constant nonsingular matrix ($\det S \neq 0$), S^{-1} is the inverse matrix of S, $\Psi_n(x)$ is a new fundamental matrix and $\vec{C}_n$ is a new arbitrary constant vector. Thus the fundamental matrix $\Psi(x)$ can be determined only up to an arbitrary nonsingular transformation

$$\Psi_n(x) = \Psi(x)S. \tag{4b}$$

The fundamental matrix $\Psi(x)$ can be determined by using any of the following methods:

1) Euler's method;
2) matrix method;
3) method of eliminations (see Sec. 4.2).

4.1.2 Homogeneous Systems. Euler's Method

In this method we assume that $\vec{y}(x) = \vec{C}\exp(\lambda x)$, where $\vec{C}$ is an arbitrary vector and λ is a certain constant. Then substituting $\vec{y}(x)$ in Eq.(2) leads to the linear system of algebraic homogeneous equations

$$(A - \lambda E)\,\vec{C} = 0. \tag{5}$$

The nontrivial solutions of this algebraic equation exists if and only if λ is chosen so as the determinant of coefficients is zero, that is, if λ is a root of the *auxiliary equation* (determinantal equation)

$$\det(A - \lambda E) = 0. \tag{6}$$

Eq.(6) is a polynomial equation in λ of degree n and hence has n roots $\lambda_1, \lambda_2, \ldots, \lambda_n$, which are called the *eigenvalues* of matrix A. Here may occur following cases:

1) all roots $\lambda_1, \ldots, \lambda_n$ are different;
2) some of the roots Eq.(6) are repeated.

4.1.3 Euler's Method. Different Eigenvalues

In this case the general solution of Eq.(2) takes the form

$$\vec{y}(x) = C_1\vec{e}_1 e^{\lambda_1 x} + C_2\vec{e}_2 e^{\lambda_2 x} + \cdots + C_n\vec{e}_n e^{\lambda_n x} \equiv \Psi(x)\vec{C}, \tag{7}$$

where C_i are arbitrary constants, which form a column vector $\vec{C}$, $\vec{e}_i$ are *eigenvectors* of matrix A, corresponding to the eigenvalues λ_i, and $\Psi(x)$ is the fundamental matrix of the system (1).

Example 1. Find the general solution of the system

$$x' = 2x + y,$$
$$y' = 3x + 4y.$$

The auxiliary equation of the given system is

$$\begin{vmatrix} 2-\lambda & 1 \\ 3 & 4-\lambda \end{vmatrix} = \lambda^2 - 6\lambda + 5 = 0.$$

Here are two roots $\lambda_1 = 1$, $\lambda_2 = 5$. Eigenvector $\vec{e}_1$ corresponding to λ_1 is a solution of the equation $A\vec{e}_1 = \lambda_1\vec{e}_1$, that is,

$$(e_1)_1 + (e_1)_2 = 0,$$
$$3(e_1)_1 + 3(e_1)_2 = 0.$$

Since $\vec{e}_1 \neq 0$ we find $(e_1)_1 = -(e_1)_2$ and

$$\vec{e}_1 = C\begin{pmatrix} -1 \\ 1 \end{pmatrix},$$

where C is an arbitrary constant. Choosing this constant equal to one we obtain $\vec{e}_1 = (-1,\ 1)^T$. In the same way we find $\vec{e}_2 = (1,\ 3)^T$. Hence the general solution of the given system is

$$\vec{y}(t) = C_1\begin{pmatrix} -1 \\ 1 \end{pmatrix} e^t + C_2\begin{pmatrix} 1 \\ 3 \end{pmatrix} e^{5t} = \begin{pmatrix} -e^t & e^{5t} \\ e^t & 3e^{5t} \end{pmatrix}\begin{pmatrix} C_1 \\ C_2 \end{pmatrix} \equiv \Psi(t)\vec{C},$$

where $\Psi(t)$ is the fundamental matrix of the given equation and $\vec{C}$ is an arbitrary constant vector.

4.1.4 Euler's Method. Repeated Eigenvalues

Let λ_1 be the unique k-fold root of the auxiliary equation and all the other roots are simple. The number k is called *algebraic multiplicity* of λ_1. It is obvious that in this case the eigenvectors corresponding to the repeated eigenvalue λ_1 we should denote by yet one index, say i. Then these eigenvectors we denote as $\vec{e}_{1i}$, where $i = 1, 2, \ldots, l$ and the number l is called *geometric multiplicity* of $\vec{e}_1$. The number l is equal to the number of arbitrary constants in the eigenvector $\vec{e}_1$ corresponding to λ_1.

In this event there are two possibilities:

1) there are $l = k$ linearly independent eigenvectors $\vec{e}_{11}, \vec{e}_{12}, \ldots, \vec{e}_{1k}$ corresponding to the eigenvalue λ_1 ;

2) there are fewer than k linearly independent eigenvectors $\vec{e}_{11}, \vec{e}_{12}, \ldots, \vec{e}_{1l}$ corresponding to the eigenvalue λ_1.

Notice that $l = k$ when $A^* \equiv \bar{A}^T = A$, where A^* is adjoint matrix of A.

1) *Algebraic and geometric multiplicities are equal.* In this case

$$\vec{y}_1(x) = (C_1\vec{e}_{11} + C_2\vec{e}_{12} + \cdots + C_k\vec{e}_{1k})e^{\lambda_1 x}, \tag{8}$$

is a particular solution of Eq.(2) corresponding to λ_1, where $\vec{e}_{1i}$ are linearly independent solutions of Eq.(5) corresponding to the eigenvalues λ_1 and C_i are arbitrary constants. The general solution of Eq.(2) then takes the form

$$\vec{y}(x) = (C_1\vec{e}_{11} + C_2\vec{e}_{12} + \cdots + C_k\vec{e}_{1k})e^{\lambda_1 x}$$

$$+C_{k+1}\vec{e}_{k+1}e^{\lambda_{k+1}x} + \cdots + C_n\vec{e}_n e^{\lambda_n x}$$

Example 2. Find the general solution of the system

$$\begin{aligned} x' &= 2x, \\ y' &= 2y, \\ z' &= 2z. \end{aligned}$$

According to Eq.(6) we find the auxiliary equation $(2-\lambda)^3 = 0$. Its root $\lambda_1 = \lambda_2 = \lambda_3 = 2$ is 3-fold. Hence $k = 3$. Eigenvectors $\vec{e}_{1i}$ corresponding to λ_1 are solutions of the system (5)

$$\begin{aligned} 2(e_{1i})_1 &= 2(e_{1i})_1, \\ 2(e_{1i})_2 &= 2(e_{1i})_2, \\ 2(e_{1i})_3 &= 2(e_{1i})_3, \end{aligned}$$

and can be determined only up to three arbitrary constant

$$\vec{e}_{1i} = C_1\begin{pmatrix}1\\0\\0\end{pmatrix} + C_2\begin{pmatrix}0\\1\\0\end{pmatrix} + C_3\begin{pmatrix}0\\0\\1\end{pmatrix}.$$

Thus $l = 3$, and here exist three eigenvectors corresponding to λ_1. Choosing these constants so that only a single component of the eigenvectors were equal to one we obtain

$$\vec{e}_{11} = \begin{pmatrix}1\\0\\0\end{pmatrix}, \quad \vec{e}_{12} = \begin{pmatrix}0\\1\\0\end{pmatrix}, \quad \vec{e}_{13} = \begin{pmatrix}0\\0\\1\end{pmatrix}.$$

Hence the general solution of the given system is

$$\begin{aligned} \vec{y}(t) = \begin{pmatrix}x\\y\\z\end{pmatrix} &= C_1\begin{pmatrix}1\\0\\0\end{pmatrix}e^{2t} + C_2\begin{pmatrix}0\\1\\0\end{pmatrix}e^{2t} + C_3\begin{pmatrix}0\\0\\1\end{pmatrix}e^{2t} \\ &= e^{2t}E\vec{C} \equiv \Psi(t)\vec{C}, \end{aligned}$$

where $\Psi(x)$ is the fundamental matrix of the given system and $\vec{C}$ is an arbitrary constant vector.

2) *Algebraic and geometric multiplicities are distinct.* In this case two methods can be used:

1) the method of associated vectors;
2) the method of undetermined coefficients.

4.1.5 Repeated Eigenvalues. Method of Associated Vectors

The solution of Eq.(2) corresponding to the repeated eigenvalue λ_1 takes the form

$$\begin{aligned} \vec{y}_1(x) = e^{\lambda_1 x}\Big(&\vec{e}_{11}P_m(x) + \vec{a}_1 P_{m-1}(x) + \vec{a}_2 P_{m-2}(x) + \cdots + \vec{a}_m P_0(x) \\ &+\vec{e}_{12}C_{m+2} + \vec{e}_{13}C_{m+3} + \cdots + \vec{e}_{1l}C_k\Big), \end{aligned} \tag{9}$$

where $\vec{e}_{1i}$ $i = 1, 2, \ldots, l \geq 1$ are linearly independent eigenvectors of matrix A corresponding to the eigenvalue λ_1; $\vec{a}_1, \vec{a}_2, \ldots, \vec{a}_m$ are vectors associated to eigenvector $\vec{e}_{11}$; $P_m(x) = C_1 + C_2 x + \ldots + C_{m+1}x^m$ is a polynomial of degree $m = k - l$ with arbitrary coefficients $C_1, C_2, \ldots, C_{m+1}$;

$$P_{i-1}(x) = P_i'(x), \qquad i = m, m-1, \ldots, 1 \tag{10}$$

and $C_{m+2}, C_{m+3}, \ldots, C_k$ are also arbitrary constants. The associated vectors $\vec{a}_i$ are determined by equations

$$(A - \lambda_1 E)\vec{a}_i = \vec{a}_{i-1}, \qquad i = 1, 2, \ldots, m, \tag{11}$$

where $\vec{a}_0 \equiv \vec{e}_{11} \neq 0$. These equations always have some solutions if the necessary conditions

$$\vec{e}_1^{\,*} \cdot \vec{a}_i = 0, \quad i = 0, 1, \ldots, m-1 \tag{12}$$

are satisfied. Here the column vector $\vec{e}_1^{\,*}$ is any solution of the equation adjoint to Eq.(5)

$$(A^* - \bar{\lambda}_1 E)\vec{e}_1^{\,*} = 0, \tag{13}$$

where matrix $A^* \equiv \bar{A}^T$ is called adjoint of matrix A and $\bar{\lambda}_1$ is the complex conjugate number of λ_1. The conditions (12) are used for choice of arbitrary constants in the eigenvector $\vec{e}_{11}$ and the associated vectors $\vec{a}_1, \vec{a}_2, \ldots, \vec{a}_m$.

Example 3. Find the general solution of the system

$$\begin{aligned} x' &= 3x - y + z, \\ y' &= x + y + z, \\ z' &= 2z. \end{aligned}$$

Proceeding as in Ex.2 we find the eigenvalues $\lambda_1 = \lambda_2 = \lambda_3 = 2$. The eigenvectors corresponding to λ_1 are solutions of Eq-s (5)

$$\begin{aligned} &(e_{1i})_1 - (e_{1i})_2 + (e_{1i})_3 = 0, \\ &(e_{1i})_1 - (e_{1i})_2 + (e_{1i})_3 = 0, \\ &0(e_{1i})_3 = 0, \end{aligned}$$

and can be determined only up to two arbitrary constants

$$\vec{e}_{1i} = C_1 \begin{pmatrix} 1 \\ 1 \\ 0 \end{pmatrix} + C_2 \begin{pmatrix} -1 \\ 0 \\ 1 \end{pmatrix}.$$

Thus $l = 2$, and here exist two eigenvectors corresponding to λ_1. We should choose the constants C_1 and C_2 so that the condition (12) is satisfied at $i = 0$ only for $\vec{e}_{11}$. According to the definition of A^* we find

$$A^* \equiv \bar{A}^T = \begin{pmatrix} 3 & 1 & 0 \\ -1 & 1 & 0 \\ 1 & 1 & 2 \end{pmatrix}.$$

It is easy to verify that $\vec{e}_1^* = D_1(1,-1,0)^T + D_2(0,0,1)^T$, where D_1 and D_2 are arbitrary constants, is a solution of Eq.(13) with obtained A^* and $\bar{\lambda} = 2$. Then conditions (12) require at $i = 0$

$$\vec{e}_1^* \cdot \vec{e}_{1i} = D_1(C_1 - C_2) - D_1 C_1 + D_2 C_2 = C_2(D_1 - D_2) = 0.$$

Since D_i are arbitrary, we should put $C_2 = 0$, and then $\vec{e}_{11} = (1,\,1,\,0)^T$. Further putting $C_2 = 1$ we find $\vec{e}_{12} = (0,1,1)^T$. We find the vector $\vec{a}_1$ associated to the eigenvector $\vec{e}_{11}$ by using Eq.(11) at $i = 1$

$$\begin{aligned} (a_1)_1 - (a_1)_2 + (a_1)_3 &= 1, \\ (a_1)_1 - (a_1)_2 + (a_1)_3 &= 1. \end{aligned}$$

This system is compatible and its solution can be determined only up to two arbitrary constants

$$\vec{a}_1 = \begin{pmatrix} 1 \\ 0 \\ 0 \end{pmatrix} + C_1 \begin{pmatrix} 1 \\ 1 \\ 0 \end{pmatrix} + C_2 \begin{pmatrix} -1 \\ 0 \\ 1 \end{pmatrix}.$$

Choosing $C_1 = C_1 = 1$ we obtain $\vec{a}_1 = (1,1,1)^T$. Now, since $m = k - l = 1$, we have $P_1(t) = C_1 + C_2 t$, where C_i are arbitrary constants. Then $P_0(t) = P_1'(t) = C_2$ and the general solution of the given system is

$$\vec{y}(t) = \left(\vec{e}_{11}(C_1 + C_2 t) + \vec{a}_1 C_2 + \vec{e}_{12} C_3\right) e^{2t},$$

or in the components

$$\begin{aligned} x &= ((C_1 + C_2 t) + C_2)\, e^{2t}, \\ y &= ((C_1 + C_2 t) + C_2 + C_3)\, e^{2t}, \\ z &= (C_2 + C_3) e^{2t}, \end{aligned}$$

where C_i are arbitrary constants. By making change of the constants $\tilde{C}_1 = C_1 - C_3$, $\tilde{C}_2 = C_2$, $\tilde{C}_3 = C_2 + C_3$ we obtain the general solution in the form

$$\vec{y}(t) = \begin{pmatrix} e^{2t} & te^{2t} & e^{2t} \\ e^{2t} & (t-1)e^{2t} & 2e^{2t} \\ 0 & 0 & e^{2t} \end{pmatrix} \begin{pmatrix} \tilde{C}_1 \\ \tilde{C}_2 \\ \tilde{C}_3 \end{pmatrix} \equiv \Psi(t)\vec{C},$$

where $\Psi(t)$ is the fundamental matrix of the given system and $\vec{C}$ is an arbitrary constant vector. Notice that the transformation with C_i to $\tilde{C}_i$ is given by the matrix

$$S^{-1} = \begin{pmatrix} 1 & 0 & -1 \\ 0 & 1 & 0 \\ 0 & 1 & 1 \end{pmatrix}.$$

Example 4. Find the general solution of the system

$$\begin{aligned} x' &= 4x - y, \\ y' &= 3x + y - z, \\ z' &= x + z. \end{aligned}$$

According to Eq.(6) we find the auxiliary equation $(2-\lambda)^3 = 0$. Its root $\lambda_1 = \lambda_2 = \lambda_3 = 2$ is 3-fold and $k = 3$. The eigenvectors $\vec{e}_{1i}$ corresponding to λ_1 are solutions of the system

$$\begin{aligned} &2(e_{1i})_1 - (e_{1i})_2 = 0, \\ &3(e_{1i})_1 - (e_{1i})_2 - (e_{1i})_3 = 0, \\ &(e_{1i})_1 - (e_{1i})_3 = 0, \end{aligned}$$

and can be determined only up to one arbitrary multiplicative constant $\vec{e}_{1i} = C(1,2,1)^T$. Thus $l = 1$ and here exist one eigenvector corresponding to λ_1. Therefore Eq. (12) should be satisfied automatically at $i = 0$. It is easy to verify, since the adjoint matrix of A is

$$A^* \equiv \bar{A}^T = \begin{pmatrix} 4 & 3 & 1 \\ -1 & 1 & 0 \\ 0 & -1 & 1 \end{pmatrix}$$

and a solution of Eq.(13) corresponding to $\bar{\lambda} = 2$ is the column vector $\vec{e}_1^* = D(-1,1,-1)^T$, where D is an arbitrary constant. It is obvious that $\vec{e}_1^* \cdot \vec{e}_{11} \equiv 0$. Then choosing $C = 1$ we obtain $\vec{e}_{11} = (1,2,1)^T$. The associated vector $\vec{a}_1$ we find by using Eq. (11) at $i = 1$

$$\begin{aligned} &2(a_1)_1 - (a_1)_2 = 1, \\ &3(a_1)_1 - (a_1)_2 - (a_1)_3 = 2, \\ &(a_1)_1 - (a_1)_3 = 1. \end{aligned}$$

This system is compatible, and its solution is $\vec{a}_1 = (C, 2C-1, C-1)^T$, where C is an arbitrary constant. According to Eq.(12) we obtain at $i = 1$, $\vec{e}_1^* \cdot \vec{a}_1 \equiv 0$ at any arbitrary constant C. Choosing $C = 1$ we obtain $\vec{a}_1 = (1,1,0)^T$. In the same manner we find associated vector $\vec{a}_2$ at $i = 2$

$$\begin{aligned} &2(a_2)_1 - (a_2)_2 = 1, \\ &3(a_2)_1 - (a_2)_2 - (a_2)_3 = 1, \\ &(a_2)_1 - (a_2)_3 = 0. \end{aligned}$$

This system is also compatible and it has the solution $\vec{a}_2 = (C, 2C-1, C)^T$, where C is an arbitrary constant. Putting $C = 1$ gives $\vec{a}_2 = (1,1,1)^T$. Further, since $m = k - l = 2$, we have

$$P_2(t) = C_1 + C_2 t + C_3 \frac{t^2}{2},$$

where C_i are arbitrary constants. Then $P_1(t) = C_2 + C_3 t$, $P_0(t) = C_3$ and the general solution of the given system is

$$\vec{y}(t) = \begin{pmatrix} x \\ y \\ z \end{pmatrix} = \begin{pmatrix} 1 \\ 2 \\ 1 \end{pmatrix} \left(C_1 + C_2 t + C_3 \frac{t^2}{2} \right) e^{2t} +$$

$$\begin{pmatrix} 1 \\ 1 \\ 0 \end{pmatrix} (C_2 + C_3 t) e^{2t} + \begin{pmatrix} 1 \\ 1 \\ 1 \end{pmatrix} C_3 e^{2t}.$$

By making change of constants $C_1 = C_1 + C_2 + C_3$, $C_2 = C_2 + C_3$, $C_3 = C_3/2$ we obtain the general solution

$$\vec{y}(t) = e^{2t} \begin{pmatrix} 1 & t & t \\ 2 & -1+2t & -2t+2t^2 \\ 1 & -1+t & 2-2t+t^2 \end{pmatrix} \begin{pmatrix} C_1 \\ C_2 \\ C_3 \end{pmatrix} \equiv \Psi(t)\vec{C},$$

where $\Psi(t)$ is the fundamental matrix and $\vec{C}$ is an arbitrary constant vector.

4.1.6 Repeated Eigenvalues. Method of Undetermined Coefficients

If λ_1 is the k-fold root of the auxiliary equation, then the solution $\vec{y}_1(x)$ can be found as

$$\begin{aligned} y_{11} &= P_{k-l}(x)e^{\lambda_1 x}, \\ y_{12} &= Q_{k-l}(x)e^{\lambda_1 x}, \\ &\vdots \\ y_{1n} &= R_{k-l}(x)e^{\lambda_1 x}, \end{aligned} \tag{14}$$

where $P_{k-l}(x)$, $Q_{k-l}(x)$, ..., $R_{k-l}(x)$ are polynomials of degree $k-l$ with at most $n \cdot (k-l+1)$ undetermined coefficients and l is the geometric multiplicities of λ_1. These coefficients can be determined by substituting the solution $\vec{y}_1(x)$ in the form of Eq.(14) in Eq. (2) and equating the coefficients of like powers of x to zero. Notice that since $l \geq 1$ we can used in Eq.(14) the polynomials of degree $k-1$ with at most $n \cdot k$ undetermined coefficients. Then the geometric multiplicities is not necessary.

Example 5. Find the general solution of the system

$$\begin{aligned} x' &= x - y + z, \\ y' &= x + y - z, \\ z' &= -y + 2z. \end{aligned}$$

The roots of the auxiliary equation are $\lambda_1 = \lambda_2 = 1$, $\lambda_3 = 2$. Since λ_1 is the 2-fold root, we find the corresponding solution $\vec{y}_1(x)$ in the form

$$\begin{aligned} x &= (a + bt)e^t, \\ y &= (c + dt)e^t, \\ z &= (e + ft)e^t, \end{aligned}$$

where a, b, ..., f are undetermined coefficients. By substituting this solution in the given system we obtain, after some minor simplifications,

$$\begin{aligned} b &= -c - dt + e + ft, \\ d &= a + bt - e - ft, \\ f &= -c - dt + e + ft. \end{aligned}$$

For these equations to be satisfied it is necessary that the coefficients of like powers of t be equal. Hence we obtain

$$\begin{aligned} b &= -c + e, \\ 0 &= -d + f, \\ d &= a - e, \\ 0 &= b - f, \\ f &= -c + e. \end{aligned}$$

Its nontrivial solution is $a = C_1$, $b = C_2$, $c = C_1 - 2C_2$, $d = C_2$, $e = C_1 - C_2$, $f = C_2$, where C_1 and C_2 are arbitrary constants. Hence the particular solution of the given system corresponding to the two-fold eigenvalue λ_1 is

$$\begin{aligned} x &= (C_1 + C_2 t)e^t, \\ y &= (C_1 - 2C_2 + C_2 t)e^t, \\ z &= (C_1 - C_2 + C_2 t)e^t. \end{aligned}$$

The solution $y_3(t)$ corresponding to $\lambda_3 = 2$ we find as has been shown above. The eigenvector $\vec{e}_3$ corresponding to λ_3 is a solution of the algebraic system

$$\begin{aligned} -(e_3)_1 - (e_3)_2 + (e_3)_3 &= 0, \\ (e_3)_1 - (e_3)_2 - (e_3)_3 &= 0, \\ -(e_3)_2 &= 0. \end{aligned}$$

It is easy to verify that $\vec{e}_3 = (1,\ 0,\ 1)^T$. Hence the general solution of the given system is

$$\begin{aligned} x &= (C_1 + C_2 t)e^t + C_3 e^{2t}, \\ y &= (C_1 - 2C_2 + C_2 t)e^t, \\ z &= (C_1 - C_2 + C_2 t)e^t + C_3 e^{2t}. \end{aligned}$$

where C_i are arbitrary constants.

4.1.7 Homogeneous Systems. Matrix Method

The general solution of the homogeneous system can be written as

$$y(x) = \exp(Ax)\vec{C}, \tag{15}$$

where $\vec{C}$ is an arbitrary constant vector and the *exponential matrix* $\exp(Ax)$ is the sum of the matrix series

$$\exp(Ax) = E + Ax + \frac{1}{2!}A^2x^2 + \frac{1}{3!}A^3x^3 + \cdots + \frac{1}{n!}A^nx^n + \cdots. \tag{16}$$

By comparing Eq.(15) and Eq.(4), we conclude that $\Psi(x) = \exp(Ax)$, however actually these matrices are similar, that is $\Psi(x) = \exp(Ax)S$, where S is a nonsingular matrix.

The exponential matrix $\exp(Ax)$ possesses the following properties:

a) if $AB = BA$, then

$$\exp(Ax + Bx) = \exp(Ax)\exp(Bx) = \exp(Bx)\exp(Ax) = \exp(Bx + Ax);$$

b) if $A = TBT^{-1}$, where a matrix T is nonsingular, then $\exp(Ax) = \exp(TBT^{-1}x) = T\exp(Bx)T^{-1}$;

c) if A is a diagonal matrix with the elements $a_{ij} = \lambda_i\delta_{ij}$, where λ_i are constants and δ_{ij} is the Kronecker delta, then $\exp(Ax)$ is also the diagonal matrix with the elements $\exp(\lambda_i x)\delta_{ij}$;

d) if a $k \times k$ matrix A is the Jordan block (the French mathematician C. Jordan (1838 - 1922))

$$A = J_k \equiv \begin{pmatrix} \lambda & 1 & 0 & \dots & 0 \\ 0 & \lambda & 1 & \dots & 0 \\ \vdots & \vdots & \vdots & \ddots & \vdots \\ 0 & 0 & 0 & \dots & 1 \\ 0 & 0 & 0 & \dots & \lambda \end{pmatrix},$$

where λ is a constant, then the exponential matrix has the *echelon form*

$$\exp(Ax) = \exp(J_k x) = e^{\lambda x}\begin{pmatrix} 1 & x & \frac{x^2}{2} & \dots & \frac{x^n}{n!} \\ 0 & 1 & x & \dots & \frac{x^{n-1}}{(n-1)!} \\ \vdots & \vdots & \vdots & \ddots & \vdots \\ 0 & 0 & 0 & \dots & x \\ 0 & 0 & 0 & \dots & 1 \end{pmatrix},$$

By using these properties the exponential matrix $\exp(Ax)$ can also be found in those cases when the direct summation in Eq.(16) is very difficult. In accordance with property b) and taking into account properties c) and d), we should find such a matrix B that takes a simplest form. As is well known such matrix is the *Jordan's canonical form* of A

$$\Lambda = T^{-1}AT, \tag{17}$$

where T is the matrix whose columns are eigenvectors $\vec{e}_1$, $\vec{e}_2$,..., $\vec{e}_n$, or eigenvectors and the corresponding associated vectors $\vec{e}_{11}$, $\vec{a}_{11}$, ..., $\vec{a}_{1m_1}$, $\vec{e}_{12}$, ..., $\vec{e}_{1l_1}$ $\vec{e}_{21}$, $\vec{a}_{21}$, ..., $\vec{a}_{2m_2}$, $\vec{e}_{22}$, ..., $\vec{e}_{2l_2}$, ... and T^{-1} is the inverse matrix of T. The matrix T is always nonsingular since the eigenvectors or the eigenvectors and the corresponding associated vectors of A are always linearly independent. Hence

$$\exp(Ax) = T\exp(\Lambda x)T^{-1}. \tag{17a}$$

Example 6. Using Eq-s(16) and (17) find the exponential matrices $\exp(Ax)$ for the following matrices A:

$$\text{a)}\quad A = \begin{pmatrix} 1 & 0 & 0 \\ 0 & 2 & 0 \\ 0 & 0 & 3 \end{pmatrix};$$

$$\text{b)}\quad A = \begin{pmatrix} 3 & -1 & 1 \\ 1 & 1 & 1 \\ 0 & 0 & 2 \end{pmatrix},$$

see Ex.3.

a). The immediate calculation gives

$$A^n = \begin{pmatrix} 1 & 0 & 0 \\ 0 & 2^n & 0 \\ 0 & 0 & 3^n \end{pmatrix}.$$

The direct summation of the matrix series (16) yields

$$\exp(Ax) = \begin{pmatrix} e^x & 0 & 0 \\ 0 & e^{2x} & 0 \\ 0 & 0 & e^{3x} \end{pmatrix};$$

b). Using eigenvectors $\vec{e}_{11}$, $\vec{e}_{12}$ and associated vector $\vec{a}_1$ found in Ex.3 we obtain

$$T = \begin{pmatrix} 1 & 1 & 0 \\ 1 & 1 & 1 \\ 0 & 1 & 1 \end{pmatrix}.$$

Then

$$T^{-1} = \begin{pmatrix} 0 & 1 & -1 \\ 1 & -1 & 1 \\ -1 & 1 & 0 \end{pmatrix}$$

and

$$\Lambda = T^{-1}AT = \begin{pmatrix} 0 & 1 & -1 \\ 1 & -1 & 1 \\ -1 & 1 & 0 \end{pmatrix} \begin{pmatrix} 3 & -1 & 0 \\ 1 & 1 & 1 \\ 0 & 0 & 2 \end{pmatrix} \begin{pmatrix} 1 & 1 & 0 \\ 1 & 1 & 1 \\ 0 & 1 & 1 \end{pmatrix}$$

$$= \begin{pmatrix} 2 & 1 & 0 \\ 0 & 2 & 0 \\ 0 & 0 & 2 \end{pmatrix}.$$

Now we find the matrix $\exp(\Lambda x)$ by using the property d) for the 2×2 Jordan's block in the two rows of Λ and the property c) for the third row of Λ.

$$\exp(\Lambda x) = e^{2x} \begin{pmatrix} 1 & x & 0 \\ 0 & 1 & 0 \\ 0 & 0 & 1 \end{pmatrix}.$$

Finally, according to Eq.(17) we obtain

$$e^{Ax} = Te^{\Lambda x}T^{-1} = \begin{pmatrix} 1 & 1 & 0 \\ 1 & 1 & 1 \\ 0 & 1 & 1 \end{pmatrix} e^{2x} \begin{pmatrix} 1 & x & 0 \\ 0 & 1 & 0 \\ 0 & 0 & 1 \end{pmatrix} \begin{pmatrix} 0 & 1 & -1 \\ 1 & -1 & 1 \\ -1 & 1 & 0 \end{pmatrix}$$

$$= e^{2x} \begin{pmatrix} 1+x & -x & x \\ x & 1-x & x \\ 0 & 0 & 1 \end{pmatrix}.$$

It is easy to verify that by the right multiplication of this matrix by the matrix

$$S = \begin{pmatrix} 1 & 0 & 1 \\ 1 & -1 & 2 \\ 0 & 0 & 1 \end{pmatrix}$$

we obtain the matrix $\Psi(x)$ in Ex.3.

Example 7. Find the general solution of the system $y' = Ay$, when

$$A = \begin{pmatrix} 2 & 1 & -1 \\ -1 & 0 & 1 \\ 1 & 1 & 0 \end{pmatrix}.$$

According to Eq-s(15) and (17) we have

$$\vec{y} = \exp(Ax)\vec{C} = T\exp(\Lambda x)T^{-1}\vec{C},$$

where $\vec{C}$ is an arbitrary constant vector, T and Λ are constant matrices. The eigenvalues of A are $\lambda_1 = 0$ and $\lambda_2 = \lambda_3 = 1$. The eigenvector corresponding to λ_1 is solution of Eq.(5)

$$\begin{aligned} &2(e_1)_1 + (e_1)_2 - (e_1)_3 = 0, \\ &-(e_1)_1 + (e_1)_3 = 0, \\ &(e_1)_1 + (e_1)_2 = 0. \end{aligned}$$

This system has a solution which is known only up to one multiplicative constant. Choosing this constant so that the first component is one, we obtain, after some minor algebraic calculations, $\vec{e}_1 = (1,\ -1,\ 1)^T$. For the eigenvectors $\vec{e}_{2i}$ corresponding to two-fold root λ_2, we have

$$\begin{aligned} &(e_{2i})_1 + (e_{2i})_2 - (e_{2i})_3 = 0, \\ &-(e_{2i})_1 - (e_{2i})_2 + (e_{2i})_3 = 0, \\ &(e_{2i})_1 + (e_{2i})_2 - (e_{2i})_3 = 0. \end{aligned}$$

Solution of this system can be determined only up to two arbitrary multiplicative constants. Hence here exist two eigenvectors corresponding to λ_2 and $l = k = 2$. Choosing these constants so that the first component of the eigenvector is one or zero we obtain $\vec{e}_{21} = (1,\ 0,\ 1)^T$ and $\vec{e}_{22} = (0,\ 1,\ 1)^T$. Hence all columns of T are eigenvectors of A and

$$T = \begin{pmatrix} 1 & 1 & 0 \\ -1 & 0 & 1 \\ 1 & 1 & 1 \end{pmatrix}.$$

By the immediate calculation we find

$$T^{-1} = \begin{pmatrix} -1 & -1 & 1 \\ 2 & 1 & -1 \\ -1 & 0 & 1 \end{pmatrix}.$$

Since $l = k$, we write automatically

$$\Lambda = T^{-1}AT = \begin{pmatrix} 0 & 0 & 0 \\ 0 & 1 & 0 \\ 0 & 0 & 1 \end{pmatrix}$$

and

$$\exp(\Lambda x) = \begin{pmatrix} 1 & 0 & 0 \\ 0 & e^x & 0 \\ 0 & 0 & e^x \end{pmatrix}.$$

Now, according to Eq-s(15) and (17) we find the general solution of the given system

$$\vec{y}(x) = T\exp(\Lambda x)T^{-1}\vec{C} = \begin{pmatrix} -C_1 - C_2 + C_3 + (2C_1 + C_2 - C_3)e^t \\ C_1 + C_2 - C_3 + (C_3 - C_1)e^t \\ -C_1 - C_2 + C_3 + (C_1 + C_2)e^t \end{pmatrix}$$

By making change of constants $\tilde{C}_1 = -C_1 - C_2 + C_3$, $\tilde{C}_2 = C_1 - C_3$, $\tilde{C}_3 = C_1 + C_2$ we obtain the general solution in the form

$$x = C_1 + (C_2 + C_3)e^t,$$
$$y = -C_1 - C_2e^t,$$
$$z = C_1 + C_3e^t,$$

where C_i are arbitrary constants and the tilde-sign is omitted.

4.1.8 Nonhomogeneous Systems

The general solution of Eq.(1) is

$$\vec{y} = \vec{y}_h + \vec{y}_p = \Psi(x)\vec{C} + \vec{y}_p, \tag{18}$$

where $\vec{y}_h$ is the general solution of the corresponding homogeneous system and $\vec{y}_p$ is a particular solution of nonhomogeneous system. We assume that the fundamental matrix $\Psi(x)$ of the corresponding homogeneous system is known. Then a particular solution $\vec{y}_p$ can be found by using any of the following methods:

1) method of variation of parameters;
2) method of undetermined coefficients;
3) matrix method.

4.1.9 Method of Variation of Parameters

Proceeding as in Sec. 2.3 we find the particular solution of Eq.(1) in the form

$$\vec{y}_p = \Psi(x)\int^x \Psi^{-1}(s)\vec{f}(s)\,ds, \tag{19}$$

where $\Psi^{-1}(s)$ is the inverse matrix of $\Psi(s)$.

Example 8. Find the general solution of the nonhomogeneous system

$$x' = 2x + y + t,$$
$$y' = 3x + 4y - 2t,$$

(see also Ex.1)

According to Ex.1, the fundamental matrix of the corresponding homogeneous system is

$$\Psi(t) = \begin{pmatrix} -e^t & e^{5t} \\ e^t & 3e^{5t} \end{pmatrix}$$

Then

$$\Psi^{-1}(t) = \frac{1}{4}\begin{pmatrix} -3e^{-t} & e^{-t} \\ e^{-5t} & e^{-5t} \end{pmatrix}.$$

Integrating we obtain, taking into account that $\vec{f}(t) = (t,\ -2t)^T$,

$$\int \Psi^{-1}(t)\vec{f}(t)\,dt = -\frac{1}{4}\int \begin{pmatrix} 5te^{-t} \\ te^{-5t} \end{pmatrix} dt = \begin{pmatrix} \frac{5}{4}(te^{-t} + e^{-t}) \\ \frac{1}{20}(te^{-5t} + \frac{1}{5}e^{-5t}) \end{pmatrix}.$$

Hence the general solution of the given system is

$$\vec{y}(t) = \begin{pmatrix} x \\ y \end{pmatrix} = \Psi(t)\vec{C} + \Psi(t)\int \Psi^{-1}(t)\vec{f}(t)dt$$

$$= C_1\begin{pmatrix} -1 \\ 1 \end{pmatrix}e^t + C_2\begin{pmatrix} 1 \\ 3 \end{pmatrix}e^{5t} + \begin{pmatrix} -\frac{6}{5}t - \frac{31}{25} \\ \frac{7}{5}t + \frac{32}{25} \end{pmatrix},$$

where C_i are arbitrary constants.

4.1.10 Method of Undetermined Coefficients

If the nonhomogeneous term $\vec{f}(x)$ in Eq.(1) is an exponential function $\exp(ax)$, or a polynomial of degree k $P_k(x) = p_0x^k + p_1x^{k-1} + \cdots + p_k$, or sinusoidal in character ($\sin(bx)$ or $\cos(bx)$), or a superposition of these functions, then particular solution of Eq.(1) can be found as a sum of these functions and all their products

$$\begin{aligned} y_{p1}(x) &= B_k(x) + e^{ax}C_k(x) + D_k(x)\cos(bx) \\ &\quad + E_k(x)\sin(bx) + e^{ax}\cos(bx)F_k(x) + e^{ax}\sin(bx)H_k(x), \\ &\vdots \\ y_{pn}(x) &= Q_k(x) + e^{ax}R_k(x) + S_k(x)\cos(bx) \\ &\quad + T_k(x)\sin(bx) + e^{ax}\cos(bx)U_k(x) + e^{ax}\sin(bx)V_k(x), \end{aligned} \tag{20}$$

where $B_k(x)$, ..., $V_k(x)$ are polynomials of degree k with at most $6n(k+1)$ undetermined coefficients. These coefficients can be determined by substituting the particular solution in the form Eq-s (20) for $\vec{y}_p(x)$ in Eq.(1) and equating the coefficients of like terms of x in both sides of system.

Example 9. Find the particular solution of the nonhomogeneous system

$$x' = 2x + y + e^{-t},$$
$$y' = 3x + 4y - 2t.$$

Since function $\vec{f}(t) = (e^{-t},\ -2t)^T = (e^{-t},\ 0)^T + (0,\ -2t)^T$, we find the particular solution in the form

$$x(t) = At + B + (at + b)e^{-t},$$
$$y(t) = Ct + D + (ct + d)e^{-t},$$

where $A, \ldots, d$ are undetermined coefficients. By substituting these expressions in the given system, we obtain

$$-(at + b - a)e^{-t} + A = (2at + 2b + ct + d + 1)e^{-t} + (2At + 2B + Ct + D),$$
$$-(ct + d - c)e^{-t} + C = (3at + 3b + 4ct + 4d)e^{-t} + (3At + 3B + 4Ct + 4D - 2t).$$

For these equations to be satisfied it is necessary that the coefficients of like terms of t be equal. Hence we have

$$\begin{aligned} -a &= 2a + c, \\ a - b &= 2b + d + 1, \\ 0 &= 2A + C, \\ A &= 2B + D, \\ -c &= 3a + 4c, \\ c - d &= 3b + 4d, \\ 0 &= 3A + 4C - 2, \\ C &= 3B + 4D. \end{aligned}$$

It is easy to verify that the solution of this system is $a = c = 0$, $b = -5/12$, $d = 1/4$, $A = -2/5$, $B = -12/25$, $C = 4/5$, $D = 14/25$. Hence the particular solution of the given system is

$$x_p(t) = -\frac{5}{12}e^{-t} - \frac{2}{5}t - \frac{12}{25},$$
$$y_p(t) = \frac{1}{4}e^{-t} + \frac{4}{5}t + \frac{14}{25}.$$

4.1.11 *Matrix Method*

The particular solution of the nonhomogeneous system can be written as

$$\vec{y}_p = \exp(Ax) \int \exp(-Ax)\vec{f}(x)\, dx, \tag{21}$$

where $\exp(Ax)$ is given by Eq. (15) and $\exp(-Ax)$ is the inverse matrix of $\exp(Ax)$.

Example 10. Find the particular solution of the nonhomogeneous system $\vec{y}'(x) = A\vec{y}(x) + \vec{f}(x)$, where the matrix A is given in Ex.7 and $\vec{f}(x) = (1,\ x,\ 1 + x)^T$.

According to Eq-s(21) and (17) we have

$$\begin{aligned}\vec{y}_p &= T\exp(\Lambda x)T^{-1}\int T\exp(-\Lambda x)T^{-1}\vec{f}(x)dx\\ &= T\exp(\Lambda x)\int \exp(-\Lambda x)T^{-1}\vec{f}(x)\,dx,\end{aligned}$$

since T and T^{-1} are constant matrices. Hence we should calculate the integral

$$\vec{u} = \int \exp(-\Lambda x)T^{-1}\vec{f}(x)\,dx$$

where matrices Λ and T^{-1} are given in Ex. 7. A substitution of these matrices and an integration then yields, after some calculations,

$$\vec{u} = (0, -e^{-x}, -(1+x)e^{-x})^T.$$

Now using the matrices T and $\exp(\Lambda x)$ found in Ex.7, we obtain the particular solution of the given system

$$\vec{y}_p = T\exp(\Lambda x)\vec{u}(x) = (-1, -(1+x), -(2+x))^T.$$

4.1.12 Initial Value Problem

The solution of the initial value problem

$$\vec{y}'(x) = A\vec{y}(x) + \vec{f}(x), \qquad \vec{y}(x_0) = \vec{y}_0 \tag{22}$$

can be written in the following form

$$\vec{y} = \exp\left(A(x-x_0)\right)\vec{y}_0 + \exp\left(Ax\right)\int_{x_0}^{x}\exp\left(-As\right)\vec{f}(s)\,ds, \tag{23}$$

or

$$\vec{y} = \Psi(x)\Psi^{-1}(x_0)\vec{y}_0 + \Psi(x)\int_{x_0}^{x}\Psi^{-1}(s)\vec{f}(s)ds, \tag{24}$$

where the exponential matrix $\exp(Ax)$ is given by Eq.(16), $\exp(-Ax)$ is the inverse matrix of $\exp(Ax)$, $\Psi(x)$ is the fundamental matrix and $\Psi^{-1}(x)$ is the inverse matrix of $\Psi(x)$.

Example 11. Solve the initial value problem

$$\begin{aligned} x' &= 2x + y + t,\\ y' &= 3x + 4y - 2t.\\ x &= 2, \quad y = -2 \quad \text{at} \quad t = 1,\end{aligned}$$

(see also Ex.8).

Using the results of Ex.8 we obtain

$$\Psi(t) = \begin{pmatrix} -e^t & e^{5t}\\ e^t & 3e^{5t}\end{pmatrix}$$

$$\Psi^{-1}(t) = \frac{1}{4}\begin{pmatrix} -3e^{-t} & e^{-t} \\ e^{-5t} & e^{-5t} \end{pmatrix}, \quad \Psi^{-1}(1) = \frac{1}{4}\begin{pmatrix} -3e^{-1} & e^{-1} \\ e^{-5} & e^{-5} \end{pmatrix}.$$

Now according to Eq.(24) we find the solution of the given initial value problem

$$\vec{y}(t) = \begin{pmatrix} x \\ y \end{pmatrix} = \frac{1}{4}\begin{pmatrix} -e^{t} & e^{5t} \\ e^{t} & 3e^{5t} \end{pmatrix} \times$$

$$\left[\begin{pmatrix} -3e^{-1} & e^{-1} \\ e^{-5} & e^{-5} \end{pmatrix}\begin{pmatrix} 2 \\ -2 \end{pmatrix} + \int_1^t \begin{pmatrix} -3e^{-s} & e^{-s} \\ e^{-5s} & e^{-5s} \end{pmatrix}\begin{pmatrix} s \\ -2s \end{pmatrix} ds\right]$$

$$= \frac{9}{2}\begin{pmatrix} 1 \\ -1 \end{pmatrix} e^{t-1} - \frac{3}{50}\begin{pmatrix} 1 \\ 3 \end{pmatrix} e^{5t-5} + \begin{pmatrix} -\frac{6}{5}t - \frac{31}{25} \\ \frac{7}{5}t + \frac{32}{25} \end{pmatrix}$$

Example 12. Solve the initial value problem

$$\vec{x}'(t) = A\vec{x}(t), \quad \vec{x}(1) = (1,\, 1,\, 1)^T,$$

where matrix A is given in Ex. 7.

According to Eq. (22) we find

$$\vec{x}(t) = \exp(A(t-1))\vec{x}(1) = T\exp(\Lambda(t-1))T^{-1}\vec{x}(1).$$

Using the results of Ex. 7 we obtain

$$\vec{x}(t) = \begin{pmatrix} 1 & 1 & 0 \\ -1 & 0 & 1 \\ 1 & 1 & 1 \end{pmatrix}\begin{pmatrix} 1 & 0 & 0 \\ 0 & e^{t-1} & 0 \\ 0 & 0 & e^{t-1} \end{pmatrix}\begin{pmatrix} -1 & -1 & 1 \\ 2 & 1 & -1 \\ -1 & 0 & 1 \end{pmatrix}\begin{pmatrix} 1 \\ 1 \\ 1 \end{pmatrix}$$

$$= 2e^{t-1}\begin{pmatrix} 1 \\ 0 \\ 1 \end{pmatrix} - \begin{pmatrix} 1 \\ -1 \\ 1 \end{pmatrix}$$

4.1.13 Laplace Transform

The Laplace transform is well adapted to solving initial value problems for linear systems with constant coefficients and nonhomogeneous terms corresponding to discontinuous or impulsive forcing functions. The Laplace transform of a vector (or a matrix) whose generic component is $y_i(x)$ (or element $a_{ij}(x)$) is a vector (or a matrix) whose corresponding component $Y_i(s)$ (or element $B_{ij}(s)$) is the transform of $y_i(x)$ (or $a_{ij}(x)$). Thus the Laplace transform of the initial value problem (22) is

$$s\vec{Y}(s) - \vec{y}_0 = A\vec{Y}(s) + \vec{F}(s)$$

or

$$(sE - A)\vec{Y}(s) = \vec{y}_0 + \vec{F}(s). \tag{25}$$

In order to solve Eq.(25) for $\vec{Y}$ we should compute $(sE-A)^{-1}$. The matrix $(sE-A)$ is nonsingular provided s is not equal to an eigenvalue of A. Hence

$$\vec{y}(x)=\mathcal{L}^{-1}\left\{(sE-A)^{-1}\left(\vec{y}_0+\vec{F}(s)\right)\right\}. \tag{26}$$

Example 13. Solve the initial value problem

$$\vec{y}'=\begin{pmatrix}2 & -1\\ 3 & -2\end{pmatrix}\vec{y}+\begin{pmatrix}1\\ -1\end{pmatrix}e^x,\quad \vec{y}(0)=\begin{pmatrix}1\\ 1\end{pmatrix}.$$

According to Eq.(26) we find, after some minor algebraic calculations,

$$(sE-A)^{-1}=\frac{1}{s^2-1}\begin{pmatrix}s+2 & -1\\ 3 & s-2\end{pmatrix},$$

$$\vec{F}(s)=\frac{1}{s-1}\begin{pmatrix}1\\ -1\end{pmatrix},$$

and

$$(sE-A)^{-1}(\vec{y}_0+\vec{F}(s))=\frac{1}{(s^2-1)(s-1)}\begin{pmatrix}s^2+s+2\\ s^2-s+4\end{pmatrix}.$$

Then by using the partial fraction expansions we rewrite this result in the form

$$(sE-A)^{-1}(\vec{y}_0+\vec{F}(s))=\frac{1}{2}\begin{pmatrix}\frac{1}{s-1}+\frac{4}{(s-1)^2}+\frac{1}{s+1}\\ -\frac{1}{s-1}+\frac{4}{(s-1)^2}+\frac{1}{s+1}\end{pmatrix}.$$

Finally, by using Tab. 2.2 in Sec. 2.4 we find the solution of the given problem

$$\vec{y}=\frac{1}{2}\begin{pmatrix}1\\ -1\end{pmatrix}e^x+\frac{1}{2}\begin{pmatrix}1\\ 3\end{pmatrix}e^{-x}+2\begin{pmatrix}1\\ 1\end{pmatrix}xe^x.$$

4.1.14 Systems of Higher Order Equations

A linear system of higher order differential equations with constant coefficients has the general form

$$\sum_{k=0}^{N}A_k\vec{y}^{(k)}(x)=\vec{f}(x), \tag{27}$$

where A_k are matrices with constant elements $a_{011}, a_{012}, \ldots, a_{0nn}, a_{111}, a_{112}, \ldots, a_{1nn}, a_{211}, \ldots, a_{Nnn}$. The number $M=k_1+k_2+\cdots+k_n\le Nn$, where k_i is the order of the highest derivative in the i-th equation, is called *formal total order* of the system (27). Notice that system (27) can be reduced to a system of M linear first order differential equations and $Nn-M$ linear algebraic equations (see nonhomogeneous systems). The general solution of Eq.(27) is a sum of a general solution of corresponding homogeneous equation and a particular solution of the nonhomogeneous equation (see Eq.(18)).

Homogeneous Systems

The general solution of the homogeneous system

$$\sum_{k=0}^{N} A_k \vec{y}^{(k)}(x) = 0$$

where the matrix A_N does not vanish, can be found by the Euler's method. By substituting $\vec{y} = \exp(\lambda x)\vec{C}$, where λ is a constant and $\vec{C}$ is an arbitrary constant vector, we find

$$\sum_{k=0}^{N} A_k \lambda^k \vec{C} = 0.$$

Since $\vec{C}$ is nontrivial, we obtain the auxiliary equation

$$\det\left|\sum_{k=0}^{N} A_k \lambda^k\right| = 0 \tag{28}$$

Further, proceeding as has been stated above, we can find all λ_i and corresponding $\vec{e}_i$, when all λ_i are distinct. In the case of repeated roots of the auxiliary equation (28), Eq.(27) should be reduced to a first order system as will be shown below.

The matrix method for these systems is beyond the scope of our text.

Example 14. Find the general solution of the system

$$x'' - 2y' + 2x = 0,$$
$$y'' + 3x' - 8y = 0.$$

According to Euler's method we assume that

$$\vec{y}(t) = \vec{C}\exp(\lambda x)$$

and by substituting this expression in the given system we obtain, after some minor algebraic simplifications,

$$(\lambda^2 + 2)C_1 - 2\lambda C_2 = 0,$$
$$3\lambda C_1 + (\lambda^2 - 8)C_2 = 0.$$

The solution of this system is nontrivial if and only if λ is a root of the auxiliary equation

$$(\lambda^2 + 2)(\lambda^2 - 8) + 6\lambda^2 = 0.$$

This equation has four roots $\lambda_1 = 2$, $\lambda_2 = -2$, $\lambda_3 = 2i$, $\lambda_4 = -2i$. It is easy to verify, that eigenvectors corresponding to these eigenvalues are

$$\vec{e}_1 = (2,\ 3)^T, \quad \vec{e}_2 = (2,\ -3)^T, \quad \vec{e}_3 = (2,\ i)^T, \quad \vec{e}_4 = (2,\ -i)^T.$$

Hence the general solution of the given system is

$$x(t) = 2C_1e^{2t} + 2C_2e^{-2t} + 2C_3\cos(2t) + 2C_4\sin(2t),$$
$$y(t) = 3C_1e^{2t} - 3C_2e^{-2t} + C_3\cos(2t) - C_4\sin(2t),$$

where C_i are arbitrary constants.

Nonhomogeneous Systems

Here is considered the method of reduction of Eq.(27) to a system of first order equations. Let $\vec{z}(x) = \vec{y}'(x)$, $\vec{u}(x) = \vec{z}'(x) = \vec{y}''(x)$, ..., $\vec{w}(x) = \vec{y}^{(N-2)}(x)$, $\vec{v}(x) = \vec{w}'(x) = \vec{y}^{(N-1)}(x)$ are new unknown vectors and $\vec{Y}(x)$ is a new vector which consists of the Nn components of the vectors $\vec{y}$, $\vec{z}$, $\vec{u}$, ..., $\vec{v}$. Then Eq. (27) together with all the definitions of these vectors can be written as a system of M first order differential and $Nn - M$ algebraic equations for components of vector $\vec{Y}(x)$. The particular solution of this system can be found by Eq.(19).

Note that the particular solution of Eq.(25) can be determined also by the method of undetermined coefficients as has been shown above.

Example 15. Find the general solution of the system

$$\begin{aligned} &x'' + 2y'' - x - 2y = 1, \\ &x' + y' - x + y = t. \end{aligned}$$

The formal total order of this system is $M = 2 + 1 = 3 < 4$. Let $z = x'$, $w = y'$ are new functions of t. Then the given system together with the definitions of x' and y' can be written as

$$\begin{aligned} &x' - z = 0, \\ &y' - w = 0, \\ &z' + 2w' - x - 2y = 1, \\ &z + w - x + y = t. \end{aligned}$$

The last equation of this system is algebraic. By elimination of $w(t)$ from this equation, we find a new system of the general form

$$\begin{aligned} &x' - z = 0, \\ &y' - x + y + z = t, \\ &2x' - 2y' - z' - x - 2y = -1. \end{aligned}$$

This system can be rewritten in the form

$$A_0\vec{y}'(t) = A_1\vec{y}(t) + \vec{f}(t),$$

where $\vec{y} = (x, y, z)^T$,

$$A_0 = \begin{pmatrix} 1 & 0 & 0 \\ 0 & 1 & 0 \\ 2 & -2 & -1 \end{pmatrix}, \quad A_1 = \begin{pmatrix} 0 & 0 & 1 \\ 1 & -1 & -1 \\ 1 & 2 & 0 \end{pmatrix},$$

and $\vec{f}(t) = (0,\ t,\ -1)^T$. Since $\det A_0 = -1$, then the inverse matrix A_0^{-1} exists. It is interesting to observe that $A_0^{-1} = A_0$. The left-hand multiplication by A_0^{-1} then gives a system in the normal form, that is $\vec{y}' = A\vec{y} + \vec{\varphi}$, where

$$A = A_0^{-1}A_1 = \begin{pmatrix} 0 & 0 & 1 \\ 1 & -1 & -1 \\ -3 & 0 & 4 \end{pmatrix}, \quad \vec{\varphi} = A_0^{-1}\vec{f} = \begin{pmatrix} 0 \\ t \\ 1 - 2t \end{pmatrix}.$$

It is easy to verify that the auxiliary equation $\det(A - \lambda E) = 0$ of the corresponding homogeneous system has three roots $\lambda_1 = -1$, $\lambda_2 = 1$, $\lambda_3 = 3$. The corresponding eigenvectors are

$$\vec{e}_1 = (0,1,1)^T, \quad \vec{e}_2 = (1,0,1)^T, \quad \vec{e}_3 = (-2,1,-6)^T.$$

Hence the fundamental and inverse matrices take the form

$$\Psi(t) = \begin{pmatrix} 0 & e^t & -2e^{3t} \\ e^{-t} & 0 & e^{3t} \\ 0 & e^t & -6e^{3t} \end{pmatrix}, \quad \Psi^{-1}(x) = \frac{1}{4}\begin{pmatrix} -e^t & 4e^t & e^t \\ 6e^{-t} & 0 & -2e^{-t} \\ -e^{-3t} & 0 & -e^{-3t} \end{pmatrix}.$$

Then according to Eq.(19) we find the particular solution of the system in the normal form

$$\vec{y}_p(t) = -\left(\frac{2}{3}t + \frac{5}{9}, \quad -\frac{1}{3}t + \frac{2}{9}, \quad \frac{2}{3}\right)^T.$$

Hence the general solution of the given system is

$$x(t) = C_2 e^t - 2C_3 e^{3t} - \frac{2}{3}t - \frac{5}{9},$$
$$y(t) = C_1 e^{-t} + C_3 e^{3t} + \frac{1}{3}t - \frac{2}{9}.$$

Notice that the particular solution of the given system can be found by the method of undetermined coefficients.

PROBLEMS

Find the general solutions of the following systems by Euler's method.

1. $\begin{cases} x' = x + y, \\ y' = -5x - y. \end{cases}$
2. $\begin{cases} x' = x - y, \\ y' = y - 4x. \end{cases}$
3. $\begin{cases} x' = x + y, \\ y' = 8x - y. \end{cases}$
4. $\begin{cases} x' = x + y, \\ y' = 3y - 2x. \end{cases}$
5. $\begin{cases} x' = 5x + 2y, \\ y' = -4x - y. \end{cases}$
6. $\begin{cases} x' = 3x + y, \\ y' = -x + y. \end{cases}$
7. $\begin{cases} x' = 3x + 2y, \\ y' = -x - y. \end{cases}$
8. $\begin{cases} x' = x - 3y, \\ y' = 3x + y. \end{cases}$
9. $\begin{cases} x' = 2x + y, \\ y' = 4y - x. \end{cases}$
10. $\begin{cases} x' = 3x - y, \\ y' = 4x - y. \end{cases}$

11. $\begin{cases} x' = 2y - 3x, \\ y' = y - 2x. \end{cases}$

12. $\begin{cases} x' = 5x + 3y, \\ y' = -3x - y. \end{cases}$

13. $\begin{cases} x' = 2x - y + z, \\ y' = x + 2y - z, \\ z' = x - y + 2z. \end{cases}$

14. $\begin{cases} x' = 2x + y, \\ y' = x + 3y - z, \\ z' = 2y + 3z - x. \end{cases}$

15. $\begin{cases} x' = 3x - y + z, \\ y' = x + y + z, \\ z' = 4x - y + 4z. \end{cases}$

16. $\begin{cases} x' = 2x + y + z, \\ y' = -2x - z, \\ z' = 2x + y + 2z. \end{cases}$

17. $\begin{cases} x' = x + z - y, \\ y' = x + y - z, \\ z' = 2x - y. \end{cases}$

18. $\begin{cases} x' = 2x + y, \\ y' = 2y + 4z, \\ z' = x - z. \end{cases}$

19. $\begin{cases} x' = 2x - y - z, \\ y' = 3x - 2y - 3z, \\ z' = 2z - x + y. \end{cases}$

20. $\begin{cases} x' = y - 2z - x, \\ y' = 4x + y, \\ z' = 2x + y - z. \end{cases}$

21. $\begin{cases} x' = 3x - 2y - z, \\ y' = 3x - 4y - 3z, \\ z' = 2x - 4y. \end{cases}$

22. $\begin{cases} x' = 2x - y - z, \\ y' = 2x - y - 2z, \\ z' = 2z - x + y. \end{cases}$

23. $\begin{cases} x' = x - 2y - z, \\ y' = y - x + z, \\ z' = x - z. \end{cases}$

Find the exponential matrix $\exp(At)$ of the following matrices.

24. $A = \begin{pmatrix} 3 & 0 \\ 0 & -2 \end{pmatrix}.$

25. $A = \begin{pmatrix} 0 & 1 \\ 1 & 0 \end{pmatrix}.$

26. $A = \begin{pmatrix} -2 & -4 \\ 1 & 2 \end{pmatrix}.$

27. $A = \begin{pmatrix} 1 & 0 \\ 0 & -1 \end{pmatrix}$.

28. $A = \begin{pmatrix} 0 & i \\ -i & 0 \end{pmatrix}$.

29. $A = \begin{pmatrix} 0 & 1 \\ -1 & 0 \end{pmatrix}$.

30. $A = \begin{pmatrix} 3 & -2 \\ 4 & -3 \end{pmatrix}$.

31. $A = \begin{pmatrix} \alpha & \beta \\ -\beta & \alpha \end{pmatrix}$.

Using the matrix method find the general solution of the linear system $\vec{x}' = A\vec{x}$ for the following matrices A.

32. $A = \begin{pmatrix} 3 & 0 \\ 0 & 3 \end{pmatrix}$.

33. $A = \begin{pmatrix} 0 & 1 \\ -1 & 0 \end{pmatrix}$.

34. $A = E + A_0$, where $A_0 = \begin{pmatrix} 0 & 1 \\ -1 & 0 \end{pmatrix}$.

35. $A = E + A_0$, where $A_0 = \begin{pmatrix} 0 & 2 \\ 0 & 0 \end{pmatrix}$.

36. $A = E + A_0$, where $A_0 = \begin{pmatrix} 0 & 3 \\ -4 & 0 \end{pmatrix}$.

37. $A = \begin{pmatrix} 2 & 3 \\ 1 & 2 \end{pmatrix}$.

38. $A = \begin{pmatrix} 1 & -2 \\ 2 & -3 \end{pmatrix}$.

39. $A = \begin{pmatrix} -2 & -4 \\ 1 & 3 \end{pmatrix}$.

40. $A = \begin{pmatrix} 3 & -2 \\ 4 & -1 \end{pmatrix}$.

41. $A = \begin{pmatrix} -3 & 2 \\ 4 & 3 \end{pmatrix}$.

42. $A = \begin{pmatrix} 0 & 0 & 1 \\ 0 & 1 & 0 \\ 1 & 0 & 0 \end{pmatrix}$.

43. $A = E + A_0$, where $A_0 = \begin{pmatrix} 0 & 0 & 1 \\ 0 & 1 & 0 \\ 1 & 0 & 0 \end{pmatrix}$.

44. $A = \begin{pmatrix} 0 & 1 & -1 \\ 1 & 0 & -1 \\ 2 & 2 & -3 \end{pmatrix}$.

45. $A = \begin{pmatrix} 2 & 0 & -1 \\ 1 & -1 & 0 \\ 3 & -1 & -1 \end{pmatrix}$.

46. $A = \begin{pmatrix} 2 & -1 & -1 \\ 1 & 0 & -1 \\ 3 & -1 & -2 \end{pmatrix}$.

47. $A = \begin{pmatrix} 1 & -2 & 2 \\ 1 & 4 & -2 \\ 1 & 5 & -3 \end{pmatrix}$.

48. $A = \begin{pmatrix} 0 & 1 & 1 \\ 1 & 1 & 0 \\ -1 & 0 & 1 \end{pmatrix}$.

49. $A = \begin{pmatrix} 3 & -3 & 1 \\ 3 & -2 & 2 \\ -1 & 2 & 0 \end{pmatrix}$.

Find the general solution of the following nonhomogeneous systems.

50. $\begin{cases} x' = x + y + \sin t, \\ y' = -x + y + \cos t. \end{cases}$

51. $\begin{cases} x' = y + 2e^t, \\ y' = x + t^2. \end{cases}$

52. $\begin{cases} x' = 2y - x, \\ y' = 4y - 3x + \frac{e^{3t}}{e^{2t}+1}. \end{cases}$

53. $\begin{cases} x' = 2x + 4y - 8, \\ y' = 3x + 6y. \end{cases}$

54. $\begin{cases} x' = 2x - 3y, \\ y' = x - 2y + 2\sin t. \end{cases}$

55. $\begin{cases} x' = x - y + 2\sin t, \\ y' = 2x - y. \end{cases}$

56. $\begin{cases} x' = 2x + y + e^t, \\ y' = -2x + 2t. \end{cases}$

57. $\begin{cases} x' = x - y + \sec t, \\ y' = 2x - y. \end{cases}$

58. $\begin{cases} x' = y + \tan^2 t - 1, \\ y' = -x + \tan t. \end{cases}$

59. $\begin{cases} x' = y + \cos t, \\ y' = -x + 1. \end{cases}$

60. $\begin{cases} x' = -5x - y + e^t, \\ y' = x - 3y + e^{2t}. \end{cases}$

61. $\begin{cases} x' = -y + \cos t, \\ y' = -x + \sin t. \end{cases}$

62. $\begin{cases} x' = 2x - 4y + 4e^{-2t}, \\ y' = 2x - 2y. \end{cases}$

63. $\begin{cases} x' = 2x + y - 2z - t + 2, \\ y' = -x + 1, \\ z' = x + y - z - t + 1. \end{cases}$

64. $\begin{cases} x' = y - z + t, \\ y' = x - z + 1, \\ z' = 2x + 2y - 3z + t + 1. \end{cases}$

Solve each of the following initial value problems.

65. $\vec{y}' = \begin{pmatrix} 1 & 1 \\ 4 & -2 \end{pmatrix} \vec{y}, \quad \vec{y}(0) = \begin{pmatrix} 0 \\ 1 \end{pmatrix}.$

66. $\vec{y}' = \begin{pmatrix} 2 & -2 \\ 4 & -2 \end{pmatrix} \vec{y}, \quad \vec{y}(0) = \begin{pmatrix} 1 \\ 0 \end{pmatrix}.$

67. $\vec{y}' = \begin{pmatrix} 3 & -2 \\ 2 & -2 \end{pmatrix} \vec{y} + \begin{pmatrix} 1 \\ 2x \end{pmatrix} e^x, \quad \vec{y}(0) = \begin{pmatrix} 1 \\ 0 \end{pmatrix}.$

68. $\vec{y}' = \begin{pmatrix} -1 & -1 \\ -2 & -1 \end{pmatrix} \vec{y}, \quad \vec{y}(0) = \begin{pmatrix} 1 \\ 0 \end{pmatrix}.$

Find the general solution of the following higher order systems.

69. $\begin{cases} x'' = 2y, \\ y'' = -2x. \end{cases}$

70. $\begin{cases} 2x' - 5y' = 4y - x, \\ 3x' - 4y' = 2x - y. \end{cases}$

71. $\begin{cases} x'' + 2y'' = x + 2y, \\ x' + y'' = x - y. \end{cases}$

72. $\begin{cases} x'' + 5x' + 2y' = -y, \\ 3x'' + y' = -5x - 3y. \end{cases}$

73. $\begin{cases} x'' = 2y + 1, \\ y' = x - y + t. \end{cases}$

74. $\begin{cases} 2x'' + 3y'' + 2x' + y' + x + y = 0, \\ x'' + 3y'' + 4x' + 2y' - x - y = 0. \end{cases}$

Find the general solution of the following problems.

75. $\begin{cases} x' = x + y - \sin t, \\ y' = -5x - y + t^2. \end{cases}$

76. $\begin{cases} x' = x - y, \\ y' = 2x + y + e^t. \end{cases}$

77. $\begin{cases} x' = y - 2y + z, \\ y' = x + y - 2z, \\ z' = -2x + y + z. \end{cases}$

78. $\begin{cases} x'' = x' - 2y + z', \\ y'' = x' + y' - 2z, \\ z'' = -2x + y' + z'. \end{cases}$

79. $\begin{cases} x' = x + y - \sin t, \\ y' = 2x - y + t^2. \end{cases}$

80. $\begin{cases} x' = y - 2y + z + 1 + t, \\ y' = x + y - 2z - t, \\ z' = -2x + y + z + e^t. \end{cases}$

81. $\begin{cases} x'' = x' - 2y + z' + 1, \\ y'' = x' + y' - 2z - t, \\ z'' = -2x + y' + z' + t^3. \end{cases}$

4.2 Linear systems

A system of n linear first order differential equations of the form

$$A(x)\vec{y}'(x) + B(x)\vec{y}(x) = \vec{\varphi}(x), \tag{1}$$

where $\vec{y}(x)$ is a vector with the components $y_1(x)$, $y_2(x)$, $\ldots$, $y_n(x)$, $A(x)$ and $B(x)$ are $n \times n$ matrices with the elements $a_{11}(x)$, $a_{12}(x)$, ..., $a_{nn}(x)$, $b_{11}(x)$, ..., $b_{nn}(x)$ and $\vec{\varphi}(x)$ is a vector with the components $\varphi_1(x)$, $\varphi_2(x)$, ..., $\varphi_n(x)$ is called system of the *general form.* If the matrix $A(x)$ is nonsingular, then by left multiplication of Eq. (1) by the inverse matrix $A^{-1}(x)$, we obtain the so called *normal form* of the linear system of first order equations

$$\vec{y}'(x) = A(x)\vec{y}(x) + \vec{f}(x), \tag{2}$$

where $A \equiv A^{-1}B$ and $\vec{f} = A^{-1}\vec{\varphi}$. The solution of Eq.(2) is a sum of the general solution of corresponding homogeneous equation

$$\vec{y}'(x) = A(x)\vec{y}(x) \tag{3}$$

and a particular solution of Eq.(2), i.e.,

$$\vec{y}(x) = \vec{y}_h(x) + \vec{y}_p(x) = \Psi(x)\vec{C} + \vec{y}_p(x), \tag{4}$$

where $\Psi(x)$ is the fundamental matrix of system (3), $\vec{C}$ is an arbitrary constant vector and $\vec{y}_p(x)$ is a particular solution of Eq.(2).

Homogeneous Linear Systems

The general solution of Eq.(2) can be determined by the following methods:
1) method of eliminations;
2) matrix method .

4.2.1 Solution by Eliminations

In this method we should first choose any equation of system (3), say the first equation. Then differentiating it with respect to x n times and substituting every time all the $y_i'(x)$ from the system (3), we obtain

$$
\begin{aligned}
y_1' &= a_{11}(x)y_1 + a_{12}(x)y_2 + \cdots + a_{1n}(x)y_n = \sum_{i=1}^{n} a_{1i}(x)y_i,\\
y_1'' &= \sum_{i=1}^{n}(a_{1i}'y_i + a_{1i}y_i') = \sum_{i=1}^{n} a_{1i}'y_i + \sum_{i,j=1}^{n} a_{1i}a_{ij}y_j \equiv \sum_{i=1}^{n} b_{1i}(x)y_i,\\
y_1''' &= \sum_{i=1}^{n}(b_{1i}'y_i + b_{1i}y_i') = \sum_{i=1}^{n} b_{1i}'y_i + \sum_{i,j=1}^{n} b_{1i}a_{ij}y_j \equiv \sum_{i=1}^{n} c_{1i}(x)y_i,\\
&\vdots\\
y_1^{(n-1)} &= \sum_{i=1}^{n}(d_{1i}'y_i + d_{1i}y_i') = \sum_{i=1}^{n} d_{1i}'y_i + \sum_{i,j=1}^{n} d_{1i}a_{ij}y_j \equiv \sum_{i=1}^{n} e_{1i}(x)y_i,\\
y_1^{(n)} &= \sum_{i=1}^{n}(e_{1i}'y_i + e_{1i}y_i') = \sum_{i=1}^{n} e_{1i}'y_i + \sum_{i,j=1}^{n} e_{1i}a_{ij}y_j \equiv \sum_{i=1}^{n} g_{1i}(x)y_i.
\end{aligned} \tag{5}
$$

The first $n-1$ linear equations of this system can be rewritten in the form

$$
\begin{aligned}
&\sum_{i=2}^{n} a_{1i}y_i = y_1' - a_{11}y_1,\\
&\sum_{i=2}^{n} b_{1i}y_i = y_1'' - b_{11}y_1,\\
&\vdots\\
&\sum_{i=2}^{n} e_{1i}y_i = y_1^{(n-1)} - e_{11}y_1,
\end{aligned} \tag{6}
$$

or in the matrix form

$$
H(x)\vec{y} = \vec{z}(x, y_1, y_1', \ldots, y_1^{(n-1)}), \tag{6a}
$$

where $H(x)$ is a $(n-1)\times(n-1)$ matrix with the rows a_{1i}, b_{1i},..., e_{1i}, $i = 2, 3, ..., n$, $\vec{y}$ is a column vector with the components y_2, y_3, ..., y_n and $\vec{z}$ is the column vector with the components $y_1' - a_{11}y_1$, $y_1'' - b_{11}y_1, \ldots, y_1^{(n-1)} - e_{11}y_1$. Thus we obtain the nonhomogeneous linear algebraic system for $y_2, y_3, \ldots, y_n$. In those points, where $\det H(x)$ exists and $\det H(x) \neq 0$, we can find its solution by using the Cramer's rule

$$
\begin{aligned}
y_2 &= \sum_{i=1}^{n-1} h_{2i}(x)y_1^{(i)},\\
y_3 &= \sum_{i=1}^{n-1} h_{3i}(x)y_1^{(i)}\\
&\vdots\\
y_n &= \sum_{i=1}^{n-1} h_{ni}(x)y_1^{(i)},
\end{aligned} \tag{7}
$$

where $h_{ij}(x)$ are elements of the inverse matrix $H^{-1}(x)$. Now, substituting these expressions in the last equation of system (5), we obtain the linear n-th order differential equation for $y_1(x)$

$$y_1^{(n)}(x) = g_{11}(x)y_1(x) + \sum_{i=2}^{n}\sum_{j=0}^{n-1} g_{1i}(x)h_{ij}(x)y_1^{(j)}(x)$$
$$\equiv -p_1(x)y^{(n-1)} - p_2(x)y^{(n-2)} - \cdots - p_n(x)y_1.$$

Solutions of this linear n-th order differential equation can be found by methods of Chapter 2. Then $y_2(x), y_3(x), \ldots, y_n(x)$ are determined in accordance with Eq-s (7).

Those points where $\det H(x) = 0$ or $\det H(x) \to \infty$, are called singular points of Eq.(3).

Example 1. Find the general solution of the homogeneous system by eliminations

$$y' = z,$$
$$z' = -2\frac{y}{x^2} + 2\frac{z}{x}.$$

Choosing the first equation and differentiating it with respect to x we obtain, after substituting z' from the second equation,

$$y' = z,$$
$$y'' = z' = -2\frac{y}{x^2} + 2\frac{z}{x}.$$

The first equation of this system gives $z = y'$. Then the second equation takes the form

$$x^2y'' - 2xy' + 2y = 0.$$

The general solution of this Euler's equation is $y(x) = C_1x + C_2x^2$, where C_i are arbitrary constants. Then we find $z = y' = C_1 + 2C_2x$.

Example 2. Reduce the homogeneous linear differential system

$$y' = \frac{z}{x^2},$$
$$z' = -2\frac{y}{x^2} + 2\frac{z}{x}$$

to a second order differential equation.

Proceeding as in Ex.1 we obtain, after some minor calculations,

$$x^4y'' + 2y = 0,$$

and $z = x^2y'$. Note that $H(x) = x^{-2}$ and $\det H(x) \to \infty$ at point $x \to 0$, $\det H(x) = 0$ at $x \to \infty$. Hence this second order differential equation for y has the singular points at $\zeta = 0$ and $\zeta \to \infty$. Its solutions can be found by using the methods of Sec. 3.1.

4.2.2 Matrix Method

The general solution of the system (2) can be written as

$$\vec{y}(x) = \exp\left(\int A(x)\,dx\right)\vec{C}, \tag{8}$$

where $\vec{C}$ is an arbitrary vector, if and only if the matrices $A(x)$ and $\int A(x)dx$ are commutative, i.e.,

$$A(x)\left(\int A(x)\,dx\right) = \left(\int A(x)\,dx\right)A(x).$$

This condition is always satisfied when the matrix $A(x)$ is a product of a constant matrix A and a scalar function $a(x)$. Then by making change of variable $t = \int a(x)\,dx$ we obtain a new linear system with constant matrix $\vec{y}'(t) = A\vec{y}(t)$. We find the solution of this system by methods of Sec. 4.1.

Example 3. Find the general solution of the system

$$\vec{y}'(x) = A(x)\vec{y}(x),$$

where

$$A(x) = \cos x \begin{pmatrix} 2 & 1 & -1 \\ -1 & 0 & 1 \\ 1 & 1 & 0 \end{pmatrix} \equiv \cos(x)A.$$

By making change of variable $t = \sin x$ we find

$$\vec{y}'(t) = A\vec{y}(t).$$

According to Ex.7 in Sec. 4.1 solution of this system is

$$y_1 = C_1 + (C_2 + C_3)e^t,$$
$$y_2 = -C_1 - C_2e^t,$$
$$y_3 = C_1 + C_3e^t.$$

Hence the solution of the given system is

$$y_1 = C_1 + (C_2 + C_3)e^{\sin x},$$
$$y_2 = -C_1 - C_2e^{\sin x},$$
$$y_3 = C_1 + C_3e^{\sin x}.$$

4.2.3 Nonhomogeneous Linear Systems

A general solution of these systems are found as has been shown in Sec. 4.1.

Example 4. Find the general solution of the nonhomogeneous linear system

$$\vec{y}' - A\frac{\vec{y}}{x} = \vec{f}(x),$$

where

$$A = \begin{pmatrix} 1 & 0 & 0 \\ 1 & 1 & 0 \\ 0 & 1 & 1 \end{pmatrix}$$

and $\vec{f}(x) = (1,\ x,\ x^{-1})^T$. Assume that $x > 0$.

Solution of the corresponding homogeneous system we find by the matrix method. A change of variable $t = -\ln x$ then yields

$$\vec{y}'(t) = A\vec{y}(t).$$

The general solution of this system is $\vec{y}(t) = \exp(At)\vec{C}$, where $\vec{C}$ is an arbitrary constant vector and $\exp(At)$ is the exponential matrix (see Sec. 4.1). By using the property d) of the exponential matrix, given in Sec.4.1, we write automatically

$$\exp(At) = \begin{pmatrix} e^t & 0 & 0 \\ te^t & e^t & 0 \\ \frac{t^2}{2}e^t & te^t & e^t \end{pmatrix} = e^t \begin{pmatrix} 1 & 0 & 0 \\ t & 1 & 0 \\ \frac{t^2}{2} & t & 1 \end{pmatrix}.$$

Hence the general solution of the corresponding homogeneous system is

$$\vec{y}(x) = \exp(-A\ln x)\vec{C} = \frac{1}{x}\begin{pmatrix} 1 & 0 & 0 \\ -\ln x & 1 & 0 \\ \frac{\ln^2 x}{2} & -\ln x & 1 \end{pmatrix}\begin{pmatrix} C_1 \\ C_2 \\ C_3 \end{pmatrix},$$

where C_i are arbitrary constants. According to Eq.(21) in Sec. 4.1 we find the particular solution of the nonhomogeneous system as

$$\vec{y}(x) = \exp(-A\ln x)\int \exp(A\ln x)\vec{f}(x)\,dx$$

$$= \frac{1}{x}\begin{pmatrix} 1 & 0 & 0 \\ -\ln x & 1 & 0 \\ \frac{\ln^2 x}{2} & -\ln x & 1 \end{pmatrix}\int x\begin{pmatrix} 1 & 0 & 0 \\ \ln x & 1 & 0 \\ \frac{\ln^2 x}{2} & \ln x & 1 \end{pmatrix}\begin{pmatrix} 1 \\ x \\ x^{-1} \end{pmatrix} dx.$$

By a multiplication, an integration and some algebraic simplifications, we obtain

$$\vec{y}(x) = \begin{pmatrix} 1 & 0 & 0 \\ -\ln x & 1 & 0 \\ \frac{\ln^2 x}{2} & -\ln x & 1 \end{pmatrix}\begin{pmatrix} \frac{x}{2} \\ \frac{x}{2}\ln x - \frac{x}{4} + \frac{x^2}{3} \\ \left(-1 + \ln x + \frac{4}{3}x\right)\frac{x}{4}\ln x - \frac{x^2}{9} + \frac{x}{8} + 1 \end{pmatrix}$$

$$= \begin{pmatrix} \frac{x}{2} \\ \frac{x^2}{3} - \frac{x}{4} \\ -\frac{x^2}{9} + \frac{x}{8} + 1 \end{pmatrix}.$$

Hence the general solution of the given system is

$$y_1 = \frac{C_1}{x} + \frac{x}{2}$$
$$y_2 = -\frac{C_1}{x}\ln x + \frac{C_2}{x} + \frac{x^2}{3} - \frac{x}{4}$$
$$y_3 = \frac{C_1}{2x}\ln^2 x - \frac{C_2}{x}\ln x + \frac{C_3}{x} - \frac{x^2}{9} + \frac{x}{8} + 1$$

Notice that this system can also be solved by the method of the undetermined coefficients as has been shown in Sec. 4.1.

4.2.4 Initial Value Problem

A solution of this problem is found as has been shown in Sec. 4.1.

Example 5. Solve the initial value problem

$$\vec{y}' + xA\vec{y} = 0, \qquad \vec{y}(0) = (1,\,0)^T,$$

where

$$A = \begin{pmatrix} 1 & 1 \\ 1 & -1 \end{pmatrix}.$$

By making change of variable $t = x^2/2$ we find the new initial value problem

$$\vec{y}'(t) = -A\vec{y}(t), \qquad \vec{y}(0) = (1,\ 0)^T.$$

According to Eq.(23) in Sec. 4.1 we obtain, formally,

$$\vec{y}(t) = \exp(-At)\vec{y}(0).$$

By the immediate calculations, we find $A^{2n} = 2^{2n}E$ and $A^{2n+1} = 2^{2n}A$, where $n = 0, 1, \ldots$ and E is the identity matrix. Hence

$$\exp(-At) = \sum_{k=0}^{\infty}(-1)^k\frac{t^k}{k!}A^k = E\sum_{n=0}^{\infty}\frac{2^n}{(2n)!}t^{2n} - A\sum_{n=0}^{\infty}\frac{2^n}{(2n+1)!}t^{2n+1}$$
$$= E\cosh(\sqrt{2}t) - A2^{-1/2}\sinh(\sqrt{2}t).$$

Thus the solution of the given initial problem is

$$\vec{y}(x) = \left(E\cosh\frac{x^2}{\sqrt{2}} - A2^{-1/2}\sinh\frac{x^2}{\sqrt{2}}\right)\vec{y}(0)$$
$$= \left(\cosh\frac{x^2}{\sqrt{2}} - 2^{-1/2}\sinh\frac{x^2}{\sqrt{2}}, -2^{-1/2}\sinh\frac{x^2}{\sqrt{2}}\right)^T$$

PROBLEMS

Determine the general solution of the following homogeneous systems.

1. $\begin{cases} xy' + y - xz = 0, \\ x^2 z' + 2y - xz = 0. \end{cases}$

2. $\begin{cases} y' - y + e^{2x} z = 0, \\ z' - e^{-2x} y + z = 0. \end{cases}$

3. $\begin{cases} x^2 y' = -y - 2z, \\ x^2 z' = 3y + 4z. \end{cases}$

4. $\begin{cases} tx' - 2x + y - z = 0, \\ ty' - x - z = 0, \\ tz' + 3x - y + 2z = 0. \end{cases}$

5. $\begin{cases} 3tx' - (t+1)x + y - tz = 0, \\ 3ty' - (t-2)x - 2y - tz = 0, \\ 3tz' - (2t-1)x - y - 2tz = 0. \end{cases}$

6. $\begin{cases} y' - \frac{y}{x-2} = 0, \\ z' - \frac{z}{x-2} = 0. \end{cases}$

7. $\begin{cases} y' - x(1-x)y - (x^3 - x^2 + x + 1)z = 0, \\ z' - (1-x)y - (x^2 - x + 1)z = 0. \end{cases}$

8. $\begin{cases} y' + \frac{z}{x} = 0, \\ z' + \frac{y}{x} = 0. \end{cases}$ 9. $\begin{cases} y' - y \sin x = 0, \\ z' - y \exp(\cos x) = 0. \end{cases}$

10. $\begin{cases} y' + \frac{y}{x} + xz = 0, \\ z' - \frac{y}{x} = 0. \end{cases}$

11. $\begin{cases} y' - \frac{z}{x^2} = 0, \\ z' - 3y - 3\frac{z}{x} = 0. \end{cases}$

12. $\begin{cases} y' - y \cos^2 x + (1 - \frac{1}{2} \sin 2x) z = 0, \\ z' - z \sin^2 x - (1 + \frac{1}{2} \sin 2x) y = 0. \end{cases}$

Determine the general solution of the following nonhomogeneous systems.

13. $\begin{cases} x^2 y' + z = x^2, \\ z' + y = x. \end{cases}$

14. $\begin{cases} xy' + 2y - 2z = x, \\ xz' + y + 5z = x^2. \end{cases}$

15. $\begin{cases} y' + (2y - z) \cot x = \sin x, \\ z' + (3y - 2z) \cot x = \cos^2 x. \end{cases}$

16. $\begin{cases} y' + (y - 2z) \tan x = \cos x, \\ z' + (y + 4z) \tan x = \sin x. \end{cases}$

17. $\begin{cases} y' + 3(y - 2z)x^2 = 2x^2, \\ z' + 3(y + 4z)x^2 = x^2. \end{cases}$

18. $\begin{cases} y' + 2\frac{y}{x} = 1, \\ z' - (1 + \frac{2}{x})y - z = -1. \end{cases}$

19. $\begin{cases} tx' + 6x - y - 3z = -5, \\ ty' + 23x - 6y - 9z = -16, \\ tz' + x + y - 2z = -3. \end{cases}$

20. $\begin{cases} x' - \frac{2}{t}x = -1, \\ y' + \frac{x}{t^3} - \frac{y}{t} = \frac{1}{t^2}, \\ z' + \frac{x}{t^4} + \frac{y}{t^2} - \frac{z}{t} = \frac{1}{t^3}. \end{cases}$

21. $\begin{cases} tx' + x - y - z = t, \\ ty' - x + y - z = t^3, \\ tz' - x - y - z = 4. \end{cases}$

Solve each of the following initial value problems.

22. $\begin{cases} y' - (2\sin x - \cos x)y - (\sin x - \cos x)z = 0, \\ z' - 2(\cos x - \sin x)y - (2\cos x - \sin x)z = 0, \\ y(\pi) = -1, \quad z(\pi) = 2. \end{cases}$

23. $\begin{cases} y' - z = 0, \\ z' - 2\frac{y}{x^2} = 0, \\ y(1) = 2, \quad z(1) = 1. \end{cases}$

24. $\begin{cases} (1+x^2)y' - xy - z = 0, \\ (1+x^2)z' + y - xz = 0, \\ y(1) = 2, \quad z(1) = 0. \end{cases}$

25. $\begin{cases} y' - xy + (x^2-1)z = 0, \\ z' - y + xz = 0, \\ y(0) = 1, \quad z(0) = 2. \end{cases}$

26. $\begin{cases} y' - 2x^3y + 2x(x^4-1)z = 0, \\ z' - 2xy + 2x^3z = 0, \\ y(-1) = 4, \quad z(-1) = 3. \end{cases}$

27. $\begin{cases} (1-x^2)y' + y\frac{1-x+x^2}{x-1} + z\frac{2x-1}{(1-x)^2} = 0, \\ (1-x^2)z' - y + z\frac{1+x-x^2}{1-x} = 0, \\ y(0) = 2, \quad z(0) = 1. \end{cases}$

28. $\begin{cases} y'(2-x) - y = 0, \\ z' - y = \frac{x-1}{x-2}, \\ y(1) = 1, \quad z(1) = 1. \end{cases}$

29. $\begin{cases} x' = 1 - \frac{2}{t}x, \\ y' = x + y + \frac{2}{t}x - 1, \\ x(3) = 1, \quad y(3) = -1. \end{cases}$

Find the general solution of the following problems.

30. $\begin{cases} x^2y' + z = x, \\ z' + y = x^2. \end{cases}$

31. $\begin{cases} y' + (y - 2z)\tan x = \sin x, \\ z' + (y + 4z)\tan x = \cos x. \end{cases}$

32. $\begin{cases} x' - \frac{2}{t}x = \frac{1}{t^3}, \\ y' + \frac{x}{t^3} - \frac{y}{t} = \frac{1}{t^2}, \\ z' + \frac{x}{t^4} + \frac{y}{t^2} - \frac{z}{t} = -1. \end{cases}$

33. $\begin{cases} y' - z = 0, \\ z' - 2\frac{y}{x^2} = 0, \\ y(1) = 1, \quad z(1) = 2. \end{cases}$

34. $\begin{cases} y' - xy + (x^2-1)z = 0, \\ z' - y + xz = 0, \\ y(0) = 2, \quad z(0) = 1. \end{cases}$

35. $\begin{cases} y'(2-x) - y = 1, \\ z' - y = \frac{x-1}{x-2}, \\ y(1) = 1, \quad z(1) = 1. \end{cases}$

36. $\begin{cases} y' - \frac{y}{x-2} = x^2, \\ z' - \frac{z}{x-2} = 1. \end{cases}$

37. $\begin{cases} y'(2-x) - y = x - 1, \\ z' - y = \frac{x-1}{x-2}, \\ y(1) = 1, \quad z(1) = 1. \end{cases}$

38. $\begin{cases} y' + \frac{y}{x-2} = x^2, \\ z' + \frac{z}{x-2} = 1. \end{cases}$

4.3 Nonlinear systems

A system of n nonlinear first order differential equations of the form

$$y_i'(x) = f_i(x, y_1, y_2, \dots, y_n), \quad i = 1, 2, \dots, n, \tag{1}$$

where $f_i(x, \vec{y})$, $i = 1, 2,\dots,n$ are given functions of $n+1$ variables is called *system of the normal form.*

A system of n nonlinear higher order differential equations of the form

$$F_i(x, y_1, y_1', \dots, y_1^{(k_1)}, y_2, y_2', \dots, y_2^{(k_2)}, \dots, y_n^{(k_n)}) = 0, \quad i = 1, 2, \dots, n, \tag{2}$$

is called *system of the general form.* Here k_i is an order of the highest derivative of y_i, F_i are given functions of $n + 1 + M$ variables, $M = k_1 + k_2 + \dots + k_n$ is the total order of system.

If Eq.(2) is solved for all the highest order derivatives, then it can be written in so called *canonical form of nonlinear system*

$$y_i^{(k_i)}(x) = g_i(x, y_1, y_1', \dots, y_1^{(k_1-1)}, y_2, y_2', \dots, y_2^{(k_2-1)}, \dots, y_n^{(k_n-1)}), \tag{3}$$

where $n = 1, 2, \dots, n$. Denoting

$$\begin{array}{llll} z_1(x) = y_1'(x), & z_2(x) = z_1' = y_1'', & \dots, & z_{k_1}(x) = z_{k_1-1}'(x) = y^{(k_1)}, \\ u_1(x) = y_2'(x), & u_2(x) = u_1' = y_2'', & \dots, & u_{k_2}(x) = u_{k_2-1}' = y_2^{(k_2)} \\ \vdots & & & \\ w_1(x) = y_n'(x), & w_2(x) = w_1' = y_n'', & \dots, & w_{k_n}(x) = w_{k_n-1}' = y_n^{(k_n)} \end{array}$$

we can write the system (3) together with this system of notations as a system of M nonlinear first order differential equations in the normal form

$$\begin{aligned}
&y_1' = z_1,\\
&z_i' = z_{i-1}, \quad i = 2,3,\dots,k_1-1,\\
&z_{k_1}' = y_1^{(k_1)} = g_1(x, z_1, z_2, \dots, z_{k_1-1}, u_1, \dots, u_{k_2-1}, \dots, w_{k_n-1}),\\
&y_2' = u_1,\\
&u_i' = u_{i-1}, \quad i = 2,3,\dots,k_2-1,\\
&u_{k_2}' = y_2^{(k_2)} = g_2(x, z_1, z_2, \dots, z_{k_1-1}, u_1, \dots, u_{k_2-1}, \dots, w_{k_n-1}),\\
&\vdots\\
&y_n' = w_1,\\
&w_i' = w_{i-1}, \quad i = 2,3,\dots,k_n-1,\\
&w_{k_n}' = y_n^{(k_n)} = g_n(x, z_1, z_2, \dots, z_{k_1-1}, u_1, \dots, u_{k_2-1}, \dots, w_{k_n-1}),
\end{aligned} \tag{4}$$

If Eq. (2) is not solved for some highest derivative then its canonical form does not exist. This case is beyond the scope of our text.

Example 1. Find the normal form of the nonlinear system

$$\begin{aligned}
2y'' + z'' + y' - 2xz &= e^x,\\
y' + z'' - z' + 2yz &= x^2.
\end{aligned}$$

This system is linear with respect to the second derivatives. Hence we can find the canonical form of this system. Subtracting the second equation from the first gives

$$\begin{aligned}
y'' &= z(x+y) - \frac{1}{2}\left(z' + x^2 - e^x\right),\\
z'' &= z' - y' - 2yz + x^2.
\end{aligned}$$

Then denoting $u = y'$ and $v = z'$ we obtain the normal form of the given system

$$\begin{aligned}
y' &= u,\\
u' &= z(x+y) - (v + x^2 - e^x)/2,\\
z' &= v,\\
v' &= u - v - 2yz + x^2.
\end{aligned}$$

Systems of Normal Form

The solution of Eq.(1) can be determined by any of the following methods:

1) method of eliminations;
2) method of integrable combinations;
3) special method for a given system.

4.3.1 Method of Eliminations

First we should choose any equation from the system (1), say $y_1' = f_1(x, \vec{y})$. Then differentiating it with respect to x n times and substituting every time the $y_i'(x)$ from the system (1), we obtain a new nonlinear system

$$\begin{aligned}
&y_1' = f_1(x,y) \equiv F_1(x,\vec{y}),\\
&y_1'' = F'_{1x} + \sum_{i=1}^{n} F'_{1y_i} y_i' = F'_{1x} + \sum_{i=1}^{n} F'_{1y_i} f_i \equiv F_2(x,\vec{y}),\\
&y_1''' = F'_{2x} + \sum_{i=1}^{n} F'_{2y_i} y_i' = F'_{2x} + \sum_{i=1}^{n} F'_{2y_i} f_i \equiv F_3(x,\vec{y}),\\
&\vdots\\
&y_1^{(n-1)} = F'_{n-2x} + \sum_{i=1}^{n} F'_{n-2y_i} y_i' = F'_{n-2x} + \sum_{i=1}^{n} F'_{n-2y_i} f_i \equiv F_{n-1}(x,\vec{y}),\\
&y_1^{(n)} = F'_{n-1x} + \sum_{i=1}^{n} F'_{n-1y_i} y_i' = F'_{n-1x} + \sum_{i=1}^{n} F'_{n-1y_i} f_i \equiv F_n(x,\vec{y}).
\end{aligned} \tag{4}$$

The first $n-1$ equations of this system form the nonhomogeneous nonlinear system for $y_2, y_3, \ldots, y_n$

$$\begin{aligned}
&F_1(x,\, y_1;\; y_2,\, y_3, \ldots, y_n) = y_1',\\
&F_2(x,\, y_1;\; y_2,\, y_3, \ldots, y_n) = y_1'',\\
&\vdots\\
&F_{n-1}(x,\, y_1;\; y_2,\, y_3, \ldots, y_n) = y_1^{(n-1)}.
\end{aligned} \tag{4a}$$

It is supposed that solution of this system can be found in the form

$$y_i = h_i(x,\, y_1,\, y_1',\, y_1'',\, \ldots,\, y_1^{(n-1)}), \qquad i = 2,\, 3,\, \ldots,\, n, \tag{5}$$

where h_i are certain functions of $n+1$ variables. Then by substituting these expressions in the last equation of system (4) we obtain a nonlinear n-th order differential equation for $y_1(x)$ of the form

$$y_1^{(n)} = F_n\left(x,\, y_1;\; \vec{h}(x,\, y_1,\, y_1',\, y_1'', \ldots, y_1^{(n-1)})\right), \tag{6}$$

where $\vec{h}$ is a vector-function with the components h_2, h_3, ..., h_n. If solution of this equation can be found as $y_1 = \psi(x;\, \vec{C})$, where ψ is a function of x and an arbitrary constant vector $\vec{C}$ with n components, then using Eq.(5), we obtain $y_2(x)$, $y_3(x), \ldots, y_n(x)$.

Example 2. Find the solution of the nonlinear system by eliminations

$$\begin{aligned}
&y' = -y^2 + \frac{1}{z},\\
&z' = yz.
\end{aligned}$$

Choosing the second equation and proceeding, as has been stated above, we obtain

$$\begin{aligned}
&z' = yz,\\
&z'' = y'z + yz' = z(-y^2 + \frac{1}{z}) + y^2 z = 1.
\end{aligned}$$

Then from the first equation we obtain $y = z'/z$. Solution of the second equation is $z = C_1 + C_2 x + x^2/2$, where C_i are arbitrary constants. Hence the general solution of the given system is

$$y(x) = \frac{C_2 + x}{C_1 + C_2 x + x^2/2},$$
$$z(x) = C_1 + C_2 x + \frac{x^2}{2}.$$

Note that this solution has a movable singular point x^*, where $z(x^*) = 0$ and $z'(x^*) \neq 0$.

Example 3. Solve the initial value problem

$$y' = \frac{1}{z - x},$$
$$z' = 1 - \frac{1}{y},$$
$$y(1) = 2, \quad z(1) = 0.$$

Choosing the first equation for elimination we find

$$y' = \frac{1}{z - x},$$
$$y'' = -\frac{z' - 1}{(z - x)^2} = \frac{1}{y(z - x)^2}.$$

Then solving the first equation of this system for z leads to $z = x + 1/y'$. A substitution of this expression in the second equation then yields

$$y'' = \frac{y'^2}{y} \quad \text{or} \quad \frac{y''}{y'} = \frac{y'}{y}.$$

A two-fold integration of this equation then gives

$$y(x) = C_2 e^{C_1 x},$$

where C_i are arbitrary constants. Hence

$$z(x) = x + \frac{e^{-C_1 x}}{C_1 C_2}.$$

By substituting these solutions into the initial conditions we obtain a system for constants C_i

$$2 = C_2 e^{C_1},$$
$$0 = 1 + \frac{e^{-C_1}}{C_1 C_2}.$$

Its solutions are $C_1 = -1/2$ and $C_2 = 2\sqrt{e}$. Hence the solution of the given initial value problem is

$$y(x) = 2 \exp\left(\frac{1 - x}{2}\right),$$
$$z(x) = x - \exp\left(\frac{x - 1}{2}\right).$$

4.3.2 Method of Integrable Combinations

In this method by some algebraic transformations of Eq. (1), namely: addition or subtraction of equations; multiplication or division of equations by some factors; some combinations of the preliminary simplifications we should find such new equation which is the exact equation for some variable. Integrating this new exact equation leads to some connection between variables. Thus the number of independent variables and number of equations can be decreased. There is the so called *symmetric form* of the system (1)

$$\frac{dy_1}{f_1(x,\vec{y})} = \frac{dy_2}{f_2(x,\vec{y})} = \cdots = \frac{dy_n}{f_n(x,\vec{y})}, \tag{7}$$

and the property of equivalence fractions:

$$\text{if} \quad \frac{a_1}{b_1} = \frac{a_2}{b_2} = \cdots = \frac{a_n}{b_n},$$
$$\text{then} \quad \frac{a_i}{b_i} = \frac{k_1 a_1 + k_2 a_2 + \cdots + k_n a_n}{k_1 b_1 + k_2 b_2 + \cdots + k_n b_n},$$

where k_i are arbitrary constants.

Example 4. Find the solution of the system

$$y' = -z + \sin y,$$
$$z' = -z \cos y,$$

by the method of integrable combinations.

Multiplying the first equation by $z \cos y$ and the second equation by $(-z + \sin y)$ and adding them, we obtain

$$y' z \cos y + (-z + \sin y) z' = 0.$$

This equation we can rewrite as

$$\frac{d}{dx}\left(z \sin y - \frac{z^2}{2}\right) = 0.$$

An integration then yields $2z \sin y - z^2 = C_1$, where C_1 is an arbitrary constant. Hence

$$y = \arcsin \frac{C_1 + z^2}{2z}.$$

Substituting $y(z)$ in the second equation of the given system, we obtain

$$z' = \pm \left(z^2 - \frac{1}{4}\left(C_1 + z^2\right)^2\right)^{1/2}.$$

A solution of this equation is the inverse function

$$x + C_2 = \pm \int \left(z^2 - \frac{1}{4}\left(C_1 + z^2\right)^2\right)^{-1/2} dz,$$

where C_2 is an arbitrary constant. Hence the solution of the given system is

$$x + C_2 = \pm \int \left(z^2 - \frac{1}{4}\left(C_1 + z^2\right)^2 \right)^{-1/2} dz, \qquad y = \arcsin \frac{C_1 + z^2}{2z}.$$

Example 5. Solve the initial value problem

$$\begin{aligned} y' &= y^2 z, \\ z' &= -yz^2 + \frac{z}{x}, \\ y &= 2, \quad z = 1 \quad \text{at} \quad x = 1. \end{aligned}$$

Multiplying the first equation by z and the second equation by y and adding them, we obtain

$$zy' + yz' = \frac{yz}{x},$$

or

$$\frac{d}{dx} yz = \frac{yz}{x}.$$

An integration then yields $yz = C_1 x$, where C_1 is an arbitrary constant. Hence $z = C_1 x / y$, and substituting z in the first equation, we obtain

$$y' = C_1 x y.$$

There is the solution $y = C_2 \exp(C_1 x^2/2)$, where C_2 is an arbitrary constant. Then

$$z = \frac{C_1}{C_2} x e^{-C_1 x^2/2}.$$

Substituting these solutions into the initial conditions gives

$$\begin{aligned} 2 &= C_2 e^{C_1/2}, \\ 1 &= \frac{C_1}{C_2} e^{-C_1/2}. \end{aligned}$$

The solution is $C_1 = 2$ and $C_2 = 2/e$. Hence the solution of the given initial value problem is

$$y = 2e^{x^2-1}, \qquad z = xe^{1-x^2}.$$

Special Methods

Here we consider only two such methods.

4.3.3 Systems of Bernoulli's Form

Consider the nonlinear system of the form (see also the Bernoulli's equation in Sec. 1.5)

$$\vec{y}'(x) = A(x)\vec{y} + F_p(x, \vec{y})\vec{y}, \tag{8}$$

where $A(x)$ is a $n \times n$ matrix with the elements $a_{11}(x), a_{12}(x), \ldots, a_{nn}(x)$ and $F_p(x, \vec{y})$ is a scalar function which is homogeneous of degree p in the arguments $y_1, y_2, \ldots, y_n$, i.e., $F_p(x, k\vec{y}) = k^p F_p(x, \vec{y})$. By substituting

$$\vec{y}(x) = \frac{\vec{z}(x)}{(u(x))^{1/p}},$$

where $\vec{z}(x)$ is an unknown vector and $u(x)$ is a scalar function, we obtain, after multiplication by $u^{1/p}$,

$$\vec{z}' - \frac{\vec{z}u'}{pu} = A\vec{z} + F_p(x, \vec{z})\frac{\vec{z}}{u}.$$

Putting $u' = -pF_p(x, \vec{z})$ then leads to

$$\vec{z}'(x) = A(x)\vec{z}(x). \tag{9}$$

A solution of this homogeneous linear first order differential system is $\vec{z}(x) = \Psi(x)\vec{b}$, where $\Psi(x)$ is the fundamental matrix of system (9) and $\vec{b}$ is an arbitrary constant vector (see Sec. 4.1, Eq. (4)). Further we obtain $u' = -pF_p(x, \Psi(x)\vec{b})$, and an integration then yields

$$u(x) = b_{n+1}^p - p\int F_p(x, \Psi(x)\vec{b})\, dx, \tag{10}$$

where b_{n+1} is an arbitrary constant. Hence the solution of Eq. (8) is

$$\vec{y}(x) = \Psi(x)\vec{b}\Big(b_{n+1}^p - p\int F_p(x, \Psi(x)\vec{b})\, dx\Big)^{-1/p}.$$

By making change of the constant vector $\vec{b} = b_{n+1}\vec{C}$, where $\vec{C}$ is a new arbitrary constant vector we obtain the general solution of Eq. (8)

$$\vec{y}(x) = \Psi(x)\vec{C}\Big(1 - p\int F_p\left(x, \Psi(x)\vec{C}\right)\, dx\Big)^{-1/p}. \tag{11}$$

When $p = 0$, approaching the limit as $p \to 0$ in Eq.(11), we obtain

$$\vec{y}(x) = \Psi(x)\vec{C}\exp\left(F_0\left(x, \Psi(x)\vec{C}\right)\, dx\right). \tag{12}$$

Example 6. Find the general solution of the system

$$x' = -\frac{x}{t} + y + t^2x^2y,$$
$$y' = -\frac{2x}{t^2} + \frac{y}{t} + t^2xy^2.$$

It is easy to verify that this system has the form of Eq.(8) with $F_p(t, \vec{y}) = xy$ and $p = 2$. Then according to Eq.(9) we obtain

$$x' = -\frac{x}{t} + y,$$
$$y' = -\frac{2x}{t^2} + \frac{y}{t},$$

Solution of this system is (see Sec. 4.2)

$$x = C_1 + C_2 t, \qquad y = 2C_2 + C_1 t^{-1},$$

where C_i are arbitrary constants. Further, according to Eq.(10) we obtain

$$u(t) = 1 - 2\int t^2(C_1 + C_2 t)(2C_2 + C_1 t^{-1})\,dt = 1 - t^2(C_1 + C_2 t)^2.$$

Hence the solution of the given system is

$$x(t) = \frac{C_1 + C_2 t}{\sqrt{1 - t^2(C_1 + C_2 t)^2}},$$
$$y(t) = \frac{2C_2 + C_1 t^{-1}}{\sqrt{1 - t^2(C_1 + C_2 t)^2}}.$$

Note that it exists on the real axes only into a closed interval of t.

4.3.4 Method of Complex Variable

In the nonlinear system of two equations

$$x'(t) = u(x, y),$$
$$y'(t) = v(x, y), \tag{13}$$

the right-hand side of which satisfies the so called *Cauchy-Riman's conditions* (The French mathematician O.L. Cauchy (1789 - 1857), the German mathematician B. Riman (1826 - 1866)).

$$u'_x(x, y) = v'_y(x, y), \qquad u'_y(x, y) = -v'_x(x, y), \tag{14}$$

can be written as a nonlinear equation in the complex plane $z(t) = x(t) + iy(t)$

$$z'(t) = u(x, y) + iv(x, y) \equiv f(z), \tag{15}$$

where $x = \mathrm{Re} z = (z + \bar{z})/2$, $y = \mathrm{Im} z = (z - \bar{z})/2i$. Solution of Eq.(15) is given as the inverse function

$$t + C = \int \frac{dz}{f(z)}, \tag{16}$$

where $C = C_1 + iC_2$ and C_i are arbitrary constants.

Example 7. Find the general solution of the system

$$x' = \frac{1}{2}(3xy^2 - x^3),$$
$$y' = \frac{1}{2}(y^3 - 3x^2y).$$

It is easy to verify that the Cauchy-Riman's conditions are satisfied. Hence we obtain

$$z'(t) = \frac{1}{2}(3xy^2 - x^3 + iy^3 - 3ix^2y) = \frac{1}{2}(-x - iy)^3 = -\frac{z^3}{2}.$$

In accordance with Eq.(16) we find the solution of this equation as

$$z(t) = \frac{1}{\sqrt{t + C_1 + iC_2}} = \frac{\pm 1}{(C_2^2 + (t + C_1)^2)^{1/4}} \exp\left(\frac{i}{2}\arctan\frac{C_2}{C_1 + t}\right),$$

where C_i are arbitrary constants. After some algebraic transformations, we find the solution for x and y

$$x(t) = \mathrm{Re} z(t) = \frac{\pm 1}{\sqrt{2}}\left[\frac{1}{\sqrt{C_2^2 + (t + C_1)^2}} + \frac{t + C_1}{C_2^2 + (t + C_1)^2}\right]^{1/2},$$

$$y(t) = \mathrm{Im} z(t) = \frac{\pm 1}{\sqrt{2}}\left[\frac{1}{\sqrt{C_2^2 + (t + C_1)^2}} - \frac{t + C_1}{C_2^2 + (t + C_1)^2}\right]^{1/2}.$$

4.3.5 Systems of Canonical Form

The nonlinear system of n higher order differential equations in the canonical form (3) can be solved immediately by the method of eliminations, or by the method of integrable combinations, or by a combination of these methods.

Example 8. Find the solution of the system

$$\varphi''(t)r(t) + 2rr'\varphi' = 0,$$
$$r'' - r\varphi'^2 - \frac{1}{r^2} = 0,$$

(so called Kepler's problem).

The first equation can be written as

$$\frac{d}{dt}\varphi' r^2 = 0.$$

Hence $r^2\varphi' = C_1$, where C_1 is an arbitrary constant. Then $\varphi' = C_1/r^2$, and substituting φ' in the second equation gives

$$r'' - \frac{C_1^2}{r^3} - \frac{1}{r^2} = 0.$$

A multiplication of this equation by the integrating factor $\mu = r'$ then yields

$$r'r'' - C_1^2\frac{r'}{r^3} - \frac{r'}{r^2} = 0.$$

or

$$\frac{d}{dt}\left(\frac{r'^2}{2} + \frac{C_1^2}{r^2} + \frac{1}{r}\right) = 0.$$

Hence

$$\frac{r'^2}{2} + \frac{C_1^2}{r^2} + \frac{1}{r} = C_2,$$

where C_2 is an arbitrary constant. Now we obtain the first order differential equation

$$r' = \pm\left(C_2 - \frac{C_1^2}{r^2} - \frac{1}{r}\right)^{1/2}.$$

Its solution exists as the inverse function

$$t + C_3 = \pm\int\left(C_2 - \frac{C_1^2}{r^2} - \frac{1}{r}\right)^{-1/2} dr,$$

where C_3 is an arbitrary constant. Obtaining from this equation r as a function of t, then we determine

$$\varphi = C_4 + C_1\int\frac{dt}{r^2(t)},$$

where C_4 is an arbitrary constant.

PROBLEMS

Find the general solution of the following system by the method of eliminations.

1. $\begin{cases} y' = z, \\ z' = \frac{z}{y}. \end{cases}$

2. $\begin{cases} y' = \frac{z}{2y}, \\ z' = y^2 + \sin x. \end{cases}$

3. $\begin{cases} y' = (1 - y^2 - z^2)y - z, \\ z' = (1 - y^2 - z^2)z - y. \end{cases}$

4. $\begin{cases} y' = \frac{x}{z}, \\ z' = -\frac{x}{y}. \end{cases}$

5. $\begin{cases} y' = \frac{y^2}{z-x}, \\ z' = 1 + y. \end{cases}$

6. $\begin{cases} x' = x + y^2, \\ y' = z - x, \\ z' = z + y^2. \end{cases}$

7. $\begin{cases} x' = x + y, \\ y' = \frac{1}{2x}\left(y^2 - x^2 + 1\right). \end{cases}$

8. $\begin{cases} x' = \frac{x}{y^2}, \\ y' = -y. \end{cases}$

9. $\begin{cases} y' = y^2 z, \\ z' = -yz^2 + \frac{z}{x}. \end{cases}$

10. $\begin{cases} x' = \frac{y}{(x-y)^2}, \\ y' = \frac{x}{(x-y)^2}. \end{cases}$

11. $\begin{cases} y'2z = y^2 - z^2 + 1, \\ z' = z + y. \end{cases}$

12. $\begin{cases} x' = 2x - 4z, \\ y' = \frac{y}{2} + \frac{z}{y}, \\ z' = 2z. \end{cases}$

13. $\begin{cases} y' = \frac{z}{x}, \\ z' = \frac{z(y+2z-1)}{x(y-1)}. \end{cases}$

14. $\begin{cases} y' = e^{x-y}, \\ z' = \frac{2z}{2x-z^2}. \end{cases}$

Find the general solution of the following system by the method of integrable combinations.

15. $\begin{cases} xy' = y + \sqrt{y^2 - x^2}, \\ z' = \frac{y+z}{z^2-x}. \end{cases}$

16. $\begin{cases} y' = \frac{z+e^y}{z+e^x}, \\ z' = \frac{z^2-e^{x+y}}{z+e^x}. \end{cases}$

17. $\dfrac{dx}{xz} = \dfrac{dy}{yz} = -\dfrac{dz}{xy}.$

18. $\begin{cases} y' = \frac{z^2+xy}{x(y-z)}, \\ z' = \frac{z(x+y)}{x(y-z)}. \end{cases}$

19. $\dfrac{dx}{\sqrt{x}} = \dfrac{dy}{\sqrt{y}} = 2dz.$

20. $\dfrac{dx}{mz - ny} = \dfrac{dy}{nx - lz} = \dfrac{dz}{ly - mx}$, where l, m, n are some constants.

21. $\begin{cases} y' = \frac{z}{(z-y)^2}, \\ z' = \frac{y}{(z-y)^2}. \end{cases}$

22. $\begin{cases} y' = 1 - \frac{1}{z}, \\ z' = \frac{1}{y-t}. \end{cases}$

23. $\dfrac{dx}{xz} = \dfrac{dy}{yz} = \dfrac{dz}{xy\sqrt{1+z^2}}.$

24. $\dfrac{dx}{x(y+z)} = \dfrac{dy}{z(z-y)} = \dfrac{dz}{y(y-z)}.$

25. $\dfrac{dx}{x(z-y)} = \dfrac{dy}{y(y-x)} = \dfrac{dz}{(y^2 - xz)}.$

26. $\dfrac{dx}{z} = \dfrac{dy}{u} = \dfrac{dz}{x} = \dfrac{du}{y}.$

27. $\dfrac{dx}{y-u} = \dfrac{dy}{z-x} = \dfrac{dz}{u-y} = \dfrac{du}{x-z}.$

28. $\dfrac{dx}{x(y^2-z^2)} = -\dfrac{dy}{y(z^2+x^2)} = \dfrac{dz}{z(x^2+y^2)}.$

29. $\dfrac{dx}{\cos y} = \dfrac{dy}{\cos x} = \dfrac{dz}{\cos y \cos x}.$

30. $\dfrac{dx}{xz} = \dfrac{dy}{yz} = \dfrac{dz}{xy}.$

Find the general solution of the following system of Bernoulli's form.

31. $\begin{cases} x' = y + xz, \\ y' = x + yz, \\ z' = x + z^2. \end{cases}$

32. $\begin{cases} x' = x + (x+y)x, \\ y' = z + (x+y)y, \\ z' = y + (x+y)z. \end{cases}$

33. $\begin{cases} x' = x - y + z + (x-z)^2 \frac{x}{y^2}, \\ y' = x + y - z + (x-z)^2 \frac{1}{y}, \\ z' = 2z - y + (x-z)^2 \frac{z}{y^2}. \end{cases}$

34. $\begin{cases} x' = 2x + y + x(2x - 3y), \\ y' = 3x + 4y + y(2x - 3y). \end{cases}$

35. $\begin{cases} x' = 2x + x^2 y e^{x/y}, \\ y' = y + xy^2 e^{x/y}. \end{cases}$

36. $\begin{cases} x' = x - y + e^{4t}\sqrt{\frac{x}{y}}, \\ y' = y - x + e^{4t}\sqrt{\frac{y}{x}}. \end{cases}$

Find the general solution of the following system by the method of complex variable.

37. $\begin{cases} x' = px + qy, \\ y' = -qx + py, \end{cases}$ where p and q are constants.

38. $\begin{cases} y' = 2xy, \\ x' = x^2 - y^2. \end{cases}$

39. $\begin{cases} x' = y + x^2 + 2xy - y^2, \\ y' = -x - x^2 + 2xy + y^2. \end{cases}$

40. $\begin{cases} x' = e^{-x} \cos y, \\ y' = -e^{-x} \sin y. \end{cases}$

41. $\begin{cases} x' = \sqrt{2}\left(x + \sqrt{x^2+y^2}\right)^{1/2}, \\ y' = \sqrt{2}\left(-x + \sqrt{x^2+y^2}\right)^{1/2}. \end{cases}$

42. $\begin{cases} x' = 1 - x^2 + y^2, \\ y' = -2xy. \end{cases}$

Find the general solution of the following system of higher order differential equations.

43. $\begin{cases} y''z' - y'z = 0, \\ y'z' - y = 0. \end{cases}$

44. $\begin{cases} x'' = x - y^2, \\ y' = \frac{1}{2y}. \end{cases}$

45. $\begin{cases} x''y' - xy'' = 0, \\ x'y' + xy = 0. \end{cases}$

46. $\begin{cases} y'' = -2zy, \\ z' = y''y'/2y^2. \end{cases}$

47. $\begin{cases} x' = 2\frac{y}{z}, \\ y' = 2\frac{x}{z}, \\ z' = x - y \end{cases}$, where $z(t)$ is a polynomial.

48. $\begin{cases} \theta'' - \varphi'^2 \sin\theta\cos\theta + \sin\theta = 0, \\ \varphi'' \sin^2\theta + \theta'\varphi' \sin 2\theta = 0. \end{cases}$

49. $\begin{cases} 2x'' - s'' \cos\alpha = 0, \\ 3s'' - x'' \cos\alpha = g\sin\alpha \end{cases}$, where α and g are some constants.

50. $\begin{cases} 2x'' - \varphi'' \cos\varphi + \varphi'^2 \sin\varphi = 0, \\ \varphi'' - x'' \cos\varphi + x'\varphi' \sin\varphi = \sin\varphi. \end{cases}$

51. $\begin{cases} \varphi'' + \psi'' \cos\theta - \psi'\theta' \sin\theta = 0, \\ \psi'' + \varphi'' \cos\theta - \varphi'\theta' \sin\theta = 0, \\ \theta'' + \psi'\varphi' \sin\theta + \sin\theta = 0. \end{cases}$

Chapter 5

PARTIAL EQUATIONS OF THE FIRST ORDER

5.1 Linear partial equations

The *linear partial differential equation of the first order*

$$P_1 \frac{\partial z}{\partial x_1} + P_2 \frac{\partial z}{\partial x_2} + \cdots + P_n \frac{\partial z}{\partial x_n} = R, \tag{1}$$

where $P_1,\ P_2,\ \ldots,\ P_n,\ R$ are functions of $x_1,\ x_2,\ \ldots,\ x_n,\ z$ may be solved from the system

$$\frac{dx_1}{P_1} = \frac{dx_2}{P_2} = \cdots = \frac{dx_n}{P_n} = \frac{dz}{R}. \tag{2}$$

If

$$u_1(x_1, \cdots, x_n, z) = a_1,\ u_2(x_1, \cdots, x_n, z) = a_2,\ \ldots,\ u_n(x_1, \cdots, x_n, z) = a_n$$

are n independent integrals of this system, the general solution of the partial differential equation is

$$\phi(u_1, u_2, \ldots, u_n) = 0, \tag{3}$$

where ϕ is an arbitrary function.

Example 1. Solve

$$y \frac{\partial z}{\partial x} - x \frac{\partial z}{\partial y} = 0.$$

From the system

$$\frac{dx}{y} = -\frac{dy}{x} = \frac{dz}{0}$$

we have the combinations

$$x dx + y dy = 0,$$

$$dz = 0$$

which, on integrating, give

$$x^2 + y^2 = a_1,$$

$$z = a_2.$$

The general solution then is

$$\phi\left(z, x^2 + y^2\right) = 0,$$

where ϕ is an arbitrary function. Or in explicit form

$$z = f\left(x^2 + y^2\right),$$

where f is an arbitrary function.

Useful equations resulting from (2) may be written down as follows. Let λ_i be any functions; then

$$\frac{dx_1}{P_1} = \frac{dx_2}{P_2} = \cdots = \frac{dx_n}{P_n} = \frac{dz}{R} = \frac{\lambda_1 dx_1 + \lambda_2 dx_2 + \cdots + \lambda_n dx_n + \lambda_{n+1} dz}{\lambda_1 P_1 + \lambda_2 P_2 + \cdots + \lambda_n P_n + \lambda_{n+1} R}.$$

The multipliers may be chosen in infinitely many ways, so that

$$\lambda_1 P_1 + \lambda_2 P_2 + \cdots + \lambda_{n+1} R = 0$$

then

$$\lambda_1 dx_1 + \lambda_2 dx_2 + \cdots + \lambda_n dx_n + \lambda_{n+1} dz = 0.$$

If it happens that the first member of this equation is the exact differential of some function $u(x_1, x_2, \ldots, z)$ then we have at once one relation

$$u\left(x_1, x_2, \ldots, z\right) = a.$$

Example 2. Solve

$$y\frac{\partial z}{\partial x} + x\frac{\partial z}{\partial y} = x - y.$$

From the system

$$\frac{dx}{y} = \frac{dy}{x} = \frac{dz}{x - y}$$

we use the multipliers 1, −1, 1

$$\frac{dx}{y} = \frac{dy}{x} = \frac{dz + dx - dy}{x - y + y - x} = \frac{dz + dx - dy}{0}$$

whence

$$xdx - ydy = 0,$$

$$dz + dx - dy = 0.$$

Integrating,

$$x^2 - y^2 = a_1,$$

$$x - y + z = a_2.$$

Then the general solution is

$$\phi(x^2 - y^2, x - y + z) = 0,$$

where ϕ is an arbitrary function.

Example 3. Solve

$$\left(x^2 + y^2\right)\frac{\partial z}{\partial x} + 2xy\frac{\partial z}{\partial y} + z^2 = 0.$$

From the system

$$\frac{dx}{x^2 + y^2} = \frac{dy}{2xy} = -\frac{dz}{z^2}$$

we have

$$-\frac{dz}{z^2} = \frac{dx + dy}{x^2 + y^2 + 2xy} = \frac{dx - dy}{x^2 + y^2 - 2xy},$$

or

$$-\frac{dz}{z^2} = \frac{d(x + y)}{(x + y)^2} = \frac{d(x - y)}{(x - y)^2}$$

whence

$$d\left(\frac{1}{z}\right) = -d\left(\frac{1}{x + y}\right),$$

$$d\left(\frac{1}{z}\right) = -d\left(\frac{1}{x - y}\right).$$

Integrating,

$$\frac{1}{z} + \frac{1}{x + y} = a_1,$$

$$\frac{1}{z} + \frac{1}{x - y} = a_2.$$

The general solution then is

$$\phi\left(\frac{1}{z} + \frac{1}{x + y}, \frac{1}{z} + \frac{1}{x - y}\right) = 0,$$

where ϕ is an arbitrary function.

Example 4. Solve

$$x\frac{\partial u}{\partial x} + y\frac{\partial u}{\partial y} + (z + u)\frac{\partial u}{\partial z} = xy.$$

We have the system

$$\frac{dx}{x} = \frac{dy}{y} = \frac{dz}{z + u} = \frac{du}{xy} \qquad (4)$$

From the first and second members,

$$\frac{dx}{x} = \frac{dy}{y}$$

whence, integrating

$$\ln|x| = \ln|y| + \ln|a_1|,$$

or taking the exponential

$$\frac{x}{y} = a_1.$$

Using the multipliers $y,\ x,\ -2$ from the first, second and forth members we have

$$\frac{ydx + xdy - 2du}{xy + xy - 2xy} = \frac{d(xy) - 2du}{0}.$$

Setting the numerator equal to zero and integrating, we have

$$xy - 2u = a_2.$$

Using the multipliers $-y,\ -x,\ 1,\ 1$ from the system (4) we have

$$\frac{dx}{x} = \frac{-ydx - xdy + dz + du}{-xy - xy + z + u + xy},$$

or

$$\frac{dx}{x} = \frac{d(-xy + z + u)}{-xy + z + u},$$

$$d(\ln|x|) = d\left[\ln|z + u - xy|\right].$$

Integrating and taking exponential

$$\frac{z + u - xy}{x} = a_3.$$

Then the general solution is

$$\phi\left(\frac{x}{y}, xy - 2u, \frac{z + u - xy}{x}\right) = 0,$$

where ϕ is an arbitrary function.

Cauchy problem.

The problem of finding an internal surface passing through a given curve

$$\Phi_1(x_1, x_2, \ldots, x_n, z) = 0,$$

$$\Phi_2(x_1, x_2, \ldots, x_n, z) = 0,$$

$$\ldots$$

$$\Phi_n(x_1, x_2, \ldots, x_n, z) = 0$$

is known as the *Cauchy problem.* This surface may be found by taking the above-mentioned function ϕ (Eq.3) not arbitrarily, but by determining the function $\Phi(a_1, a_2, \ldots, a_n)$ via elimination of $x_1,\ x_2,\ \ldots,\ x_n,\ z$ from the equations

$$\Phi_1(x_1, x_2, \ldots, x_n, z) = 0,$$

$$\Phi_2(x_1, x_2, \ldots, x_n, z) = 0,$$

$$\ldots$$

$$\Phi_n(x_1, x_2, \ldots, x_n, z) = 0,$$

$$u_1(x_1, x_2, \ldots, x_n, z) = a_1,$$

$$u_2(x_1, x_2, \ldots, x_n, z) = a_2,$$

$$\ldots$$

$$u_n(x_1, x_2, \ldots, x_n, z) = a_n$$

as a result of which we obtain the equation $\Phi(a_1, a_2, \ldots, a_n) = 0$, and the desired integral will be $\Phi(u_1, u_2, \ldots, u_n) = 0$.

Example 5. Find the solution of

$$x\frac{\partial z}{\partial x} - y\frac{\partial z}{\partial y} = 0.$$

Satisfying the condition $z = 2x$ when $y = 1$.

From the system

$$\frac{dx}{x} = -\frac{dy}{y} = \frac{dz}{0}$$

we have the combinations

$$\frac{dx}{x} = -\frac{dy}{y},$$

$$dz = 0.$$

Integrating and taking exponential we have

$$xy = a_1,$$

$$z = a_2$$

whence the general solution is

$$z = \phi(xy),$$

where ϕ is the function to be determined from the condition

$$z(x, y = 1) = 2x = \phi(x).$$

Eliminating x, y and z from the equations

$$xy = a_1,$$

$$z = a_2,$$

$$z = 2x,$$

$$y = 1$$

we have

$$a_2 = 2a_1$$

hence the solution is

$$z = 2xy.$$

Example 6. Find the solution of

$$\frac{\partial u}{\partial x} + \frac{\partial u}{\partial y} + 2\frac{\partial u}{\partial z} = 0$$

satisfying the condition $u = yz$ when $x = 1$.

We have the system

$$\frac{dx}{1} = \frac{dy}{1} = \frac{dz}{2} = \frac{du}{0}.$$

From these

$$x - y = a_1,$$

$$2x - z = a_2,$$

$$u = a_3.$$

The general solution is then

$$u = \phi(x - y, 2x - z),$$

where the function $\phi(t_1, t_2)$ must be found from the condition

$$u(x = 1, y) = yz = \phi(-y + 1, 2 - z).$$

Eliminating x, y, z, u from the equations

$$x - y = a_1$$

$$2x - z = a_2,$$

$$u = a_3,$$

$$x = 1,$$

$$u = yz$$

we have

$$a_3 = (2 - a_2)(1 - a_1),$$

whence the solution of the Cauchy problem is

$$u = (2 - 2x + z)(1 - x + y).$$

PROBLEMS

Solve

1. $x\dfrac{\partial u}{\partial x} + y\dfrac{\partial u}{\partial y} + z\dfrac{\partial u}{\partial z} = 0.$
2. $(x+2y)\dfrac{\partial z}{\partial x} - y\dfrac{\partial z}{\partial y} = 0.$
3. $(x-z)\dfrac{\partial u}{\partial x} + (y-z)\dfrac{\partial u}{\partial y} + 2z\dfrac{\partial u}{\partial z} = 0.$
4. $xz\dfrac{\partial z}{\partial x} + yz\dfrac{\partial z}{\partial y} = x.$
5. $x^2\dfrac{\partial z}{\partial x} - xy\dfrac{\partial z}{\partial y} + y^2 = 0.$
6. $x\dfrac{\partial z}{\partial x} + y\dfrac{\partial z}{\partial y} = xy + z.$
7. $x\dfrac{\partial z}{\partial x} + y\dfrac{\partial z}{\partial y} = 2xy\sqrt{a^2 - z^2}.$
8. $(x^2+y^2)\dfrac{\partial z}{\partial x} = y^2 + z^2.$
9. $x\dfrac{\partial z}{\partial x} + y\dfrac{\partial z}{\partial y} = z - x^2 - y^2.$
10. $(y+x)\dfrac{\partial z}{\partial x} + (y-x)\dfrac{\partial z}{\partial y} = z.$
11. $(z+y-x)\dfrac{\partial z}{\partial x} + (z+x-y)\dfrac{\partial z}{\partial y} = x + y - z.$
12. $(z+e^x)\dfrac{\partial z}{\partial x} + (z+e^y)\dfrac{\partial z}{\partial y} = z^2 - e^{x+y}.$
13. $x\dfrac{\partial u}{\partial x} + (z+u)\dfrac{\partial u}{\partial y} + (y+u)\dfrac{\partial u}{\partial z} = y + z.$
14. $\dfrac{\partial u}{\partial x} + \dfrac{\partial u}{\partial y} + \dfrac{\partial u}{\partial z} = xyz.$
15. $x\dfrac{\partial u}{\partial x} + y\dfrac{\partial u}{\partial y} + z\dfrac{\partial u}{\partial z} = u + \dfrac{xy}{z}.$
16. $x\dfrac{\partial u}{\partial x} + y\dfrac{\partial u}{\partial y} + z\dfrac{\partial u}{\partial z} = x^2 + 2u.$
17. $xz\dfrac{\partial z}{\partial x} + yz\dfrac{\partial z}{\partial y} = xy.$
18. $\dfrac{\partial z}{\partial x} - a\dfrac{\partial z}{\partial y} = e^{mx}\cos py.$

19. $xy\frac{\partial z}{\partial x} + yz\frac{\partial z}{\partial y} + xy = 0,\ xy = a^2$ when $z = h$.

20. $xy\frac{\partial z}{\partial x} - y^2\frac{\partial z}{\partial y} = x,\ 2ayz = a^2 + 2$ when $x = a$.

21. $(x^2 + y^2)\frac{\partial z}{\partial x} + 2xy\frac{\partial z}{\partial y} = xz,\ y^2 + z^2 = a^2$ when $x = a$.

22. $z\frac{\partial z}{\partial x} - z\frac{\partial z}{\partial y} = y - x,\ z = y^2$ when $x = 1$.

23. $2xz\frac{\partial z}{\partial x} + 2yz\frac{\partial z}{\partial y} = z^2 - x^2 - y^2,\ x^2 + y^2 + z^2 = a^2$ when $x + y + z = 0$.

24. $x\frac{\partial z}{\partial x} - y\frac{\partial z}{\partial y} = z,\ z = x^3$ when $y = x$.

25. $yz\frac{\partial z}{\partial x} + z\frac{\partial z}{\partial y} = xy;\ x = a,\ y^2 + z^2 = a^2$.

26. $z\frac{\partial z}{\partial x} - xy\frac{\partial z}{\partial y} = 2xz;\ x + y = 2,\ yz = 1$.

27. $z\frac{\partial z}{\partial x} + (z^2 - x^2)\frac{\partial z}{\partial y} + x = 0;\ y = x^2,\ z = 2x$.

28. $(y - z)\frac{\partial z}{\partial x} + (z - x)\frac{\partial z}{\partial y} = x - y;\ z = y = -x$.

29. $x\frac{\partial z}{\partial x} + (xz + y)\frac{\partial z}{\partial y} = z;\ x + y = 2z,\ xz = 1$.

30. $y^2\frac{\partial z}{\partial x} + yz\frac{\partial z}{\partial y} + z^2 = 0;\ x - y = 0,\ x - yz = 1$.

5.2 Pfaffian equation

The differential equation

$$P(x,y,z)dz + Q(x,y,z)dy + R(x,y,z)dz = 0 \tag{1}$$

is called *Pfaffian* after the German mathematician J.F.Pfaff(1765 - 1825).

It may happen that (1) is the exact differential of a function or can be made so by the introduction of a suitable factor. The Pfaffian is then called integrable. The necessary condition that the Pfaffian be integrable is

$$P\left(\frac{\partial R}{\partial y} - \frac{\partial Q}{\partial z}\right) + Q\left(\frac{\partial P}{\partial z} - \frac{\partial R}{\partial x}\right) + R\left(\frac{\partial Q}{\partial x} - \frac{\partial P}{\partial y}\right) \equiv 0 \tag{2}$$

If the Pfaffian is integrable then it determines the family of surfaces $U(x,y,z) = C$ that are orthogonal to the vector lines of the field $\vec{F} = P(x,y,z)\vec{\imath} + Q(x,y,z)\vec{\jmath} + R(z,y,z)\vec{k}$ where $\vec{\imath},\ \vec{\jmath},\ \vec{k}$ are unit vectors directed along the coordinate axis.

The rule for finding the integral is as follows. Hold z constant and integrate the resulting equation in x and y, replacing the constant of integration by a function

$f(z)$. Differentiate the result with respect to all the variables, and compare with the original equation. Here an equation will result for the determination of $f(z)$. (Due to symmetry of (1) we may put constant x or y).

Example 1. Solve the equation

$$3yzdx + 2xzdy + xydz = 0.$$

It is readily found that the equation is integrable

$$3yz(x - 2x) + 2xz(3y - y) + xy(2z - 3z) \equiv 0.$$

Holding z constant,

$$3ydx + 2xdy = 0$$

and integrating, we get

$$2\frac{dy}{y} = -3\frac{dx}{x},$$

or

$$2\ln|y| = -3\ln|x| + \ln|f(z)|.$$

Taking exponential of both sides yields

$$y^2x^3 = f(z).$$

Whence differentiating,

$$3x^2y^2dx + 2yx^3dy = f'dz$$

and comparing with the original equation multiplied by x^2y/z,

$$3x^2y^2dx + 2x^3ydy + \frac{x^3y^2}{z}dz = 0$$

we require that

$$f'(z) = -\frac{x^3y^2}{z} = -\frac{f(z)}{z}.$$

From this we get

$$f(z) = \frac{C}{z}.$$

The solution is then

$$x^3y^2z = C.$$

Line Integrals

If the equation (1) is already exact and hence

$$\frac{\partial P}{\partial y} = \frac{\partial Q}{\partial x},$$

$$\frac{\partial P}{\partial z} = \frac{\partial R}{\partial x},$$
$$\frac{\partial Q}{\partial z} = \frac{\partial R}{\partial y},$$

we can write the solution as a *line integral.* We have the solution $U = C$, where

$$U(x,y,z) = \int_{a,b,c}^{x,y,z} [P(x',y',z')dx' + Q(x',y',z')dy' + R(x',y',z')dz']$$

the integral, which is independent of the path, being taken along some curve connecting (a,b,c) and (x,y,z).

Example 2. Solve the equation

$$(y+z)dx + (x+z)dy + (x+y)dz = 0.$$

This equation is exact. We integrate along the straight line from the origin to (x,y,z)

$$x' = xt, \quad y' = yt, \quad z' = zt$$

getting the solution

$$U(x,y,z) = \int_o^1 [(y+z)txdt + (x+z)tydt + (x+y)tzdt]$$

$$= [x(y+z) + (x+z)y + (x+y)z] \frac{1}{2}t^2 \Big|_0^1 = xy + yz + zx = C.$$

5.2.1 *Mayer's Method*

We take a family of planes parallel to the z axis through the point (x_0, y_0, z_0)

$$y = k(x - x_0) + y_0. \tag{3}$$

Here $dy = kdx$ and we eliminate y from (1). We solve the resulting differential equation of the first order in z and x, and get the particular solution such that $z = z_0$ when $x = x_0$. This solution together with (3) gives, with varying k, a family of integral curves through the point. By eliminating k we have the equation of the integral surface through the point.

By taking z_0, say, as arbitrary, we get a family of surfaces. We would choose x_0 and y_0 so as to simplify the work as much as possible ($x_0 = y_0 = 0$, for example), provided the resulting line on which the point moves is not objectionable. This method is known as Mayer's method.

Example 3. Solve the equation

$$zydx + xzdy + (xy+1)dz = 0.$$

It is readily found that the equation is integrable. Putting

$$y = kx, \quad dy = kdx$$

we have

$$2zxkdx + (kx^2 + 1)dz = 0,$$

or

$$\frac{dz}{z} = -k\frac{2x}{kx^2+1}dx.$$

Integrating both sides yields

$$z = \frac{z_0}{kx^2+1}.$$

Here $z = z_0$ when $x = x_0 = 0$. Eliminating k by putting $k = y/x$, we have the final result

$$z = \frac{z_0}{xy+1},$$

or

$$xyz + z = C.$$

If the condition (2) is not fulfilled, then there does not exist a family of surface $U(x,y,z) = C$ that are orthogonal to the vector lines of the field $\vec{F}(x,y,z)$. But the lines that have the same properties may be determined as follows. One can arbitrary specify equation

$$U_1(x,y,z) = 0 \tag{6}$$

and with the aid of (6) eliminate from (1) one of the variables, say z, then we get a differential equation of the form

$$M(x,y)dx + N(x,y)dy = 0$$

integrating this equation we find the desired lines on an arbitrary chosen surface $U_1(x,y,z) = 0$.

PROBLEMS

Find the solutions of the Pfaffians.

1. $2yzdx + zxdy + 3xydz = 0$.
2. $(2xz^2 + y)\,dx + xdy + (2x^2z + 3z^2)\,dz = 0$.
3. $(yz - 1)dx + (z - x)xdy + (1 - xy)dz = 0$.
4. $(2yz + 3x)dx + xzdy + xydz = 0$.
5. $(2x^2 + 2xz + 2xy^2 - 1)\,dx - 2ydy - dz = 0$.
6. $(2x^2 + 2xy + 2xz^2 + 1)\,dx + dy + 2zdz = 0$.
7. $(yz - z)dx - xzdy + xydz = 0$.
8. $(z-y)^3(z - 2x + y)dx + (x-z)^3(x - 2y + z)dy + (y-x)^3(y - 2z + x)dz = 0$.
9. $z\,(1 - z^2)\,dx + zdy - (x + y + xz^2)\,dz = 0$.
10. $(3x^2 + yz)\,ydx - x^2dy + (x + 2z)y^2dz = 0$.
11. $(1 - 4x)dx + (1 + 4y)dy - 4zdz = 0,\ x = y = z = 1$.

12. $dx + (y+z)dy + zdz = 0.$
13. $(1+yz)dx + x(z-x)dy - (1+xy)dz = 0.$
14. $(y^2+yz+z^2)\,dx + (z^2+xz+x^2)\,dy + (x^2+xy+y^2)\,dz = 0.$
15. $zdx + \left((x-y^2)^2 - 2yz\right)dy + (y^2-x)\,dz = 0.$
16. $yzdx - xzdy + x^2dz = 0.$
17. $yzdx + (e^{xy}+xz)dy - dz = 0.$
18. $(x^2+2z-x^3)\,dx + x^3dy - xdz = 0.$
19. $\dfrac{2(x+y)}{z}dx + \dfrac{2(x+y)}{z}dy - \left(\dfrac{x+y}{z}\right)^2 dz = 0.$
20. $(y+a)^2dx + zdy - (y+a)dz = 0.$
21. $(y^2+yz)\,dx + (zx+z^2)\,dy + (y^2-xy)\,dz = 0.$
22. $(x^2y-y^3-y^2z)\,dx + (xy^2-x^3-x^2z)dy + (xy^2+x^2y)\,dz = 0.$
23. $(2x+y^2+2xz)\,dx + 2xydy + x^2dz = du.$
24. $yzdx = xzdy + y^2dz.$

5.3 Nonlinear partial equations

Given an equation of the first order

$$F(x,y,z,p,q) = 0, \tag{1}$$

where

$$p = \frac{\partial z}{\partial x},\ q = \frac{\partial z}{\partial y}.$$

A solution containing two arbitrary constants $\Phi(x,y,z,a,b)$ is called a *complete solution* or *integral.* If any one-parameter subsystem of surfaces $\Phi(x,y,z,a,\phi(a)) = 0$ is taken and its envelope is formed by eliminating a from the equations $\Phi = 0,\ \Phi_a = 0$ a *general integral* is obtained. If the envelope of the complete integral (obtained by eliminating $a,\ b$ from the equations $\Phi = 0,\ \Phi_a = 0,\ \Phi_b = 0$) exists, a *singular integral* of the equation is obtained.

Example 1. Find general and singular integral of the equation

$$z^2(1+p^2+q^2) = 1$$

if its complete integral is

$$\Phi(x,y,z,a,b) = (x-a)^2 + (y-b)^2 + z^2 - 1.$$

If $b = \phi(a)$ then we have for the general solution

$$\Phi_a = -2(x-a) - 2\left[y - \phi(a)\right]\phi' = 0,$$

$$(x-a)^2 + \left[y-\phi(a)\right]^2 + z^2 - 1 = 0.$$

Similarly for the singular integral we have

$$\Phi_a = -2(x-a) = 0, \; a = x,$$

$$\Phi_b = -2(y-b) = 0, \; b = y$$

and from the complete integral we get the singular solution

$$z^2 - 1 = 0 \text{ or } z = \pm 1.$$

Simple cases

Consider some simple cases.

(a) $F(p,q) = 0$. Then $p, \; q$ are supposed to be constant.

Example 2. Solve

$$p^2 - q - 1 = 0.$$

Let $p = a = const$, then

$$z = ax + f(y)$$

and

$$\frac{\partial z}{\partial y} = f'_y.$$

From the given equation we have

$$f'_y = q = p^2 - 1 = a^2 - 1$$

whence

$$f(y) = (a^2 - 1)y + b.$$

The resulting complete integral is then

$$z = ax + (a^2 - 1)y + b.$$

(b) $F(p,q,z) = 0$. Then we try a solution of the form $z = f(x + ay)$

Example 3. Find complete integral

$$zp^2 = q.$$

Putting $z = f(t)$, where $t = x + ay$ we obtain

$$p = f', \; q = af'.$$

On substituting these values into the given equation we get

$$ff'^2 = af',$$

or

$$ff' = a.$$

Integrating both sides with respect to t gives

$$f^2 = 2at + b.$$

Thus the required complete solution is

$$z^2 = 2a(x + ay) + b.$$

(c) $F_1(x,p) = F_2(y,q)$. We solve the problem by putting $F_1(x,p)$ and $F_2(y,q)$ constant.

Example 4. Find complete integral

$$xp = y + q.$$

Setting $xp = a$, we have

$$x\frac{\partial z}{\partial x} = a,$$

or

$$\frac{\partial z}{\partial x} = \frac{a}{x}.$$

Integrating with respect to x we obtain

$$z = a \ln |x| + f(y).$$

Putting these into the original equation we get

$$f' + y = a.$$

Integrating with respect to y we find

$$f = -\frac{1}{2}y^2 + ay + b.$$

The resulting complete integral is

$$z = a \ln |x| - \frac{1}{2}y^2 + ay + b.$$

5.3.1 Lagrange - Charpit's Method

We define a *strip* as a curve endowed with a planar element at each point, provided the planar element passes through the tangent to the curve and the orientation of the planar element varies continuously as we proceed along the curve. A strip is a *characteristic strip* of (1) if at each point the curve is tangent to a ruling of the conical element along this ruling. (This definition is inadequate if the conical element degenerates into a line.)

THEOREM . *Each solution of the system*

$$\frac{dx}{dt} = \frac{\partial F}{\partial p},$$

$$\frac{dy}{dt} = \frac{\partial F}{\partial q},$$

$$\frac{dz}{dt} = p\frac{\partial F}{\partial p} + q\frac{\partial F}{\partial q}, \tag{2}$$

$$\frac{dp}{dt} = -\left(\frac{\partial F}{\partial x} + p\frac{\partial F}{\partial z}\right),$$

$$\frac{dq}{dt} = -\left(\frac{\partial F}{\partial y} + q\frac{\partial F}{\partial z}\right)$$

is a strip along which $F(x, y, z, p, q) = const.$ *A solution is a characteristic strip if and only if* $C = 0$.

It is of greatest importance, where possible, to find a complete solution. The other various solutions may be obtained by setting up functional relations between the parameters and by elimination.

To find a complete solution of (1) we must find integral of Eq.-s(2) written in the form

$$\frac{dx}{\frac{\partial F}{\partial p}} = \frac{dy}{\frac{\partial F}{\partial q}} = \frac{dz}{p\frac{\partial F}{\partial p} + q\frac{\partial F}{\partial q}} = \frac{dp}{-\frac{\partial F}{\partial x} - p\frac{\partial F}{\partial z}} = \frac{dq}{-\frac{\partial F}{\partial y} - q\frac{\partial F}{\partial z}} \tag{3}$$

These equations are called the *subsidiary equations* of (1).

Then from this integral together with the given differential equation (1) we find p and q. Substituting these values of p and q into

$$dz = pdx + qdy \tag{3}$$

we have integrable equation. The method just described is known as *Lagrange - Charpit's method.*

Example 5. Solve the differential equation

$$z = x\frac{\partial z}{\partial x} + y\frac{\partial z}{\partial y} + \frac{\partial z}{\partial x}\frac{\partial z}{\partial y}.$$

We have

$$px + qy + pq - z = 0.$$

The subsidiary equations are

$$\frac{dx}{x+q} = \frac{dy}{y+p} = \frac{dz}{px+qy+2pq} = \frac{dp}{0} = \frac{dq}{0}.$$

One integral is

$$q = a.$$

From the given equation

$$p = \frac{z - ay}{x + a}$$

and we have to integrate

$$dz = \frac{z - ay}{x + a}dx + ady, \tag{4}$$

which is integrable. Since

$$P = \frac{z - ay}{x + a}, \quad Q = a, \quad R = -1,$$

$$P\left(\frac{\partial R}{\partial y} - \frac{\partial Q}{\partial z}\right) + Q\left(\frac{\partial P}{\partial z} - \frac{\partial R}{\partial x}\right) + R\left(\frac{\partial Q}{\partial x} - \frac{\partial P}{\partial y}\right) \equiv 0.$$

Holding x constant in (4)

$$dz = ady$$

and integrating, we get

$$z = ay + f(x)$$

whence differentiating,

$$dz = ady + f'dx$$

and comparing with (4)

$$f' = \frac{z - ay}{x + a}, \quad f' = \frac{f}{x + a},$$

or

$$f = b(x + a).$$

Hence the complete solution of the given equation is

$$z = ay + bx + ab.$$

Example 6. Solve the differential equation

$$\frac{\partial z}{\partial x}\frac{\partial z}{\partial y} = 9z^2.$$

We have

$$pq - 9z^2 = 0.$$

The subsidiary equations are

$$\frac{dx}{q} = \frac{dy}{p} = \frac{dz}{pq + pq} = \frac{dp}{18pz} = \frac{dq}{18qz}.$$

From the forth and fifth members

$$\frac{dp}{p} = \frac{dq}{q}$$

whence, integrating we have one integral

$$p = aq.$$

From this and the given equation we get

$$q = \frac{3z}{\sqrt{a}},$$

$$p = 3z\sqrt{a}$$

and we have to integrate

$$dz = 3z\sqrt{a}dx + \frac{3z}{\sqrt{a}}dy.$$

Writing this

$$\frac{dz}{z} = \frac{3}{\sqrt{a}}(adx + dy)$$

we have

$$\ln z = \frac{3}{\sqrt{a}}(ax + y) + \ln b.$$

Hence taking exponential of both sides, we obtain a complete solution of given equation

$$z = b\exp\left[\frac{3(ax+y)}{\sqrt{a}}\right].$$

PROBLEMS

Find complete solutions $\left(p = \frac{\partial z}{\partial x},\ q = \frac{\partial z}{\partial y},\ p_i = \frac{\partial z}{\partial x_i}\right)$.

1. $zp^2 + q^2 = 4$.
2. $p^2 + q^2 = x - y$.
3. $z = xp + yq + f(p, q)$.
4. $p[q^2 + 4] + q(k - 2z) = 0$.
5. $(4p^2 + q^2)\,y = zq$.
6. $p^2 + q^2 - xp - yq + \frac{1}{2}xy = 0$.
7. $pq = x^2y^2$.
8. $z = xp + yq + p^3q^3$.
9. $p = \sin q$.
10. $q = 3z^2p^2$.
11. $p = 2zq^2$.
12. $z = xp + yq + p^2$.

13. $z^2(1+p^2+q^2) = R^2,\ R = const.$
14. $z = xp + yq - p^2q.$
15. $p = 2q^2 + 1.$
16. $p = e^q.$
17. $p^2q^3 = 1.$
18. $4z = pq.$
19. $z^2 = 1 + p^2 + q^2.$
20. $q^2 = z^2p^2(1-p^2).$
21. $p^3 + q^3 = 27z.$
22. $p^2 = zq.$
23. $p^2 = q + x.$
24. $pq = xy.$
25. $yp = 2xy + \ln q$
26. $q = xyp^2.$
27. $pe^y = qe^x.$
28. $q(p - \cos x) = \cos y.$
29. $z = px + qy - p^2q.$
30. $px + qy - (px+z)^2q = 0.$
31. $p^{1/2} - q^{1/3} = 3x - 2y.$
32. $p^3 + q^3 - 3pqz = 0.$
33. $p_1^2 + p_2^2 + p_3^2 = 4z.$
34. $p^2 + 6p + 2q + 4 = 0.$
35. $p^2z^2 + q^2 = p^2q.$
36. $2z + p^2 + qy + 2y^2 = 0.$
37. $yzp^2 = q.$
38. $pxy + pq + qy = yz.$
39. $2x(z^2q^2+1) = pz.$
40. $q = 3p^2.$
41. $z^2(p^2z^2 + q^2) = 1.$
42. $p - 3x^2 = q^2 - y.$
43. $z = px + qy + p^2 + q^2.$

Find singular solution.

44. $z = px + qy + \ln pq.$
45. $z = px + qy + p^2 + pq + q^2.$
46. $z = px + qy + \frac{1}{2}p^2q^2$
47. $z = px + qy + \frac{p}{q}.$
48. $4z = pq.$
49. $z^2 = 1 + p^2 + q^2$
50. $p^3 + q^3 = 27z.$

Chapter 6

NONLINEAR EQUATIONS AND STABILITY

Consider the two simultaneous differential equations

$$\frac{dx}{dt} = F(x, y), \quad \frac{dy}{dt} = G(x, y), \tag{1}$$

in which the independent variable t does not appear explicitly. Such a system is said to be *autonomous*. We will assume that the functions F and G are continuous and have continuous partial derivatives in some domain D in the xy plane. Then if (x_0, y_0) is a point in this domain, there exists a unique solution $x = \phi(t)$, $y = \psi(t)$ of the system (1) satisfying the initial conditions

$$x(t_0) = x_0, \quad y(t_0) = y_0. \tag{2}$$

The solution is defined in some interval $\alpha < t < \beta$ which contains the point t_0.

It is very helpful to interpret a solution $x = \phi(t)$, $y = \psi(t)$ of the system (1) as a parametric representation of a curve in the xy plane, usually referred to as the *phase plane.* A curve in the phase plane that is described parametrically by a solution of the system (1) is called a *trajectory* of the system.

In determining the trajectories of the system (1), it is often convenient to eliminate the parameter t from the solution $x = \phi(t)$, $y = \psi(t)$, obtaining a relation between x and y. Alternatively, in a region where $F(x, y) \neq 0$, we have from Eq.-s(1)

$$\frac{dy}{dx} = \frac{\frac{dy}{dt}}{\frac{dx}{dt}} = \frac{G(x, y)}{F(x, y)} \tag{3}$$

which is a first order differential equation for $y = f(x)$. The one parameter family of solutions of Eq.(3) are the trajectories of the system (1).

A point (x_0, y_0) at which both $F(x, y)$ and $G(x, y)$ vanish presents special difficulties and has a special significance. Such points are called *critical points.* If the point (x_0, y_0) is a critical point of the system (1) then $x = x_0$, $y = y_0$ is a solution of the system (1). The trajectory of this solution is, of course, simply the single point (x_0, y_0). The particle at (x_0, y_0) is then often said to be at rest or in equilibrium. A trajectory, represented by the solution $x = \phi(t)$, $y = \psi(t)$, is said to approach the critical point (x_0, y_0) as $t \to \infty$ if $\lim_{t\to\infty} \phi(t) = x_0$ and $\lim_{t\to\infty} \psi(t) = y_0$.

Example 1. Sketch the trajectory corresponding to the solution satisfying the specified initial conditions, and indicate direction of motion for increasing t

$$\frac{dx}{dt} = -x, \quad \frac{dy}{dt} = -2y, \quad x(0) = 4, \quad y(0) = 2.$$

The solution of the given system is

$$x = 4e^{-t}, \quad y = 2e^{-2t}.$$

By eliminating t we have

$$y = \frac{x^2}{8}.$$

As $t \to \infty$ then $x \to 0$, $y \to 0$. The path followed by the particle is sketched in Fig.1

Example 2. a). Determine the critical points for the following system

$$\frac{dx}{dt} = x - x^2 - xy$$

$$\frac{dy}{dt} = \frac{1}{2}y - \frac{1}{4}y^2 - \frac{3}{4}xy.$$

The critical points of the given system are the solutions of the nonlinear algebraic equations

$$x - x^2 - xy = 0,$$

$$\frac{1}{2}y - \frac{1}{4}y^2 - \frac{3}{4}xy = 0,$$

or

$$x(1 - x - y) = 0,$$

$$y\left(\frac{1}{2} - \frac{1}{4}y - \frac{3}{4}x\right) = 0.$$

From this system we get the critical points $(0,0)$; $(0,2)$; $(1,0)$ and $\left(\frac{1}{2}, \frac{1}{2}\right)$.

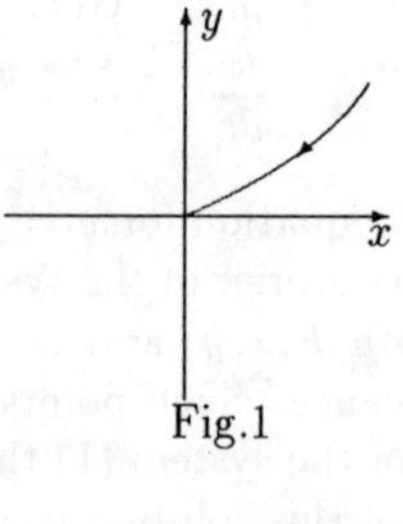

Fig.1

b). Determine the critical points for the following system

$$\frac{dx}{dt} = x - x^2 + xy$$

$$\frac{dy}{dt} = \frac{1}{2}y + \frac{1}{4}y^2 + \frac{3}{4}xy.$$

The critical points of the given system are the solutions of the nonlinear algebraic equations

$$x - x^2 + xy = 0,$$

$$\frac{1}{2}y + \frac{1}{4}y^2 + \frac{3}{4}xy = 0,$$

or

$$x(1 - x + y) = 0,$$

$$y\left(\frac{1}{2} + \frac{1}{4}y + \frac{3}{4}x\right) = 0.$$

From this system we get the critical points $(0,0)$, $(0,-2)$, $(1,0)$ and $\left(-\frac{1}{4}, \frac{5}{4}\right)$.

6.1 Phase plane. Linear systems

A critical point (x_0, y_0) is an *isolated critical point* of the system (1) if there is some circle about the critical point inside which there are no other critical points.

We assume that the origin is an isolated critical point and that in the neighborhood of $(0,0)$ the functions F and G have the form

$$F(x,y) = ax + by + F_1(x,y),$$

$$G(x,y) = cx + dy + G_1(x,y)$$

where $ad - bc \neq 0$, and the functions F_1 and G_1 are continuous, have continuous first partial derivatives, and are small in the sense that

$$\frac{F_1(x,y)}{r} \to 0, \quad \frac{G_1(x,y)}{r} \to 0 \text{ as } r \to 0$$

where $r = (x^2 + y^2)^{1/2}$. Such a system is often referred to as an *almost linear system* in the neighborhood of the critical point $(0,0)$.

A consequence of the assumption that $F_1(x,y)$ and $G_1(x,y)$ are small compared to the linear terms $ax + by$ and $cx + dy$ near the origin is that in many cases (but not all) the trajectories of the linear autonomous system

$$\frac{dx}{dt} = ax + by,$$

$$\frac{dy}{dt} = cx + dy \tag{4}$$

are good approximations to those of the almost linear system, in the neighborhood of the critical point $(0,0)$.

It is possible to solve the system (4) by the methods of Chapter 4. The solutions are of the form $x = Ae^{\lambda t}$, $y = Be^{\lambda t}$ where λ is a root of the auxiliary equation

$$\lambda^2 - (a+d)\lambda + ad - bc = 0. \tag{5}$$

Notice that $\lambda = 0$ cannot be a root of Eq.(5). If it were, then $ad - bc$ would necessarily be zero, implying that the equations $ax + by = 0$, $cx + dy = 0$ would have nontrivial solutions. These nontrivial solutions would be determined only up to a multiplicative constant and would lie on the line passing through the origin. Thus other critical points would be located arbitrarily close to the origin, violating the assumption that the origin is an isolated critical point.

There are a number of different cases that must be considered depending on whether the roots of Eq.(5) are both negative, both positive, one positive and one negative, complex with positive real parts, complex with negative real parts, or pure imaginary.

The trajectories of the system exhibit one of the following three types of behavior.

a) All the trajectories approach the critical point at $t \to \infty$ ($Re\lambda_{1,2} < 0$).

b) The trajectories neither approach the critical point nor tend to infinity as $t \to \infty$ ($Re\lambda_{1,2} = 0$).

c) At least one (possibly all) of the trajectories tends to infinity as $t \to \infty$. This will be the case if at least one of the roots of the auxiliary equation (5) is positive or if the roots have positive real parts.

These types of behavior are summarized in Table 6.1.

A critical point $x = x_0$, $y = y_0$ (an equilibrium solution $x = x_0$, $y = y_0$) of the autonomous system (1) is said to be a *stable critical point* if, given any $\epsilon > 0$, it is possible to find $\delta > 0$ such that every solution $x = \phi(t)$, $y = \psi(t)$ of the system (1), which at $t = 0$ satisfies

$$\left\{[\phi(0) - x_0]^2 + [\psi(0) - y_0]^2\right\}^{1/2} < \delta$$

exists and satisfies

$$\left\{[\phi(t) - x_0]^2 + [\psi(t) - y_0]^2\right\}^{1/2} < \epsilon$$

for all $t \geq 0$.

All solutions that start "sufficiently close" to (x_0, y_0) stay "close" to (x_0, y_0). However, the trajectory of the solution does not have to approach the initial point (x_0, y_0).

Table 6.1.

$\frac{dx}{dt} = ax + by,\ \frac{dy}{dt} = cx + dy$		$\lambda^2 - (a+d)\lambda + (ad - bc) = 0$, $ad - bc \neq 0$
Roots of the auxiliary equation	Type of critical point	Stability
$\lambda_1 > \lambda_2 > 0$	Improper node	Unstable
$\lambda_1 < \lambda_2 < 0$		Asymptotically stable
$\lambda_2 < 0 < \lambda_1$	Saddle point	Unstable
$\lambda_1 = \lambda_2 > 0$	Proper or improper	Unstable
$\lambda_1 = \lambda_2 < 0$	node	Asymptotically stable
$\lambda_1, \lambda_2 = \nu \pm i\mu$	Spiral point	
$\nu > 0$		Unstable
$\nu < 0$		Asymptotically stable
$\lambda_1 = i\mu,\ \lambda_2 = -i\mu$	Center	Stable

Types of critical points

A critical point (x_0, y_0) is said to be *asymptotically stable*, if it is stable and if there exists a $0 < \delta_0 < \delta$, such that if a solution $x = \phi(t)$, $y = \psi(t)$ satisfies

$$\left\{[\phi(0) - x_0]^2 + [\psi(0) - y_0]^2\right\}^{1/2} < \delta_0$$

then

$$\lim_{t\to\infty} \phi(t) = x_0, \qquad \lim_{t\to\infty} \psi(t) = y_0.$$

Trajectories that start "sufficiently close" to (x_0, y_0) must not only stay "close" but must eventually approach (x_0, y_0) as t approaches infinity. Note that asymptotic stability is a stronger requirement of stability.

A critical point that is not stable is said to be *unstable.*

The stability characteristics are summarized in the following theorem.

THEOREM 1. *The critical point* $(0,0)$ *of the linear system (4) is asymptotically stable if the roots of the auxiliary equation (5) are real and negative or have negative real parts. It is stable, but not asymptotically stable, if the roots are pure imaginary. If either of the roots is real and positive or if they have positive real parts, it is unstable.*

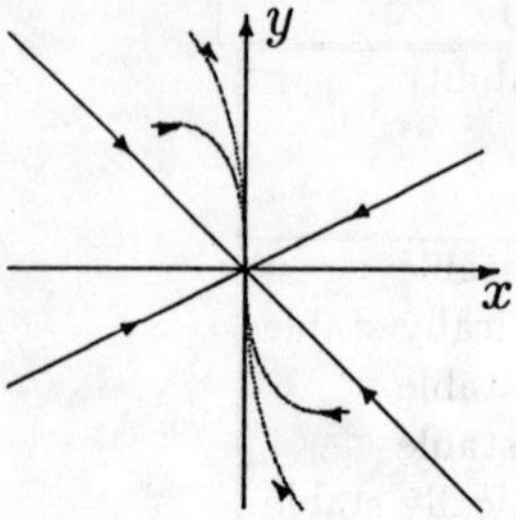

Fig.2

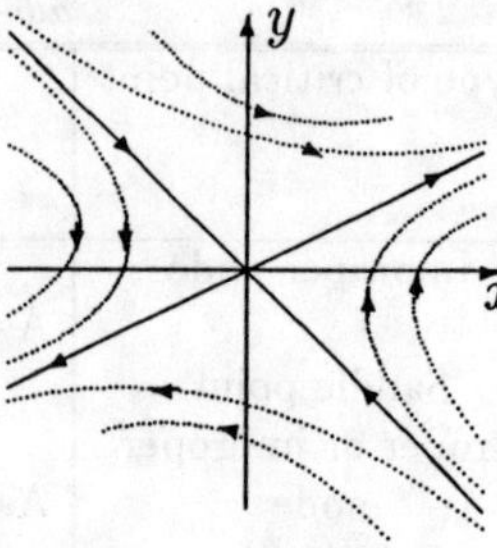

Fig.3

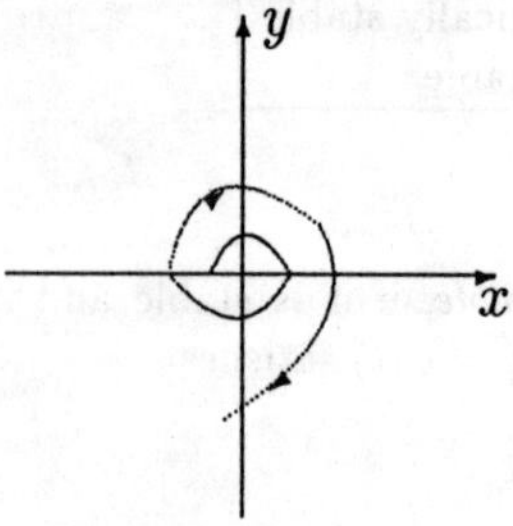

Fig.4

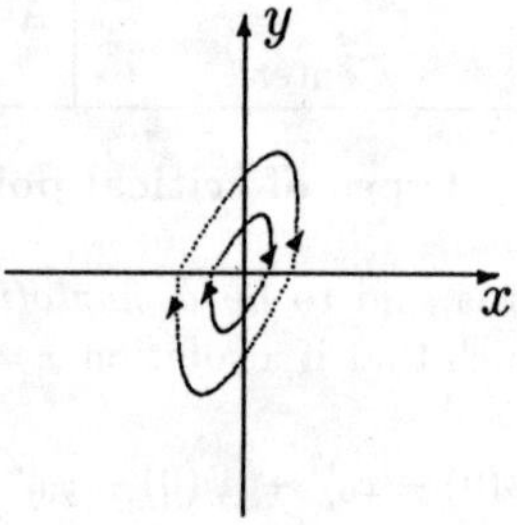

Fig.5

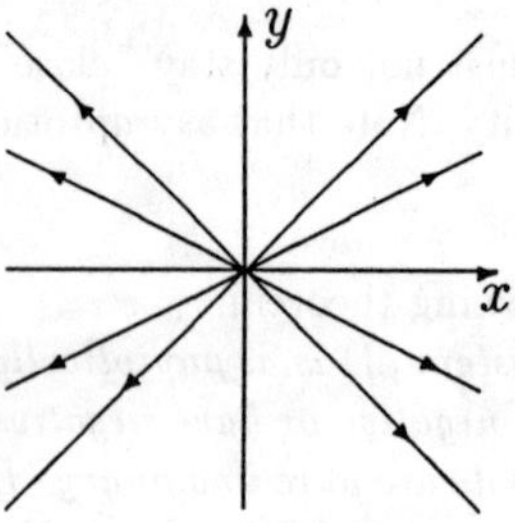

Fig.6

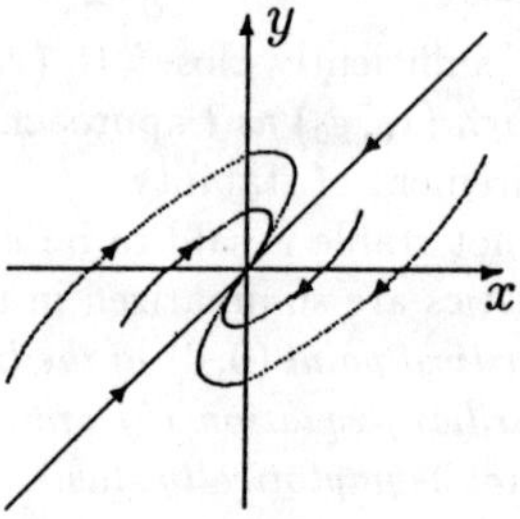

Fig.7

Example 3. Classify the critical point (0,0) and determine whether it is stable, asymptotically stable or unstable. Sketch the trajectory and indicate the direction of motion with increasing t.

$$\frac{dx}{dt} = 2y - 3x, \quad \frac{dy}{dt} = x - 4y.$$

Rewrite this system in the matrix form

$$\begin{pmatrix} \dot{x} \\ \dot{y} \end{pmatrix} = \begin{pmatrix} -3 & 2 \\ 1 & -4 \end{pmatrix} \begin{pmatrix} x \\ y \end{pmatrix}$$

The auxiliary equation is

$$\begin{vmatrix} -3-\lambda & 2 \\ 1 & -4-\lambda \end{vmatrix} = 0,$$

or

$$\lambda^2 + 7\lambda + 10 = 0.$$

The roots are $\lambda_1 = -2$, $\lambda_2 = -5$. The general solution is

$$\begin{pmatrix} x \\ y \end{pmatrix} = \begin{pmatrix} z_1 \\ z_2 \end{pmatrix} e^{-2t} + \begin{pmatrix} z_3 \\ z_4 \end{pmatrix} e^{-5t} \equiv C_1 \vec{a}_1 e^{-2t} + C_2 \vec{a}_2 e^{-5t},$$

where C_1 and C_2 are arbitrary constants, $\vec{a}_1$, $\vec{a}_2$ are eigenvectors.

Since $\lambda_2 < \lambda_1 < 0$ the critical point is an asymptotically stable improper node. Every solution of the given system approaches the solution $x = 0$, $y = 0$ exponentially as $t \to \infty$. To sketch the trajectory first we must find eigenvectors corresponding to eigenvalues λ_1 and λ_2. For λ_1 we have

$$\begin{pmatrix} -3 & 2 \\ 1 & -4 \end{pmatrix} \begin{pmatrix} z_1 \\ z_2 \end{pmatrix} = -2 \begin{pmatrix} z_1 \\ z_2 \end{pmatrix},$$

or

$$-z_1 + 2z_2 = 0.$$

Whence the first eigenvector is $\vec{a}_1 = \begin{pmatrix} 1 \\ 1/2 \end{pmatrix}$. Similarly, for λ_2 we get $\vec{a}_1 = \begin{pmatrix} 1 \\ -1 \end{pmatrix}$. Then we draw the straight lines passing through (0,0) which are parallel to these vectors and sketch trajectories tangent in the origin to the line corresponding to $\vec{a}_1$ (since $|\lambda_1| < |\lambda_2|$). The trajectory is sketched in Fig.2. **Example 4.** The same problem for the following system

$$\frac{dx}{dt} = 3x + 4y, \quad \frac{dy}{dt} = 2x + y,$$

or

$$\begin{pmatrix} \dot{x} \\ \dot{y} \end{pmatrix} = \begin{pmatrix} 3 & 4 \\ 2 & 1 \end{pmatrix} \begin{pmatrix} x \\ y \end{pmatrix}$$

The auxiliary equation is

$$\begin{vmatrix} 3-\lambda & 4 \\ 2 & 1-\lambda \end{vmatrix} = 0,$$

or

$$\lambda^2 - 4\lambda - 5 = 0.$$

The roots are $\lambda_1 = -1$, $\lambda_2 = 5$. The general solution is

$$\begin{pmatrix} x \\ y \end{pmatrix} = \begin{pmatrix} z_1 \\ z_2 \end{pmatrix} e^{-t} + \begin{pmatrix} z_3 \\ z_4 \end{pmatrix} e^{5t} \equiv C_1\vec{a}_1 e^{-t} + C_2\vec{a}_2 e^{5t}.$$

Critical point is an unstable saddle point. We now find eigenvectors corresponding to eigenvalues λ_1 and λ_2. We have for λ_1

$$\begin{pmatrix} 3 & 4 \\ 2 & 1 \end{pmatrix}\begin{pmatrix} z_1 \\ z_2 \end{pmatrix} = -\begin{pmatrix} z_1 \\ z_2 \end{pmatrix},$$

or

$$z_1 + z_2 = 0.$$

Whence $\vec{a}_1 = \begin{pmatrix} 1 \\ -1 \end{pmatrix}$. Similarly, for λ_2 we find $\vec{a}_2 = \begin{pmatrix} 1 \\ 1/2 \end{pmatrix}$.

Consider the direction field near the critical point. We first draw the straight lines passing through the origin which are parallel to the eigenvectors. The integral curves are tangent to these straight lines.

It is now possible for the direction of motion to be toward the critical point on some trajectories and away from the critical point on other trajectories. For any initial point on the line $y = -x$ $(C_1 \neq 0,\ C_2 = 0)$ we have $x \to 0$, $y \to 0$ as $t \to \infty$ (since $\lambda_1 = -1 < 0$); while for an initial point on the line $y = \dfrac{1}{2}x$ we have $x, y \to \pm\infty$ as $t \to \infty$ (since $\lambda_2 = 5 > 0$). It is important to note that only the trajectories on the line corresponding to the root $\lambda < 0$ approach the critical point; all the others approach infinity as t increases. The trajectory is sketched in Fig.3.

Example 5. The same problem for the following system

$$\frac{dx}{dt} = y, \quad \frac{dy}{dt} = y - 2x,$$

or

$$\begin{pmatrix} \dot{x} \\ \dot{y} \end{pmatrix} = \begin{pmatrix} 0 & 1 \\ -2 & 1 \end{pmatrix}\begin{pmatrix} x \\ y \end{pmatrix}$$

The auxiliary equation is

$$\begin{vmatrix} -\lambda & 1 \\ -2 & 1-\lambda \end{vmatrix} = 0,$$

or

$$\lambda^2 - \lambda + 2 = 0.$$

The roots are $\lambda_1 = 1/2 + i\sqrt{3}/2$, $\lambda_2 = 1/2 - i\sqrt{3}/2$. The general solution is

$$\begin{pmatrix} x \\ y \end{pmatrix} = e^{t/2}\left(C_1\vec{a}_1 e^{it\sqrt{3}/2} + C_2\vec{a}_2 e^{-it\sqrt{3}/2}\right)$$

and critical point is unstable spiral point. Since $Re\lambda = 1/2 > 0$, the direction of motion is away from the critical point $(0,0)$. To find the direction of spiral motion (clockwise or anticlockwise) we must calculate the velocity vector $\begin{pmatrix} \dot{x} \\ \dot{y} \end{pmatrix}$ in any convenient point (x, y) for example, setting $x = 1$, $y = 0$ we obtain $\dot{x} = 0$, $\dot{y} = -2$ and motion along the trajectory is clockwise. The trajectory is sketched in Fig.4.

Example 6. The same problem for the following system

$$\frac{dx}{dt} = x - y,$$

$$\frac{dy}{dt} = 2x - y,$$

or in the matrix form

$$\begin{pmatrix} \dot{x} \\ \dot{y} \end{pmatrix} = \begin{pmatrix} 1 & -1 \\ 2 & -1 \end{pmatrix}\begin{pmatrix} x \\ y \end{pmatrix}.$$

The auxiliary equation is

$$\begin{vmatrix} 1-\lambda & -1 \\ 2 & -1-\lambda \end{vmatrix} = 0,$$

or

$$\lambda^2 + 1 = 0.$$

The roots are $\lambda_1 = i$, $\lambda_2 = -i$. Since $Re\lambda = 0$, critical point is stable center. The trajectories are a family of ellipses centered at the origin. We find the direction of motion by setting $x = y = 1$, then $\dot{x} = 0$, $\dot{y} = 1$ and we have anticlockwise motion. The trajectories are sketched in Fig.5

Example 7. The same problem for the equation

$$\frac{dy}{dx} = \frac{y}{x}.$$

This equation is equivalent to the system

$$\frac{dx}{dt} = x, \quad \frac{dy}{dt} = y$$

or in the matrix form

$$\begin{pmatrix} \dot{x} \\ \dot{y} \end{pmatrix} = \begin{pmatrix} 1 & 0 \\ 0 & 1 \end{pmatrix}\begin{pmatrix} x \\ y \end{pmatrix}.$$

The auxiliary equation is $(\lambda - 1)^2 = 0$ and the roots are $\lambda_1 = \lambda_2 = 1$. The general solution is

$$x = Ae^t, \quad y = Be^t$$

or

$$\begin{pmatrix} x \\ y \end{pmatrix} = A\begin{pmatrix} 1 \\ 0 \end{pmatrix} e^t + B\begin{pmatrix} 0 \\ 1 \end{pmatrix} e^t.$$

Since $\lambda_1 = \lambda_2 > 0$, the critical point is unstable proper node. For $A \neq 0$, $B \neq 0$ the trajectories are the straight lines . The trajectories are sketched in Fig.6.

Example 8. The same problem for the equation

$$\frac{dy}{dx} = \frac{y - 2x}{2y - 3x}.$$

This equation is equivalent to the system

$$\begin{pmatrix} \dot{x} \\ \dot{y} \end{pmatrix} = \begin{pmatrix} -3 & 2 \\ -2 & 1 \end{pmatrix} \begin{pmatrix} x \\ y \end{pmatrix}.$$

The auxiliary equation is

$$\begin{vmatrix} -3 - \lambda & 2 \\ -2 & 1 - \lambda \end{vmatrix} = 0,$$

or $\lambda^2 + 2\lambda + 1 = 0$. The roots are $\lambda_1 = \lambda_2 = -1$. For this value of λ equation

$$(-3 - \lambda)z_1 + 2z_2 = 0$$

implies that $z_1 = z_2$. Hence there is only the one eigenvector $\vec{a}_1$ given by $\begin{pmatrix} 1 \\ 1 \end{pmatrix}$ corresponding to the double eigenvalue. Thus one solution of the original system is

$$\begin{pmatrix} x \\ y \end{pmatrix} = \begin{pmatrix} 1 \\ 1 \end{pmatrix} e^{-t},$$

but there is no second solution of the form

$$\begin{pmatrix} x \\ y \end{pmatrix} = \vec{a}e^{-t}.$$

We attempt to find a second solution of the form

$$\begin{pmatrix} x \\ y \end{pmatrix} = \vec{a}_2 te^{-t} + \vec{a}_3 e^{-t},$$

where $\vec{a}_2, \vec{a}_3$ are constant vectors. Substituting for $\begin{pmatrix} x \\ y \end{pmatrix}$ in the original system gives

$$-\vec{a}_2 te^{-t} + (\vec{a}_2 - \vec{a}_3)\, e^{-t} = \begin{pmatrix} -3 & 2 \\ -2 & 1 \end{pmatrix} (\vec{a}_2 t + \vec{a}_3)e^{-t}.$$

In order for this equation to be satisfied for all t it is necessary for the coefficients of te^{-t} and e^{-t} each to be zero. Equating coefficients of te^{-t} and e^{-t} on each side gives the conditions

$$\begin{pmatrix} -3 & 2 \\ -2 & 1 \end{pmatrix} \vec{a}_2 = -\vec{a}_2$$

and

$$\begin{pmatrix} -3 & 2 \\ -2 & 1 \end{pmatrix} \vec{a}_3 = \vec{a}_2 - \vec{a}_3$$

for determination of $\vec{a}_2$ and $\vec{a}_3$. The first equation is satisfied if $\vec{a}_2$ is the eigenvector of the matrix $\begin{pmatrix} -3 & 2 \\ -2 & 1 \end{pmatrix}$ corresponding to the eigenvalue $\lambda = -1$, that is $\vec{a}_2 = \vec{a}_1 = \begin{pmatrix} 1 \\ 1 \end{pmatrix}$. The solution of the second equation is not unique since an arbitrary multiple of $\vec{a}_1$ can be included. Upon solving equation

$$\begin{pmatrix} -3 & 2 \\ -2 & 1 \end{pmatrix} \begin{pmatrix} z_1 \\ z_2 \end{pmatrix} = \begin{pmatrix} 1 \\ 1 \end{pmatrix} - \begin{pmatrix} z_1 \\ z_2 \end{pmatrix}$$

we find that

$$-2z_1 + 2z_2 = 1,$$

or

$$z_2 = \frac{1}{2}(1 + 2z_1)$$

whence

$$\vec{a}_3 = C\vec{a}_1 + \begin{pmatrix} 0 \\ 1/2 \end{pmatrix},$$

where C is an arbitrary constant. Substituting for $\vec{a}_2$ and $\vec{a}_3$ gives the second solution of the original system of the form

$$\begin{pmatrix} x \\ y \end{pmatrix} = \begin{pmatrix} 1 \\ 1 \end{pmatrix} te^{-t} + \left[C \begin{pmatrix} 1 \\ 1 \end{pmatrix} + \begin{pmatrix} 0 \\ 1/2 \end{pmatrix} \right] e^{-t}.$$

The general solution is

$$\begin{pmatrix} x \\ y \end{pmatrix} = C_1 \begin{pmatrix} 1 \\ 1 \end{pmatrix} e^{-t} + C_2 \left[\begin{pmatrix} 1 \\ 1 \end{pmatrix} te^{-t} + \begin{pmatrix} 0 \\ 1/2 \end{pmatrix} e^{-t} \right].$$

It is clear from the solution obtained that all of the trajectories approach the origin as $t \to \infty$. The critical point is stable improper node. The trajectories are tangent to the straight line passing through the origin which is parallel to the vector $\vec{a}_1$. The trajectories are sketched in Fig.7.

PROBLEMS

Classify the critical point $(0,0)$ and determine whether it is stable, asymptotically stable or unstable. Sketch the trajectory and indicate the direction of motion with increasing t.

1. $\dot{x} = 3x + y, \quad \dot{y} = -2x + y.$

2. $\dot{x} = -x + 2y, \quad \dot{y} = x + y.$
3. $\dot{x} = -x + 3y, \quad \dot{y} = -x + y.$
4. $\dot{x} = -2x - y, \quad \dot{y} = 3x - y.$
5. $\dot{x} = 3x - y, \quad \dot{y} = x + y.$
6. $\dot{x} = 3x, \quad \dot{y} = 3y.$
7. $\dot{x} = 3x, \quad \dot{y} = 2x + y.$
8. $\dot{x} = 2x - y, \quad \dot{y} = x.$
9. $\dot{x} = x + 3y, \quad \dot{y} = -6x - 5y.$
10. $\dot{x} = x, \quad \dot{y} = 2x - y.$
11. $\dot{x} = -2x - 5y, \quad \dot{y} = 2x + 2y.$
12. $\dot{x} = 3x + y, \quad \dot{y} = y - x.$
13. $\dot{x} = 3x - 2y, \quad \dot{y} = -6x + 4y.$
14. $\dot{x} = -2x + y, \quad \dot{y} = -4x + 2y.$
15. $\dot{x} = x + 2y, \quad \dot{y} = -3x + y.$
16. $\dot{x} = -2x + \frac{y}{3}, \quad \dot{y} = -2x + \frac{1}{2}y.$
17. $\dot{x} = -x + 3y, \quad \dot{y} = -x + 2y.$
18. $\dot{x} = -y, \quad \dot{y} = x - 2y.$
19. $\dot{x} = -6x - 5y, \quad \dot{y} = -2x - 5y.$
20. $\dot{x} = -x + 2y, \quad \dot{y} = -2x - 5y.$

Investigate the zero solution of the following equations for stability.

21. $\dot{x} = y, \quad \dot{y} = 5ax - a^2y.$
22. $\dot{x} = a^2x - 3y, \quad \dot{y} = ax + 4y.$
23. $\dot{x} = ax - y, \quad \dot{y} = ay - z \quad \dot{z} = az - x.$
24. $\dot{x} = x - 3y + 4z, \quad \dot{y} = 4x - 7y + 8z, \quad \dot{z} = 6x - 7y + 7z.$
25. $\dot{x} = 2x - y + 2z, \quad \dot{y} = 5x - 3y + 3z, \quad \dot{z} = -x - 2z.$
26. $x'' + 2ax' + x = 0.$
27. $x'' - (1 - a)x' - ax = 0.$
28. $x'' - 2a^2x' + 4x = 0.$
29. $x'' + ax' + (a - 1)x = 0.$
30. $x'' + 2(1 - a)x' - 2ax = 0.$

6.2 Almost linear systems

The question of the stability of the nonlinear system

$$\frac{dx}{dt} = ax + by + F_1(x, y),$$

$$\frac{dy}{dt} = cx + dy + G_1(x, y) \tag{6}$$

is answered by the following theorem, which is due to Russian mathematician A.M.Liapunov (1857-1918).

THEOREM 2. *If the critical point* $(0,0)$ *of the almost linear system (6) is an asymptotically stable critical point of the linear system (4), then it is an asymptotically stable critical point of the almost liner system (6). If it is an unstable critical point of the linear system (4), then it is an unstable critical point of the almost linear system (6).*

The case not covered by Theorem 2 is that of a critical point which is a center of the linear system (4). Then λ_1 and λ_2 are pure imaginary, and the "small" changes introduced by the nonlinear terms may lead to complex value for λ_1 and λ_2 with nonzero real parts. Depending on the real parts of λ_1 and λ_2 the center of the linear system (4) may become a stable or unstable spiral point, or it may even remain a center of the almost linear system.

The critical point of the almost linear system (6) is of the same type as that of the linear system (4) if (a) λ_1, λ_2 are real, unequal, and of the same sign (improper node); (b) λ_1, λ_2 are real and of opposite sign (saddle point); (c) λ_1, λ_2 are complex with real parts not equal to zero (spiral point). The two cases not mentioned are λ_1 and λ_2 pure imaginary and λ_1 and λ_2 real and equal. For the former, as mentioned earlier, nothing can be inferred about the critical point of the almost linear system. For the latter ($\lambda_1 = \lambda_2$), while the asymptotic stability or instability does not change, the nature of the critical point for the almost linear system may be altered. The reason is that if the coefficients in an auxiliary equation that has equal roots are slightly perturbed, it is possible that (1) λ_1 and λ_2 remain equal, (2) λ_1 and λ_2 remain real but become unequal (improper node), or (3) λ_1 and λ_2 split apart to form two complex roots whose real parts are only slightly changed from the value of $\lambda_1 = \lambda_2$ but with small nonzero imaginary parts (spiral point).

Even though the critical point is of the same type in cases (a), (b), and (c) as that of the linear system, the trajectories of the almost linear system may be considerably different in appearance from those of the corresponding linear system.

Example 9. Determine all real critical points of the following system of equations and discuss their type and stability

$$\dot{x} = x^2 - y,$$

$$\dot{y} = x^2 - (y-2)^2. \tag{7}$$

The critical points of the system (7) are the solutions of the nonlinear algebraic equations

$$x^2 - y = 0,$$

$$x^2 - (y-2)^2 = 0,$$

or

$$y = x^2,$$

$$x^2 - (x^2-2)^2 = -x^4 + 5x^2 - 4 = 0.$$

From the last we obtain $x^2 = 1$, $x^2 = 4$. Hence we have four critical points in the xy: $(1,1); (-1,1); (2,4); (-2,4)$.

$x = 1,\ y = 1$. To examine the nature of this critical point we let $x = 1 + u$, $y = 1 + v$. Substituting for x and y in Eq.-s(7) we obtain

$$\frac{du}{dt} = 1 + 2u + u^2 - 1 - v = 2u - v + u^2,$$

$$\frac{dv}{dt} = 1 + 2u + u^2 - 1 - v^2 + 2v = 2u + 2v + u^2 - v^2 \qquad (8)$$

The system (8) is an almost linear system. The corresponding linear system is

$$\begin{pmatrix} \dot{u} \\ \dot{v} \end{pmatrix} = \begin{pmatrix} 2 & -1 \\ 2 & 2 \end{pmatrix} \begin{pmatrix} u \\ v \end{pmatrix}$$

and the roots of the auxiliary equation

$$\begin{vmatrix} 2-\lambda & -1 \\ 2 & 2-\lambda \end{vmatrix} = \lambda^2 - 4\lambda + 6 = 0$$

are

$$\lambda_1 = 2 + \sqrt{2}i, \quad \lambda_2 = 2 - \sqrt{2}i$$

since $Re\lambda > 0$ the critical point is unstable spiral point.

$x = -1,\ y = 1$. To examine this critical point, it is convenient to translate it to the origin by letting $x = -1 + u$, $y = 1 + v$. Substituting for x and y in Eq.-s (7) and simplifying we obtain

$$\frac{du}{dt} = -2u - v + u^2,$$

$$\frac{dv}{dt} = -2u + 2v + u^2 - v^2.$$

The corresponding linear system is

$$\begin{pmatrix} \dot{u} \\ \dot{v} \end{pmatrix} = \begin{pmatrix} -2 & -1 \\ -2 & 2 \end{pmatrix} \begin{pmatrix} u \\ v \end{pmatrix}$$

the roots of the auxiliary equation

$$\begin{vmatrix} -2-\lambda & -1 \\ -2 & 2-\lambda \end{vmatrix} = \lambda^2 - 6 = 0$$

are

$$\lambda_1 = \sqrt{6}, \quad \lambda_2 = -\sqrt{6}.$$

Thus $x = -1$, $y = 1$ is an unstable saddle point.

$x = 2,\ y = 4$. Letting $x = 2 + u$, $y = 4 + v$ and substituting in Eq.-s (7) we obtain an almost linear system

$$\frac{du}{dt} = 4u - v + u^2,$$

$$\frac{dv}{dt} = 4u - 4v + u^2 - v^2.$$

The corresponding linear system is

$$\begin{pmatrix} \dot{u} \\ \dot{v} \end{pmatrix} = \begin{pmatrix} -4 & -1 \\ 4 & -4 \end{pmatrix} \begin{pmatrix} u \\ v \end{pmatrix}$$

the roots of the auxiliary equation

$$\begin{vmatrix} 4-\lambda & -1 \\ 4 & -4-\lambda \end{vmatrix} = \lambda^2 - 12 = 0$$

are

$$\lambda_1 = \sqrt{12}, \quad \lambda_2 = -\sqrt{12}.$$

Thus $x = 2,\ y = 4$ is an unstable saddle point.

$x = -2,\ y = 4$. Letting $x = -2 + u,\ y = 4 + v$ and substituting in Eq.-s (7) we obtain an almost linear system

$$\frac{du}{dt} = -4u - v + u^2,$$

$$\frac{dv}{dt} = -4u - 4v + u^2 - v^2.$$

The corresponding linear system is

$$\begin{pmatrix} \dot{u} \\ \dot{v} \end{pmatrix} = \begin{pmatrix} -4 & -1 \\ -4 & -4 \end{pmatrix} \begin{pmatrix} u \\ v \end{pmatrix}.$$

The roots of the auxiliary equation

$$\begin{vmatrix} -4-\lambda & -1 \\ -4 & -4-\lambda \end{vmatrix} = \lambda^2 + 8\lambda + 12 = 0$$

are

$$\lambda_1 = -2, \quad \lambda_2 = -6.$$

Thus $x = -2,\ y = 4$ is an asymptotically stable improper node.

If auxiliary equation has a high degree, then its solution is complicated; for this reason, very important are methods which permit establishing (without solving the equation) whether all its roots have a negative real part or not.

THEOREM 3. *A necessary and sufficient condition for the negativity of the real parts of all the roots of the polynomial*

$$z^n + a_1 z^{n-1} + \ldots + a_{n-1} z + a_n \equiv f(z) \tag{8}$$

with real coefficients is the positivity of all the principal diagonals of the minors of the Hurwitz matrix

$$\begin{pmatrix} a_1 & 1 & 0 & 0 & \cdots & 0 \\ a_3 & a_2 & a_1 & 1 & \cdots & 0 \\ a_5 & a_4 & a_3 & a_2 & \cdots & 0 \\ a_7 & a_6 & a_5 & a_4 & \cdots & 0 \\ \cdots & & & & & \\ 0 & 0 & 0 & 0 & \cdots & a_n \end{pmatrix}.$$

This Theorem is known as *Hurwitz' theorem* after the German mathematician A. Hurwitz (1859 - 1919).

Denote the principal diagonal minors of the Hurwitz matrix:

$$\Delta_1 = a_1,$$

$$\Delta_2 = \begin{vmatrix} a_1 & 1 \\ a_3 & a_2 \end{vmatrix},$$

$$\Delta_3 = \begin{vmatrix} a_1 & 1 & 0 \\ a_3 & a_2 & a_1 \\ a_5 & a_4 & a_3 \end{vmatrix}.$$

Example 10. Determine the stability for the following equation

$$y^{iv} + 2y''' + 4y'' + 3y' + 2y = 0.$$

The auxiliary equation and Hurwitz matrix are

$$\lambda^4 + 2\lambda^3 + 4\lambda^2 + 3\lambda + 2 = 0,$$

$$\begin{pmatrix} 2 & 1 & 0 & 0 \\ 3 & 4 & 2 & 1 \\ 0 & 2 & 3 & 4 \\ 0 & 0 & 0 & 2 \end{pmatrix}.$$

The Hurwitz conditions reduce to

$$\Delta_1 = 2 > 0,$$

$$\Delta_2 = \begin{vmatrix} 2 & 1 \\ 3 & 4 \end{vmatrix} = 5 > 0,$$

$$\Delta_3 = \begin{vmatrix} 2 & 1 & 0 \\ 3 & 4 & 2 \\ 0 & 2 & 3 \end{vmatrix} = 24 - 8 - 9 = 7 > 0,$$

$$\Delta_4 = \begin{vmatrix} 2 & 1 & 0 & 0 \\ 3 & 4 & 2 & 1 \\ 0 & 2 & 3 & 4 \\ 0 & 0 & 0 & 2 \end{vmatrix} = 14 > 0.$$

Then the solution is asymptotically stable.

Example 11. For what values of the parameters a and b is the trivial solution of the equation

$$y''' + ay'' + by' + 2y = 0$$

asymptotically stable.

The auxiliary equation and Hurwitz matrix are

$$\lambda^3 + a\lambda^2 + b\lambda + 2 = 0,$$

$$\begin{pmatrix} a & 1 & 0 \\ 2 & b & a \\ 0 & 0 & 2 \end{pmatrix}.$$

By the Hurwitz criterion,

$$\Delta_1 = a > 0,$$

$$\Delta_2 = \begin{vmatrix} a & 1 \\ 2 & b \end{vmatrix} = ab - 2 > 0,$$

$$\Delta_3 = \begin{vmatrix} a & 1 & 0 \\ 2 & b & a \\ 0 & 0 & 2 \end{vmatrix} = 2ab - 4 > 0.$$

In the given case these conditions reduce to $a > 0$, $ab > 2$.

The Mikhailov criterion

For the zero solution $y \equiv 0$ of equation

$$a_0 y^{(n)} + a_1 y^{(n-1)} + \cdots + a_n y = 0 \tag{9}$$

to be stable it is necessary and sufficient that

(i) for the change of ω from 0 to $+\infty$ the vector $f(i\omega)$ should rotate through an angle $\phi = n\frac{\pi}{2}$, i.e. it should make $\frac{\pi}{4}$ rotations counterclockwise;

(ii) the locus of $f(i\omega)$ should not pass through the origin $(0,0)$ as ω changes from 0 to $+\infty$.

It follows that for solution of equation (9) to be stable it is necessary and sufficient that $a_n a_{n-1} > 0$, and that all roots of the polynomials

$$p(\theta) = a_n - a_{n-2}\theta + a_{n-4}\theta^2 - \ldots,$$

$$q(\eta) = a_{n-1} - a_{n-3}\eta + a_{n-5}\eta^2 - \ldots$$

should be positive and alternative, i.e. between any two roots of one equation there must be a root of the other equation

$$0 < \theta_1 < \eta_1 < \theta_2 < \eta_2 < \cdots.$$

Example 12. Determine the stability for the following equation

$$y^{v} + 4y^{iv} + 16y''' + 25y'' + 13y' + 9y = 0.$$

Here $a_n = 9,\ a_{n-1} = 13$, and the polynomials

$$p(\theta) = 9 - 25\theta + 4\theta^2 = 0,$$

$$q(\eta) = 13 - 16\eta + \eta^2 = 0$$

have the roots $0 < \theta_1 = \left(25 - \sqrt{481}\right)/8 < \eta_1 = 8 - \sqrt{51} < \theta_2 = \left(25 + \sqrt{481}\right)/8 < \eta_2 = 8 + \sqrt{51}$.

By the Mikhailov criterion the solution is asymptotically stable.

PROBLEMS

Determine all real critical points of the following systems of equations and discuss their type and stability.

1. $\dot{x} = -\sin y, \quad \dot{y} = 2x + \sqrt{1 - 3x - \sin y}$.
2. $\dot{x} = y, \quad \dot{y} = \sin(x + y)$.
3. $\dot{x} = \ln(y^2 - x), \quad \dot{y} = x - y - 1$.
4. $\dot{x} = 3 - \sqrt{4 + x^2 + y}, \quad \dot{y} = \ln(x^2 - 3)$.

Determine the stability of the critical point $x \equiv 0,\ y \equiv 0$ of the following systems.

5. $\dot{x} = -x + 3y + x^3 \sin y, \quad \dot{y} = -x - 4y + 1 - \cos y^4$.
6. $\dot{x} = -2x + 4\sin^2 y, \quad \dot{y} = x - 3y + 3x^2$.
7. $\dot{x} = 3x - 22\sin y + 2x^2 - 4y^3, \quad \dot{y} = \sin x - 5y + e^{x^3} - 1$.
8. $\dot{x} = -10x + 4e^y - 4\cos y^2 + x^2, \quad \dot{y} = 2e^x - 2 - y + x^3$.
9. $\dot{x} = 7x + 2\sin y - y^3, \quad \dot{y} = e^x - 3y - 1 + 2x^2$.
10. $\dot{x} = \frac{1}{4}(e^x - 1) - 9y + x^3, \quad \dot{y} = \frac{1}{5}x - \sin y + 2y^3$;
11. $\dot{x} = 5x + y\cos y - \frac{x^2}{3}, \quad \dot{y} = 3x + 2y + x^3 - 3y^2 e^y$.
12. $\dot{x} = 4y - 3x^3, \quad \dot{y} = -3x - 7y^3$.
13. $\dot{x} = -2y - 3x^4, \quad \dot{y} = 2x - 2y^4$.
14. $\dot{x} = -\frac{3}{2}x + \sin y \cos y, \quad \dot{y} = -2x - y + \sin x^3$.
15. $\dot{x} = \ln(4y + e^{-3x}) + x^3, \quad \dot{y} = 2y - 1 + \sqrt[3]{1 - 6x + y^2}$.
16. $\dot{x} = e^{x+2y} - \cos 3x + xy, \quad \dot{y} = \sqrt{4 + 8x} - 2e^y - y^2$.

Determine the stability for the following equations.

17. $y''' - 3y' + 2y = 0$.
18. $y^{iv} + 4y''' + 7y'' + 6y' + 2y = 0$.
19. $y''' + 5y'' + 9y' + 5y = 0$.
20. $y^{iv} - 2y''' + y'' + 2y' - 2y = 0$.
21. $y^{iv} + 7y''' + 17y'' + 17y' + 6y = 0$.
22. $y''' - 3y'' + 12y' - 10y = 0$.
23. $y^{iv} + 5y''' + 18y'' + 34y' + 20y = 0$.

24. $y^{iv} + 7y''' + 19y'' + 23y' + 10y = 0$.
25. $y^{iv} + 11y''' + 41y'' + 61y' + 30y = 0$.
26. $y^{v} + 3y^{iv} - 5y''' - 15y'' + 4y' + 12y = 0$.
27. $y^{v} + 7y^{iv} + 33y''' + 88y'' + 122y' + 60y = 0$.

Determine values of the parameters a and b for which the trivial solution of the following equations is asymptotically stable.

28. $y''' + ay'' + by' + y = 0$.
29. $y^{iv} + 3y''' + ay'' + 2y' + by = 0$.
30. $y''' + 3y'' + ay' + by = 0$.
31. $y^{iv} + ay''' + y'' + 2y' + y = 0$.
32. $y^{iv} + y''' + ay'' + y' + by = 0$.
33. $y^{iv} + 2y''' + 3y'' + 2y' + ay = 0$.
34. $y^{iv} + ay''' + 4y'' + 2y' + by = 0$.

Using the Mikhailov criterion determine the stability for the following equations.

35. $2y''' + 7y'' + 7y' + 2y = 0$.
36. $y''' + 2y'' + 2y' + y = 0$.
37. $2y^{iv} + 13y''' + 28y'' + 23y' + 6y = 0$.
38. $3y^{iv} + 13y''' + 19y'' + 11y' + 2y = 0$.
39. $2y^{iv} + 6y''' + 9y'' + 6y' + 2y = 0$.
40. $y^{iv} + 4y''' + 16y'' + 24y' + 20y = 0$.
41. $y^{v} + 13y^{iv} + 43y''' + 51y'' + 40y' + 12y = 0$.
42. $y^{v} + 3y^{iv} + 2y''' + y'' + 3y' + 2y = 0$.
43. $y^{vi} + y^{v} + 3y^{iv} + 2y''' + 4y'' + 2y' + 2y = 0$.

6.3 Liapunov's second method

In Section 2 we studied how the stability of a critical point of an almost linear system can usually be determined from a study of the corresponding linear system. However, no conclusion can be drawn when the critical point is a center of the corresponding linear system. In this section we will study another approach, known as *Liapunov's second method* or *direct* method.

Let $V(x, y)$ be defined on some domain D containing the origin. Then V is said to be *positive definite* on D if $V(0,0) = 0$ and $V(x, y) > 0$ for all other points in D. Similarly, V is said to be *negative definite* on D if $V(0,0) = 0$ and $V(x, y) > 0$ for all other points in D. If the inequalities $>$ and $<$ are replaced by $\geq$ and $\leq$, then V is said to be *positive semidefinite* and *negative semidefinite*, respectively.

THEOREM 4. *Suppose that the autonomous system (4) has an isolated critical point at origin. If there exists a function V that is continuous and has continuous first partial derivatives, is positive definite, and for which the function $\dot{V}$ is given by*

$$\dot{V}(x, y) = V_x(x, y)F(x, y) + V_y(x, y)G(x, y) \tag{9}$$

is negative definite on some domain D in the xy plane containing $(0,0)$, then the origin is an asymptotically stable critical point. If $\dot{V}$ is negative semidefinite, then the origin is a stable critical point.

We note that the difficulty in using Theorem 4 is that it tells us nothing about how to construct the function V, called a *Liapunov function*, assuming that one exists.

Example 13. Show that the critical point $(0,0)$ of the autonomous system

$$\dot{x} = y - x + xy, \quad \dot{y} = x - y - x^2 - y^3$$

is asymptotically stable.

We try to construct a Liapunov function of the form

$$V(x,y) = ax^2 + bxy + cy^2.$$

Then

$$V_x = 2ax + by, \quad V_y = bx + 2cy$$

so

$$\begin{aligned}\dot{V}(x,y) &= (2ax+by)(y-x+xy) + (bx+2cy)(x-y-x^2-y^3)\\ &= x^2(b-2a) + y^2(b-2c) + xy(2a+2c-2b) + 2x^2y(a-c)\\ &\quad + by^2x - bx^3 - bxy^3 - 2cy^4.\end{aligned}$$

If we chose $b = 0$, and a and c to be positive and equal numbers, then

$$\dot{V} = -2a(x^2+y^2+2xy) - 2ay^4 = -2a\left[y^4 + (x+y)^2\right]$$

is negative definite and $V = a(x^2+y^2)$ is positive definite. Thus by Theorem 4 the origin is an asymptotically stable critical point.

THEOREM 5. *Let the origin be an isolated critical point of the autonomous system (4). Let V be a function that is continuous and has continuous first partial derivatives, suppose that $V(0,0) = 0$ and that in every neighborhood of the origin there is at least one point at which V is positive (negative). Then if there exists a domain D containing the origin such that the function $\dot{V}$ as given by (9) is positive (negative) on D, then the origin is an unstable critical point.*

Example 14. Show that the critical point $(0,0)$ of the autonomous system

$$\dot{x} = x^3 - y, \quad \dot{y} = x + y^3$$

is unstable.

We consider a function $V(x,y) = x^2 + y^2$ which is positive definite, and

$$\dot{V} = 2x\left(x^3 - y\right) + 2y\left(x + y^4\right) = 2\left(x^4 + y^4\right)$$

is positive definite. Thus by Theorem 5 the origin is an unstable critical point.

THEOREM 6. *If there is a differentiable function $v(x_1, x_2, \ldots, x_n)$ that satisfies the following conditions in a certain closed h - neighborhood of the coordinate origin (1) in an arbitrary small neighborhood U of the origin there exists a region $(v > 0)$ in which $v > 0$, and $v = 0$ on a part of the boundary of the region $(v > 0)$ lying in U; (2) in the region $(v > 0)$ the derivative*

$$\frac{dV}{dt} = \sum_{i=1}^{n} \frac{\partial V}{\partial x_i} f_i(x_1, x_2, \ldots, x_n, t) > 0$$

and in the region $(v \geq \alpha)$, $\alpha > 0$, the derivative $\frac{dV}{dt} \geq \beta > 0$, then the critical point $x_i \equiv 0$ $(i = 1, 2, \ldots, n)$ of the system

$$\frac{dx_i}{dt} = f_i(x_1, x_2, \ldots, x_n, t) \quad (i = 1, 2, \ldots, n)$$

is unstable.

This Theorem is known as *Chetaev's instability theorem* after the Russian mathematician N.G.Chetaev (1902 - 1959).

Example 15. Test the stability of the point $x \equiv 0$, $y \equiv 0$ of the system of equations

$$\frac{dx}{dt} = x^5 - y^3,$$

$$\frac{dy}{dt} = -x^3 + y^5$$

The function $V = x^4 - y^4$ satisfies the conditions of Theorem 6:

$$1) V > 0, \ \text{for} |x| > |y|,$$

$$2) \frac{dV}{dt} = 4x^3\left(x^5 - y^3\right) - 4y^3\left(-x^3 + y^5\right) = 4\left(x^8 - y^8\right) > 0$$

for $|x| > |y|$; and for $V \geq \alpha > 0$, $\frac{dV}{dt} \geq \beta > 0$. Hence, the critical point $x \equiv 0$, $y \equiv 0$ is unstable.

As we mentioned above in the case when the real parts of all roots of the auxiliary equation are nonpositive, with the real part of at least one root being zero, investigation for stability by the linearized system is in general impossible. We illustrate this by the next examples.

Example 16. Show that the plane autonomous system

$$\frac{dx}{dt} = y - x^3,$$

$$\frac{dy}{dt} = -x^3$$

is stable, thought its linearization is unstable.

We consider the function $V(x,y) = x^4 + 2y^2$, which is positive definite, and

$$\frac{dV}{dt} = 4x^3\left(y - x^3\right) + 4y\left(-x^3\right) = -4x^6 \leq 0.$$

Thus, by the Theorem 4 the origin is an asymptotically stable critical point. But the linearized system

$$\dot{x} = y,$$

$$\dot{y} = 0$$

is unstable, since its solution is $y = const = a, x = at + b$ $(x \to \infty$ when $t \to \infty)$.

Example 17. Test for stability the point $x \equiv 0,\ y \equiv 0$ of the systems

$$\frac{dx}{dt} = 2y - 3x^5,$$

$$\frac{dy}{dt} = -2x - 3y^5 \qquad (10)$$

and

$$\frac{dx}{dt} = 2y + 3x^5,$$

$$\frac{dy}{dt} = -2x + 3y^5. \qquad (11)$$

The critical point $x \equiv 0,\ y \equiv 0$ of the system (10) is asymptotically stable, since for this system the Liapunov function $V = x^2 + y^2$ satisfies

$$\frac{dV}{dt} = 2x\left(2y - 3x^5\right) + 2y\left(-2x - 3y^5\right) = -6\left(x^6 + y^6\right) \leq 0.$$

At the same time the critical point $x \equiv 0,\ y \equiv 0$ of the system (11) is unstable by the Theorem 5; on taking $V = x^2 + y^2$ we have

$$\frac{dV}{dt} = 2x\left(2y + 3x^5\right) + 2y\left(-2x + 3y^5\right) = 6\left(x^6 + y^6\right) \geq 0.$$

Systems (10) and (11) have the same linearized system

$$\frac{dx}{dt} = 2y,$$

$$\frac{dy}{dt} = -2x.$$

The auxiliary equation for this system is

$$\begin{vmatrix} -\lambda & 2 \\ -2 & -\lambda \end{vmatrix} = \lambda^2 + 4 = 0$$

and the roots are $\lambda_{1,2} = \pm 2i$.

Thus in the case when $\mathrm{Re}\lambda = 0$ nonlinear terms can influence the stability of a critical point.

PROBLEMS

By constructing suitable Liapunov functions of the form $ax^2 + by^2 + cx^4 + dy^4$ where a, b, c and d are to be determined, show that for each of the following systems the critical point at the origin is of the indicated type.

1. $\dot{x} = -x^3 + xy^2, \quad \dot{y} = -2x^2y - y^3$; asymptotically stable.
2. $\dot{x} = -\frac{1}{2}x^3 + 2xy^2, \quad \dot{y} = -y^3$; asymptotically stable.
3. $\dot{x} = -x^3 + 2y^3, \quad \dot{y} = -2xy^2$; stable (at least).
4. $\dot{x} = x^3 - y^3, \quad \dot{y} = 2xy^2 + 4x^2y + 2y^3$; unstable.
5. $\dot{x} = 2y^3 - x^5, \quad \dot{y} = -x - y^3 + y^5$; asymptotically stable.
6. $\dot{x} = xy - x^3 + y^3, \quad \dot{y} = x^2 - y^3$; unstable.
7. $\dot{x} = -3y - 2x^3, \quad \dot{y} = 2x - 3y^3$; asymptotically stable.
8. $\dot{x} = -xy^4, \quad \dot{y} = x^4y$; stable.
9. $\dot{x} = x + 2xy^2, \quad \dot{y} = -2y + 4x^2y^3$; unstable.
10. $\dot{x} = -y - \frac{x}{2} - \frac{x^3}{4}, \quad \dot{y} = x - \frac{y}{2} - \frac{1}{4}y^3$; asymptotically stable.
11. $\dot{x} = y + x^3, \quad \dot{y} = -x + y^3$; unstable.
12. $\dot{x} = y + x^2y^2 - \frac{1}{4}x^5, \quad \dot{y} = -2x - x^3y - \frac{1}{2}y^3$; stable.
13. $\dot{x} = -x, \quad \dot{y} = -y$; asymptotically stable.
14. $\dot{x} = x + x^3, \quad \dot{y} = -y - y^3$; unstable.
15. $\dot{x} = xy^4 - 2x^3 - y, \quad \dot{y} = 2x^2y^3 - y^7 + 2x$; stable.
16. $\dot{x} = -2x - 3y, \quad \dot{y} = x - y$;asymptotically stable.
17. $\dot{x} = x^5 + y^3, \quad \dot{y} = x^3 - y^5$; unstable.
18. $\dot{x} = xy - x^3 + y, \quad \dot{y} = x^4 - x^2y - x^3$; stable.
19. $\dot{x} = -x - y - x^3 - y^2, \quad \dot{y} = x - y + xy$; stable.
20. $\dot{x} = -y + x^5, \quad \dot{y} = x + y^5$; unstable.
21. $\dot{x} = -2x + 4xy^2, \quad \dot{y} = y + 2x^2y$; unstable.
22. $\dot{x} = y - x, \quad \dot{y} = -x - 2y^3$; asymptotically stable.
23. $\dot{x} = x(2 + \cos x), \quad \dot{y} = -y$; unstable.
24. $\dot{x} = -\frac{1}{2}x - \frac{1}{2}xy^2, \quad \dot{y} = -\frac{3}{4}y + 3xz^3,$

    $\dot{z} = -\frac{2}{3}z - 2xyz^2$; stable.

Chapter 7

CALCULUS OF VARIATIONS

Functionals are variable quantities whose values are determined by the choice of one or several functions.

The calculus of variations investigates methods that permit finding of maximal and minimal values of functionals.

7.1 Euler's equation

The curve $y = y(x)$ which extremize the functional

$$v\,[y(x)] = \int_{x_0}^{x_1} F\,[x, y(x), y'(x)]\,dx \tag{1}$$

where the boundary point is being fixed: $y(x_0) = y_0$ and $y(x_1) = y_1$ is a solution of the second-order differential equation

$$F_y - \frac{d}{dx}F_{y'} = 0, \tag{2}$$

or in expanded form

$$\frac{\partial F}{\partial y} - \frac{\partial^2 F}{\partial x \partial y'} - \frac{\partial^2 F}{\partial y \partial y'}y' - \frac{\partial^2 F}{\partial y'^2}y'' = 0. \tag{3}$$

The equation (2) is called *Euler's equation* (it was first published in 1744).

The boundary value problem

$$F_y - \frac{d}{dx}F_{y'} = 0,\ y(x_0) = y_0,\ y(x_1) = y_1$$

does not always have a solution and if the solution exists, it may not be unique.

The integral curves of Euler's equation $y = y(x, C_1, C_2)$ are called *extremals.* Only on extremals that satisfy boundary conditions $y(x_0) = y_0,\ y(x_1) = y_1$ can the functional be extremized. However, in order to establish whether indeed an extremum is achieved on them, one has to take the *sufficient conditions.* In the case of $F(x, y, y')$ this sufficient condition takes the form $F_{y'y'} > 0$ on the extremal for weak minimum and $F_{y'y'} < 0$ for weak maximum and is called the *Legendre condition.*

Example 1. Test for an extremum the functional

$$v[y(x)] = \int_0^1 (y'^3 + x)dx, \; y(0) = 0, \; y(1) = 2.$$

The Euler equation is of the form

$$\frac{d}{dx}\left(3y'^2\right) = 0,$$

or

$$y'y'' = 0.$$

Whence either $y' = 0$ or $y'' = 0$ and respectively

$$y = \begin{cases} const \\ ax + b \end{cases}.$$

The first solution does not satisfy the boundary conditions. The second one gives $y = 2x$. On the extremal $y = 2x$ the second derivative $F_{y'y'} = 6y' = 12 > 0$. Hence, the straight line $y = 2x$ achieves a weak minimum.

Example 2. Find the extremals of the functional

$$v\left[y(x)\right] = \int_{x_0}^{x_1} \left(y'^2 + 2xy'\right) dx.$$

The Euler equation is of the form

$$\frac{d}{dx}(2y' + 2x) = 0,$$

or

$$y'' + 1 = 0$$

whence the extremals are

$$y = -\frac{x^2}{2} + C_1x + C_2.$$

Example 3. Find the extremals of the functional

$$v[y(x)] = \int_0^1 \left(y^2 + 2xyy'\right) dx, \; y(0) = 0, \; y(1) = 1.$$

The Euler equation is of the form

$$2y - \frac{d}{dx}(2xy) = 2y - (2y + 2xy') = -2xy' = 0$$

whence

$$y = const$$

This solution does not satisfy the boundary conditions. An extremum is not achieved in the class of continuous functions.

Example 4. What curve extremize the functional

$$v[y(x)] = \int_0^1 \left(y'^2 + xy\right) dx, \; y(0) = 0, \; y(1) = 1?$$

The Euler equation is of the form

$$x - \frac{d}{dx}(2y') = x - 2y'' = 0$$

whence

$$y = \frac{x^3}{12} + C_1 x + C_2.$$

Using the boundary conditions, we get

$$y = \frac{x^3}{12} + \frac{11}{12}x.$$

PROBLEMS

Find extremals of the following functionals.

1. $v(y) = \int_0^1 \left(y'^2 + xy\right) dx; \; y(0) = y(1) = 0.$
2. $v(y) = \int_1^e \left(2y - x^2 y'^2\right) dx; \; y(0) = 0, \; y(e) = e.$
3. $v(y) = \int_0^1 \left(y'^2 + yy' + 12xy\right) dx; \; y(0) = y(1) = 0.$
4. $v(y) = \int_0^1 (e^y + xy')dx; \; y(0) = 0, \; y(1) = 1.$
5. $v(y) = \int_0^1 \left(y'^2 + y^2 + xy\right) dx; \; y(0) = y(1) = 0.$
6. $v(y) = \int_0^1 \left(y'^2 + y^2 + 2ye^x\right); \; y(0) = 0, \; y(1) = \frac{1}{2e}.$
7. $v(y) = \int_0^{\ln 2} \left(y'^2 + 3y^2\right) e^{2x}; \; y(0) = 0, \; y(\ln 2) = 15/8.$
8. $v(y) = \int_0^1 \left(y'^2 - y^2 - y\right) e^{2x} dx; \; y(0) = 0, \; y(1) = e^{-1}.$
9. $v(y) = \int_0^1 (e^y + xy')dx; \; y(0) = 0, \; y(1) = a.$
10. $v(y) = \int_0^1 \left(x + y'^2\right) dx; \; y(0) = 1, \; y(1) = 2.$
11. $v(y) = \int_0^1 \left(y^2 + y'^2\right) dx; \; y(0) = 0, \; y(1) = 1.$
12. $v(y) = \int_0^1 \left(y'^2 + 4y^2\right) dx; \; y(0) = e^2, \; y(1) = 1.$
13. $v(y) = \int_0^1 \left(2e^y + y^2\right) dx; \; y(0) = 1, \; y(1) = e.$

14. $v(y) = \int_a^b \left(xy' + y'^2\right) dx.$

15. $v(y) = \int_a^b \left(y + \frac{y^3}{3}\right) dx.$

16. $v(y) = \int_0^1 \left(e^{x+y} - y - \sin x\right) dx;\ y(0) = 0,\ y(1) = -1.$

17. $v(y) = \int_0^1 y'^2 dx;\ y(0) = 0;\ y(1) = 1.$

18. $v(y) = \int_{-1}^1 \frac{\sqrt{1+y'^2}}{y'} dx;\ y(-1) = 0,\ y(1) = 1\ .$

19. $v(y) = \int_0^{3/2} \left(y'^3 + 2x\right) dx;\ y(0) = 0,\ y(3/2) = 1.$

20. $v(y) = \int_{-1}^1 \left(xy' + y'^2\right) dx;\ y(-1) = 1,\ y(1) = 0.$

21. $v(y) = \int_0^1 \left(y - y'^2\right) dx;\ y(0) = y(1) = 0.$

Now we consider the functionals of the form

$$\int_{x_0}^{x_1} F\left(x, y_1, y_2, ..., y_n, y_1', y_2', ..., y_n'\right) dx.$$

The extremizing functions must satisfy the system of *Euler's equations*

$$F_{y_i} - \frac{d}{dx} F_{y_i'} = 0\ , i = 1,\ 2,\ \ldots,\ n \tag{4}$$

Example 5. Find the extremals of the functional

$$v[y(x), z(x)] = \int_{x_0}^{x_1} \left(2yz + y'^2 - z'^2\right) dx.$$

The system of Euler's equations is of the form

$$2z - \frac{d}{dx}(2y') = 0,$$

$$2y - \frac{d}{dx}(-2z') = 0,$$

or

$$y'' - z = 0,$$
$$z'' + y = 0.$$

Eliminating one of the unknown functions, say z, we get

$$y^{iv} + y = 0.$$

This is a linear differential equation with constant coefficients. The auxiliary equation is

$$\lambda^4 + 1 = 0$$

and the roots are

$$\lambda = \exp\left(\frac{i\pi + 2\pi k}{4}\right)$$

where $k = 0, 1, 2, 3$, then

$$\lambda = \pm\left(\frac{\sqrt{2}}{2} \pm i\frac{\sqrt{2}}{2}\right).$$

Hence

$$y = \left(C_1 \cos\frac{\sqrt{2}}{2}x + C_2 \sin\frac{\sqrt{2}}{2}x\right) e^{x\sqrt{2}/2}$$

$$+ \left(C_3 \cos\frac{\sqrt{2}}{2}x + C_4 \sin\frac{\sqrt{2}}{2}x\right) e^{-x\,\sqrt{2}/2},$$

$$z = y''.$$

PROBLEMS

Find extremals.

22. $v(y,z) = \int_0^{\pi/4} \left(2z - 4y^2 + y'^2 - z'^2\right) dx;$
$y(0) = 0,\ y\left(\frac{\pi}{4}\right) = 1,\ z(0) = 0,\ z\left(\frac{\pi}{4}\right) = 1.$

23. $v(y,z) = \int_{-1}^{1} \left(2xy - y'^2 + \frac{z'^3}{3}\right);$
$y(1) = 0,\ y(-1) = 2,\ z(1) = 1,\ z(-1) = -1.$

24. $v(y,z) = \int_0^{\pi/2} \left(y'^2 + z'^2 - 2yz\right) dx;$
$y(0) = 0,\ y\left(\frac{\pi}{2}\right) = 1,\ z(0) = 0,\ z\left(\frac{\pi}{2}\right) = 1.$

25. $v(y,z) = \int_0^{1} \left(y'^2 + z'^2 2y\right) dx;$
$y(0) = 1,\ y(1) = \frac{3}{2},\ z(0) = 0,\ z(1) = 1.$

26. $v(y_1,y_2) = \int_0^{\pi/2} \left(y_1'^2 + y_2'^2 + 2y_1y_2\right) dx;$
$y_1(0) = y_2(0) = 0,\ y_1\left(\frac{\pi}{2}\right) = 1,\ y_2\left(\frac{\pi}{2}\right) = -1.$

27. $v(y_1,y_2) = \int_0^{1} \left(y_1'^2 + y_2'^2 - 2y_1y_2\right) dx;$
$y_1(0) = y_2(0) = 0,\ y_1(1) = \sinh 1,\ y_2(1) = -\sinh 1.$

28. $v(y_1,y_2) = \int_0^{\pi/2} (y_1'y_2' - y_1y_2) dx;$
$y_1(0) = y_2(0) = 0,\ y_1\left(\frac{\pi}{2}\right) = 1,\ y_2\left(\frac{\pi}{2}\right) = -1.$

29. $v(y_1,y_2) = \int_0^{1} (y_1'y_2' + y_1y_2)\, dx;$
$y_1(0) = y_2(0) = 0,\ y_1(1) = e,\ y_2(1) = e^{-1}.$

30. $v(y_1, y_2) = \int_0^1 \left(y_1'y_2' + 6xy_1 + 12x^2y_2\right)dx;$
$y_1(0) = y_2(0) = 0,\ y_1(1) = y_2(1) = 1.$
31. $v(y_1, y_2) = \int_1^3 \left(xy_1'^2 + y_2'^2 + xy_1y_2\right)dx;$
$y_1(1) = 1,\ y_2(1) = y_2(3) = 0,\ y_1(3) = \ln 3 + 1.$
32. $v(y_1, y_2) = \int_0^\pi \left(y_1'^2 - y_2'^2 + 2y_1'y_2' + 2y_1 \cos x + 2y_2^2\right)dx;$
$y_1(0) = -1,\ y_2(0) = y_2(\pi) = 0,\ y_1(\pi) = 1 + \pi.$
33. $v(y_1, y_2) = \int_0^1 \left(y_1'^2 + y_2'^2 + 2y_1\right)dx;$
$y_1(0) = y_2(0) = 1,\ y_1(1) = \frac{3}{2},\ y_2(1) = 1.$

7.2 Conditional extremum

In this section we consider the problem of investigating the functional for an extremum:

$$v(y) = \int_{x_0}^{x_1} F(x, y_1, y_2, \dots, y_n, y_1', y_2', \dots, y_n')\, dx, \tag{1}$$

$$y_j(x_0) = y_{j0},\ y_j(x_1) = y_{j1},\ j = 1,\ 2,\ \dots,\ n$$

under the conditions

$$\phi_i(x, y_1, y_2, \dots, y_n) = 0,\ i = 1,\ 2,\ \dots,\ m < n \tag{2}$$

In mechanics, constraints of the type (2) are called *holonomic*, while constraints of the type

$$\phi_i(x, y_1, y_2, \dots, y_n, y_1', y_2', \dots, y_n') = 0,\ i = 1,\ 2,\ \dots,\ m < n \tag{3}$$

are called *nonholonomic*.

The functions $y_1,\ y_2,\ \dots,\ y_n$ that extremize the functional (1) under the conditions (2) satisfy the Euler equations formed for the *auxiliary functional*

$$L = \int_{x_0}^{x_1} \left[F + \sum_{i=1}^{m} \lambda_i \phi_i\right] dx$$

in the case of an appropriate choice of so-called *Lagrange multipliers* $\lambda_i(x)$.

Example 1. Test for an extremum the functional

$$v = \int_0^1 \left(y_1'^2 + y_2'^2 + x^2\right)dx,$$

$$y_1(0) = y_2(0) = 0,\ y_1(1) = y_2(1) = 1,\ y_1 - y_2 = 0.$$

First we set up the auxiliary functional

$$L = \int_0^1 \left[y_1'^2 + y_2'^2 + x^2 + \lambda(y_1 - y_2)\right] dx.$$

The Euler equations are of the form

$$\lambda - \frac{d}{dx}(2y_1') = \lambda - 2y_1'' = 0,$$

$$-\lambda - \frac{d}{dx}(2y_2') = -\lambda - 2y_2'' = 0.$$

The general solution of this system of equations is

$$y_1 = \frac{\lambda}{4}x^2 + C_1 x + C_2,$$

$$y_2 = -\frac{\lambda}{4}x^2 + C_3 x + C_4.$$

From the boundary conditions and subsidiary condition $y_1 - y_2 = 0$ we get

$$y_1(0) = C_2 = 0,$$

$$y_2(0) = C_4 = 0,$$

$$y_1(1) = \frac{\lambda}{4} + C_1 = 1,$$

$$y_2(1) = -\frac{\lambda}{4} + C_3 = 1,$$

$$y_1 - y_2 = \frac{\lambda}{2}x^2 + (C_1 - C_3)x = 0$$

whence $C_1 = C_3 = 1,\ \lambda = 0$ and the desired extremal is

$$y_1 = y_2 = x.$$

Example 2. Find the functions $y_1(x)$ and $y_2(x)$ that render the functional

$$v = \int_0^1 \left(y_1'^2 + y_2'^2 + 1\right) dx,$$

$$y_1(0) = y_2(0) = y_2(1) = 0,\ y_1(1) = 2,\ y_1 + y_2 - 2x^2 = 0$$

a conditional extremum.

The auxiliary functional is

$$L = \int_0^1 \left[y_1'^2 + y_2'^2 + 1 + \lambda\left(y_1 + y_2 - 2x^2\right)\right] dx$$

The Euler equations are of the form

$$\lambda - \frac{d}{dx}(2y_1') = \lambda - 2y_1'' = 0,$$

$$\lambda - \frac{d}{dx}(2y_2') = \lambda - 2y_2'' = 0$$

whence

$$y_1 = \frac{\lambda}{4}x^2 + C_1 x + C_2,$$

$$y_2 = \frac{\lambda}{4}x^2 + C_3 x + C_4.$$

Integration constants and undetermined multiplier λ can be determined by the boundary conditions

$$y_1(0) = C_2 = 0,$$

$$y_2(0) = C_4 = 0,$$

$$y_1(1) = \frac{\lambda}{4} + C_1 = 2,$$

$$y_2(1) = \frac{\lambda}{4} + C_3 = 0$$

and the subsidiary condition

$$\frac{1}{2}\lambda x^2 + (C_1 + C_3)\,x - 2x^2 = 0.$$

From these equations we find $\lambda = 4,\ C_1 = -C_3 = 1$.

Thus, the required extremizing functions are

$$y_1 = x^2 + x,$$

$$y_2 = x^2 - x.$$

Now we consider the nonholonomic constraints.

The functions $y_1(x),\ y_2(x),\ \ldots,\ y_n(x)$ that render the functional (1) a conditional extremum with the constraints of the type (3) must satisfy the system of $m+n$ equations

$$F^*_{y_j} - \frac{d}{dx}F^*_{y_j'} = 0,\ j = 1,\ 2, \ldots,\ n, \tag{4}$$

$$\phi_i(x, y_1, y_2, \ldots, y_n, y_1', y_2', \ldots, y_n') = 0,\ i = 1,\ 2,\ \ldots,\ m, \tag{5}$$

where

$$F^* = F + \sum_{i=1}^{m} \lambda_i(x)\phi_i$$

and auxiliary functional

$$L = \int_{x_0}^{x_1} F^* dx$$

is regarded as a functional dependent on the $m + n$ functions $y_1, y_2, \ldots, y_n, \lambda_1, \lambda_2, \ldots, \lambda_m$.

Example 4. Find the extremals of the functional

$$v(y,z) = \int_0^1 \left(y'^2 + z^2\right) dx,$$

$$y(0) = z(1) = 0, \; y(1) = z(0) = 1, \; y' - z = 0.$$

The auxiliary functional is of the form

$$L = \int_0^1 \left[y'^2 + z^2 + \lambda\left(y' - z\right)\right] dx.$$

The system of equations (4), (5) is

$$-\frac{d}{dx}\left[2y' + \lambda\right] = 0,$$

$$2z - \lambda = 0,$$

$$y' - z = 0,$$

whence

$$2y' + \lambda = const,$$

$$\lambda = 2z,$$

$$z = y'$$

Hence, the extremals are

$$y = C_1 x + C_2,$$

$$z = C_1.$$

The boundary conditions yield a system of linear equations for determining C_1 and C_2:

$$y(0) = C_2 = 0,$$

$$y(1) = C_1 + C_2 = 1,$$

$$z(0) = C_1 = 1,$$

$$z(1) = C_1 = 0.$$

This system of equations does not have a solution, therefore the given variational problem does not have a solution as well.

PROBLEMS

Find the extremals of the following functionals.

1. $v(y_1, y_2) = \int_0^{\pi/2} \left(y_1^2 + y_2^2 - y_1'^2 - y_2'^2 + \cos x\right) dx;$
$y_1(0) = y_2(0) = y\left(\frac{\pi}{2}\right) = 1,\ y_2\left(\frac{\pi}{2}\right) = -1;$
$y_1 - y_2 - \sin x = 0.$
2. $v(y_1, y_2) = \int_0^1 \left(y_1'^2 + 2y_2'^2 + y_2^2\right) dx;$
$y_1(0) - 2,\ y_2(0) = 1,\ y_1(1) = -e^{-1},\ y_2(1) = 0;$
$y_1 - y_2' = 0.$
3. $v(y_1, y_2) = \int_0^1 \left(y_1'^2 + y_2'^2 + x^3\right) dx;$
$y_1(0) = y_2(1) = 2,\ y_2(0) = y_1(1) = 1;$
$y_1 - 2y_2 + 3x = 0.$
4. $v(y_1, y_2) = \int_0^1 \left(y_1'^2 + y_2'^2 + 1\right) dx;$
$y_1(0) = y_2(0) = y_2(1) = 0,\ y_1(1) = 2;$
$y_1 + y_2 - 2x^2 = 0.$
5. $v(y_1, y_2) = \int_0^1 \left(y_1'^2 + y_2^2\right) dx;$
$y_1(0) = y_2(1) = 0,\ y_2(0) = y_1(1) = 1,$
$y_1' - y_2 = 0.$
6. $v(y_1, y_2) = \int_0^{\pi} \left(y_1'^2 - y_2'^2\right) dx;$
$y_1(0) = y_2(0) = y_1(\pi) = 0,\ y_2(\pi) = \frac{\pi}{2};$
$y_1' - y_2 + \cos x = 0.$
7. $v(y_1, y_2) = \int_0^{\pi/2} \left(y_1'^2 + y_2'^2 + 2y_1y_2\right) dx;$
$y_1(0) = 1,\ y_2(0) = -1,\ y_1\left(\frac{\pi}{2}\right) = \frac{\pi^2}{4} + 1,\ y_2\left(\frac{\pi}{2}\right) = \frac{\pi^2}{4} - 1;$
$y_1' + y_2' - 4x = 0.$
8. $v(y_1, y_2) = \int_1^2 \sqrt{1 + y_1'^2 + y_2'^2}dx;$
$y_1(1) = -1,\ y_1(2) = 1,\ y_2(1) = 0,\ y_2(2) = -1;\ 15x - 7y_1 + y_2 = 22.$

7.2.1 *Isoperimetric Problem*

For an *isoperimetric problem*

$$v(y) = \int_{x_0}^{x_1} F(x, y, y')\, dx = min,$$

$$k(y) = \int_{x_0}^{x_1} G(x, y, y')\, dx = l,\ y(x_0) = y_0,\ y(x_1) = y_1$$

the Euler equation becomes

$$F_y^* - \frac{d}{dx} F_{y'}^* = 0,$$

where

$$F^* = F + \lambda G$$

with *Lagrange multiplier* λ.

Two integration constants contained in a general solution of this differential equation of the second order and an undetermined constant λ can be determined, for instance, by the boundary conditions $y(x_0) = y_0$, $y(x_1) = y_1$ and a subsidiary condition $k(y) = l$.

Example 5. Find the extremum of the functional

$$v(y) = \int_0^1 \left(x^2 + y'^2\right) dx$$

subject to the conditions

$$k(y) = \int_0^1 y^2 dx = 2, \ y(0) = 0, \ y(1) = 0.$$

First we set up the auxiliary functional

$$L = \int_0^1 (F + \lambda G)\, dx = \int_0^1 \left(x^2 + y'^2 + \lambda y^2\right) dx$$

and then write Euler's equation for it

$$F^*_y - \frac{d}{dx} F^*_{y'} = 2\lambda y - \frac{d}{dx}(2y') = 2\lambda y - 2y'' = 0,$$

or

$$y'' - \lambda y = 0.$$

The general solution of this equation is

$$y = C_1 e^{\sqrt{\lambda}x} + C_2 e^{-\sqrt{\lambda}x}$$

From the boundary conditions and subsidiary condition we get

$$C_1 + C_2 = 0,$$

$$C_1 e^{\sqrt{\lambda}} + C_2 e^{-\sqrt{\lambda}} = 0,$$

$$\int_0^1 \left(C_1 e^{\sqrt{\lambda}x} + C_2 e^{-\sqrt{\lambda}x}\right)^2 dx$$

$$= \frac{1}{2\sqrt{\lambda}} \left[C_1^2 \left(e^{2\sqrt{\lambda}} - 1\right) + C_2^2 \left(e^{-2\sqrt{\lambda}} - 1\right)\right] + 2C_1 C_2 = 2,$$

or

$$C_1 = -C_2,$$

$$e^{2\sqrt{\lambda}} = 1 = e^{i2\pi k},$$

$$C_1 C_2 = 1,$$

whence $C_1 = -C_2 = \pm i$, $\lambda = -\pi^2 k^2$, $k = 0, \pm 1, \pm 2, \ldots$.

Hence, the required extremals are

$$y = \pm i \left(e^{i\pi kx} - e^{-i\pi kx} \right) = \pm 2 \sin \pi kx.$$

To determine the extremum of the functional

$$v(y) = \int_{x_0}^{x_1} F(x, y_1, y_2, \ldots, y_n, y_1', y_2', \ldots, y_n')\, dx$$

under the conditions

$$\int_{x_0}^{x_1} G_i(x, y_1, y_2, \ldots, y_n)\, dx = l_i, \ i = 1, 2, \ldots, m,$$

where l_i are constants, it is necessary to set up an *auxiliary functional*

$$L = \int_{x_0}^{x_1} \left(F + \sum_{i=1}^{m} \lambda_i G_i \right) dx$$

and to write the Euler equations for it.

Example 6. Find extremals of the functional

$$v(y_1, y_2) = \int_0^1 y_1' y_2' dx$$

$$y_1(0) = y_2(0) = y_1(1) = 0,\ y_2(1) = 1,\ \int_0^1 x y_1 dx = 0,\ \int_0^1 x y_2 dx = 0.$$

The auxiliary functional is of the form

$$L = \int_0^1 (y_1' y_2' + \lambda_1 x y_1 + \lambda_2 x y_2)\, dx$$

and the Euler equations are

$$\lambda_1 x - \frac{d}{dx}(y_2') = \lambda_1 x - y_2'' = 0,$$

$$\lambda_2 x - \frac{d}{dx}(y_1') = \lambda_2 x - y_1'' = 0.$$

The general solution is

$$y_1 = \lambda_1 \frac{x^3}{6} + C_1 x + C_2,$$

$$y_2 = \lambda_2 \frac{x^3}{6} + C_3 x + C_4.$$

From the boundary conditions we have

$$y_1(0) = C_2 = 0,$$

$$y_2(0) = C_4 = 0,$$

$$y_1(1) = \frac{\lambda_1}{6} + C_1 = 0,$$

$$y_2(1) = \frac{\lambda_2}{6} + C_3 = 1.$$

From the subsidiary condition we get

$$\int_0^1 \left(C_1 x^3 + C_1 x\right) x dx = -\frac{2}{15} C_1 = 0,$$

$$\int_0^1 \left[(C_3 - 1)x^3 + C_3 x\right] x dx = (C_3 - 1)\frac{1}{5} + \frac{1}{3} C_3 = 0.$$

Whence $C_1 = 0,\ C_3 = -3/2,\ \lambda_1 = 0,\ \lambda_2 = 15.$

Hence, the desired extremal is

$$y_1 = 0,\ y_2 = \frac{5}{2}x^2 - \frac{3}{2}x.$$

PROBLEMS

Find the extremals of the following functionals.

1. $v(y) = \int_0^1 y'^2 dx;$

   $y(0) = 0,\ y(1) = 1;\ \int_0^1 xy dx = 0.$

2. $v(y) = \int_0^1 y'^2 dx;$

   $y(0) = y(1) = 0;\ \int_0^1 y dx = 1,\ \int_0^1 xy dx = 0.$

3. $v(y) = \int_0^\pi y'^2 dx;$

   $y(0) = 0,\ y(\pi) = 1;\ \int_0^\pi y \sin x dx = 0.$

4. $v(y) = \int_0^\pi y \sin x dx;$

   $y(0) = 0,\ y(\pi) = \pi;\ \int_0^\pi y'^2 dx = \frac{3}{2}\pi.$

5. $v(y) = \int_0^1 \left(y'^2 + y^2\right) dx;$

   $y(0) = 0,\ y(1) = e^{-1};\ \int_0^1 e^{-x} y dx = \frac{1}{4}\left(1 - 3e^{-2}\right).$

6. $v(y) = \int_0^1 y'^2 dx;$

$y(0) = 1,\ y(1) = 6;\ \int_0^1 y dx = 3.$

7. $v(y) = \int_0^1 y'^2 dx;$

$y(0) = 0,\ y(1) = \frac{1}{4};\ \int_0^1 \left(y - y'^2\right) dx = \frac{1}{12}.$

8. $v(y) = \int_0^\pi y'^2 dx;$

$y(0) = y(\pi) = 0,\ \int_0^\pi y^2 dx = 1.$

9. $v(y_1, y_2) = \int_0^1 y_1' y_2' dx;$

$y_1(0) = y_2(0) = y_1(1) = 0,\ y_2(1) = 1;\ \int_0^1 y_1 dx = 1,\ \int_0^1 y_2 dx = 0.$

10. $v(y_1, y_2) = \int_0^1 (y_1 + y_2) dx;$

$y_1(0) = y_2(0) = 0,\ y_1(1) = 1,\ y_2(1) = -3;\ \int_0^1 y_1' y_2' dx = 0.$

11. $v(y_1, y_2) = \int_0^1 \left(y_1'^2 + y_2'^2\right) dx;$

$y_1(0) = y_2(0) = y_1(1) = y_2(1) = 0;\ \int_0^1 y_1 y_2 dx = -2.$

12. $v(y_1, y_2) = \int_0^1 x\,(y_1 - y_2) dx;$

$y_1(0) = y_2(0) = y_2(1) = 0,\ y_1(1) = 2;\ \int_0^1 y_1' y_2' dx = -\frac{2}{5}.$

13. $v(y, z) = \int_0^1 \left(y'^{\,2} + z'^{\,2} - 4xz' - 4z\right) dx;$

$y(0) = z(0) = 0,\ y(1) = z(1) = 1,\ (y'^{\,2} - xy' - z'^{\,2})\, dx = 2.$

7.3 Movable end points

First we consider an elementary problem with moving boundaries. In this case the endpoints (x_0, y_0) and (x_1, y_1) of the argument function $y = y(x)$ of the functional

$$v(y) = \int_{x_0}^{x_1} F(x, y, y') dx \tag{1}$$

must lie on the given curves $\Phi_0(x_0, y_0) = 0$ and $\Phi_1(x_1, y_1) = 0$. Then the extremizing function must satisfy the Euler equation and the *condition of transversality*

$$\left[F - \left(\frac{\Phi'_{0x}}{\Phi'_{0y}} + y'\right) F_{y'}\right]_{x=x_0} = 0,$$

$$\left[F - \left(\frac{\Phi'_{1x}}{\Phi'_{1y}} + y'\right) F_{y'}\right]_{x=x_1} = 0, \tag{2}$$

or if the curves are given in the explicit form $y_0 = \phi_0(x_0)$ and $y_1 = \phi_1(x_1)$

$$[F + (\phi_0' - y')F_{y'}]_{x=x_0} = 0,$$
$$[F + (\phi_1' - y')F_{y'}]_{x=x_1} = 0. \qquad (3)$$

If the boundary condition $y(x_0)$ or $y(x_1)$ is absent, we must write the so-called *natural* boundary condition

$$F_{y'}|_{x=x_0} = 0,$$

or

$$F_{y'}|_{x=x_1} = 0 \qquad (4)$$

Example 1. Find the extremals of the functional

$$v(y) = \int_0^{x_1} y'^2 dx;\ y(0) = 0,\ y(x_1) = \frac{2}{1-x_1}.$$

The Euler equation is of the form

$$\frac{d}{dx}(2y') = 2y'' = 0,$$

whence the general solution is

$$y = C_1 x + C_2.$$

The two integration constants and the unknown endpoint x_1 can be determined by the boundary conditions

$$y(0) = C_2 = 0,$$

$$y(x_1) = C_1 x_1 = \frac{2}{1-x_1}$$

and the condition of transversality (3)

$$\left[y'^2 + \left(\frac{2}{(1-x)^2} - y'\right) 2y'\right]_{x=x_1} = C_1^2 + \left(\frac{2}{(1-x_1)^2} - C_1\right) 2C_1 = 0,$$

or

$$C_1\left(\frac{4}{(1-x_1)^2} - C_1\right) = 0$$

whence either $C_1 = 0$ or $C_1 = \dfrac{4}{(1-x_1)^2}$.

From this and the second boundary condition we have the system of equations

$$C_1 x_1 = \frac{2}{1-x_1},$$

$$C_1 = \frac{4}{(1-x_1)^2}$$

for determining the endpoint x_1 and constant C_1. Solving this system we find $x_1 = \frac{1}{3}$, and $C_1 = 9$.

Thus the desired extremal is

$$y = 9x, \; x_1 = \frac{1}{3}.$$

Example 2. Find the extremals of the functional

$$v(y) = \int_0^{\pi/6} \left(y'^2 - y^2\right) dx, \; y(0) = 1.$$

In this problem the second boundary condition is absent, therefore we use the natural boundary condition (4)

$$F_{y'}|_{x=\pi/6} = 0. \tag{5}$$

The Euler equation is of the form

$$-2y - \frac{d}{dx}(2y') = -2y - 2y'' = 0,$$

or

$$y'' + y = 0.$$

The general solution is

$$y = C_1 \sin x + C_2 \cos x.$$

From the boundary condition $y(0) = 1$ we have $C_2 = 1$ and from the natural boundary condition (5) we get

$$C_1 \cos\frac{\pi}{6} - \sin\frac{\pi}{6} = 0$$

whence $C_1 = \sqrt{3}/3$ and the desired extremal is

$$y = \cos x + \frac{\sqrt{3}}{3} \sin x.$$

Now we consider the functional dependent on several functions

$$v(y, z) = \int_{x_0}^{x_1} F(x, y, z, y', z')dx, \tag{6}$$

where

$$y(x_0) = \phi_0(x_0), \; y(x_1) = \phi_1(x_1), \; z(x_0) = \psi_0(x_0), \; z(x_1) = \psi_1(x_1).$$

The extremum of the functional (6) can only be attained on integral curves of the system of Euler equations

$$F_y - \frac{d}{dx}F_{y'} = 0,$$

$$F_z - \frac{d}{dx}F_{z'} = 0,$$

and the condition of transversality in this case is

$$[F + (\phi_0' - y')F_{y'} + (\psi_0' - z')F_{z'}]_{x=x_0} = 0,$$

$$[F + (\phi_1' - y')F_{y'} + (\psi_1' - z')F_{z'}]_{x=x_1} = 0. \quad (7)$$

Example 3. Find the extremals of the functional

$$v(y,z) = \int_{x_0}^{x_1} \left(y'^2 + z'^2 + 2y\right) dx,$$

$$y(x_0) = x_0,\ y(x_1) = \frac{1}{2},\ z(x_0) = 1,\ z(x_1) = x_1.$$

The system of Euler equations is

$$2 - \frac{d}{dx}(2y') = 2 - 2y'' = 0,$$

$$\frac{d}{dx}(2z') = 2z'' = 0,$$

or

$$y'' = 1,$$

$$z'' = 0.$$

Hence we have the general solution of the form

$$y = \frac{x^2}{2} + C_1 x + C_2,$$

$$z = C_3 x + C_4.$$

From the boundary conditions and condition of transversality we get the system of equations for determining $C_1,\ C_2,\ C_3,\ C_4,\ x_0$ and x_1

$$\frac{x_0^2}{2} + C_1 x_0 + C_2 = x_0,$$

$$\frac{x_1^2}{2} + C_1 x_1 + C_2 = \frac{1}{2},$$

$$C_3 x_0 + C_4 = 1,$$

$$C_3 x_1 + C_4 = x_1,$$

$$\Big[C_3^2 + (x + C_1)^2 + x^2 + 2C_1 x + 2C_2$$

$$+2(x + C_1)(1 - x - C_1) - C_3 2C_3\Big]_{x=x_0} = 0,$$

$$\Big[C_3^2 + (x + C_1)^2 + x^2 + 2C_1x + 2C_2$$
$$-2(x + C_1)(x + C_1) + (1 - C_3)2C_3\Big]_{x=x_1} = 0.$$

Solving this system we find $x_0 = 0,\ x_1 = 1,\ C_1 = C_2 = C_3 = 0,\ C_4 = 1$ and the required extremal is

$$y = \frac{x^2}{2},\ z = 1.$$

Example 4. Find the shortest distance between the point $M(1,1)$ and the circle $x^2 + y^2 = 4$.

The distance between point and curve is given by the formula

$$l = \int_{x_0}^{x_1} \sqrt{1 + y'^2} dx.$$

Here we have to find the minimum of l provided $x_1^2 + y_1^2 = 4$ and $x_0 = 1,\ y(x_0) = 1$.

The Euler equation is of the form

$$\frac{d}{dx}\left(\frac{y'}{\sqrt{1 + y'^2}}\right) = 0,$$

or

$$\frac{y'}{\sqrt{1 + y'^2}} = const,$$

whence

$$y' = const = C_1,$$

and the general solution is

$$y = C_1x + C_2.$$

From the boundary conditions we get

$$C_1 + C_2 = 1,$$

$$(C_1x_1 + C_2)^2 + x_1^2 = 4.$$

The condition of transversality gives

$$\sqrt{1 + C_1^2} - \left(\frac{x_1}{C_1x_1 + C_2} + C_1\right)\frac{C_1}{\sqrt{1 + C_1^2}} = 0.$$

Thus we have the system of equations for determining $C_1,\ C_2,\ x_1$

$$C_2 = 1 - C_1,$$

$$(C_1x_1 + 1 - C_1)^2 + x_1^2 = 4,$$

$$\sqrt{1+C_1^2}-\left(\frac{x_1}{C_1x_1+1-C_1}+C_1\right)\frac{C_1}{\sqrt{1+C_1^2}}=0,$$

whence $C_1 = 1,\ C_2 = 0,\ x_1 = \pm\sqrt{2}$. The endpoint $x_1 = \sqrt{2}$ corresponds to the shortest distance, which is equal to

$$l=\int_1^{\sqrt{2}}\sqrt{2}dx=2-\sqrt{2}.$$

Example 5. Find the shortest distance between the circle $x^2 + y^2 = 9$ and the straight line $y = x + 5$.

The distance between circle and straight line is given by the same formula as in Example 4

$$l=\int_{x_0}^{x_1}\sqrt{1+y'^2}dx$$

where the endpoint x_0 lies on the straight line $y = x + 5$ and the endpoint x_1 lies on the circle $x^2 + y^2 = 9$.

The general solution of the Euler equation is (see Example 4)

$$y = C_1x + C_2.$$

The boundary conditions and condition of transversality are

$$C_1x_0 + C_2 = x_0 + 5,$$

$$x_1^2 + (C_1x_1 + C_2)^2 = 9,$$

$$\sqrt{1+C_1^2}+(1-C_1)\frac{C_1}{\sqrt{1+C_1^2}}=0,$$

$$\sqrt{1+C_1^2}-\left(\frac{x_1}{C_1x_1+C_2}+C_1\right)\frac{C_1}{\sqrt{1+C_1^2}}=0,$$

whence $C_1 = -1,\ C_2 = 0,\ x_0 = -\frac{5}{2},\ x_1 = -3\sqrt{2}/2$

Thus the desired distance is

$$l=\int_{-5/2}^{-3\sqrt{2}/2}\sqrt{2}dx=\frac{5\sqrt{2}}{2}-3.$$

If the boundary points $M_0(x_0, y_0, z_0)$ and $M_1(x_1, y_1, z_1)$ can move along some surfaces $z = \phi_0(x, y)$ and $z = \phi_1(x, y)$, then the condition of transversality will be

$$[F - y'F_{y'} + (\phi'_{0\,x} - z')F_{z'}]_{x=x_0} = 0,$$

$$\left[F_{y'} + F_{z'}\phi'_{0\,y}\right]_{x=x_0} = 0,$$

$$\left[F - y'F_{y'} + (\phi'_{1\,x} - z')F_{z'}\right]_{x=x_1} = 0,$$
$$\left[F_{y'} + F_{z'}\phi'_{1\,y}\right]_{x=x_1} = 0.$$

PROBLEMS

Find extremals of the following functionals.

1. $v(y) = \int_0^{x_1} y'^2 dx;$
 $y(0) = 0,\ y(x_1) = -x_1 - 1.$
2. $v(y) = \int_0^{x_1} \sqrt{1 + y'^2} dx;$
 $y(0) = 0,\ y(x_1) = x_1^{-2}.$
3. $v(y) = \int_{x_0}^{x_1} \sqrt{1 + y'^2} dx;$
 $y(x_0) = x_0^2,\ y(x_1) = x_1 - 5.$
4. $v(y) = \int_0^{1/2} \left(y - y'^2\right) dx;\ y(0) = 0.$
5. $v(y) = \int_0^{1} \left(y'^2 + y\right) dx;\ y(1) = 0.$
6. $v(y) = \int_0^{\pi/2} \left(y'^2 - y^2\right) dx;\ y(0) = 0.$
7. $v(y) = \int_0^{x_1} \left(y'^2 + y^2\right) dx;$
 $y(0) = 0,\ y(x_1) = 1.$
8. $v(y) = \int_0^{x_1} \frac{\sqrt{1 + y'^2}}{y} dx;$
 $y(0) = 1,\ y(x_1) = x_1 - 1.$
9. $v(y) = \int_0^{\pi/4} \left(y'^2 - y^2 + 4y\cos x\right) dx;$
 $y(0) = 0.$
10. $v(y) = \int_0^{1} \left(\frac{1}{2}y'^2 + yy' + y' + y\right) dx.$
11. Find the shortest distance between the circle $x^2 + y^2 = a^2$, $z = 0$ and the hyperbola $z^2 - y^2 = c^2$, $x = 0$.
12. Find the shortest distance between the point (1,0) and the ellipse $\frac{x^2}{9} + \frac{y^2}{4} = 1$.
13. Find the shortest distance between the point (-1,5) and the parabola $y^2 = x$.
14. Find the shortest distance between the circle $x^2 + y^2 = 1$ and the straight line $x + y = 4$.
15. Find the shortest distance between the point (-1,3) and the straight line $y = 1 - 3x$.
16. Find the shortest distance from the point (0,0,3) to the surface $z = x^2 + y^2$.
17. Find the shortest distance from the point (2,0,5) to the surface $z = x^2 + y^2$.
18. Find the shortest distance between the surfaces $x^2 + y^2 + z^2 = 4$ and $\frac{x^2}{25} + \frac{y^2}{16} + \frac{z^2}{9} = 1$.
19. Find the shortest distance from the point $(1,1,1)$ to the sphere $x^2+y^2+z^2 = 1$.

7.4 Bolza problem

The problem of determining the function $y(x)$ minimizing the functional

$$v(y) = \int_a^b F(x, y, y')dx + f\left[(y(a), y(b)\right]$$

where F and f are known functions can be reduced to the solution of the Euler equation

$$F_y - \frac{d}{dx}F_{y'} = 0$$

subject to the condition of transversality, which in this case assumes the form

$$\left[F_{y'} - \frac{\partial f}{\partial y(a)}\right]_{x=a} = 0,$$

$$\left[F_{y'} - \frac{\partial f}{\partial y(b)}\right]_{x=b} = 0.$$

The problem of this type is called *Bolza problem* after Oskar Bolza (1857 – 1942).

Example 6. Find the extremals of the functional

$$v(y) = \int_0^3 4y'^2y^2dx + y^4(0) - 8y(3).$$

The Euler equation has the form

$$8y'^2y - 4\frac{d}{dx}\left(2y'y^2\right) = 8y'^2y - 8y^2y'' - 16yy'^2 = 0,$$

or

$$y''y + y'^2 = 0, \ y \neq 0,$$

whence on substitution $y' = z(y)$ we have

$$y'' = \frac{d}{dx}z(y) = \left(\frac{d}{dy}z(y)\right)\frac{dy}{dx} = z'z,$$

then we get the separable equation

$$z'zy + z^2 = 0,$$

or

$$\frac{dz}{z} = -\frac{dy}{y}, \ z \neq 0.$$

Integrating yields

$$z = \frac{C_1}{y}.$$

Returning to the original variables, we have the equation

$$y' = \frac{C_1}{y}.$$

Its solution is readily found to be

$$y = \sqrt{C_1 x + C_2}.$$

Two integration constants can be determined by the condition of transversality

$$\left[8y'y^2 - 4y^3(0)\right]_{x=0} = 0,$$

$$\left[8y'y^2 - 8\right]_{x=3} = 0,$$

or

$$(C_1 - C_2)\, C_2 = 0,$$

$$\frac{1}{2}\frac{C_1}{\sqrt{C_2 + 3C_1}}(3C_1 + C_2) - 1 = 0.$$

Solving this system we find two solutions

$$C_1 = \sqrt[3]{\frac{4}{3}},\; C_2 = 0$$

and

$$C_1 = 1,\; C_2 = 1.$$

Thus, an extremum can only be attained on the curves

$$y_1 = \sqrt[6]{\frac{4}{3}x^2},$$

$$y_2 = \sqrt{x+1}.$$

PROBLEMS

Find extremals of the following functionals.

1. $v(y) = \int_0^1 y'^2 dx + y^2(0) - 2y^2(1).$
2. $v(y) = \int_0^1 \left(y'^2 + y^2\right) dx - 2y(1)\sinh 1.$
3. $v(y) = \int_0^\pi \left(y'^2 + y^2 - 4y \sin x\right) dx + 2y^2(0) + 2y(\pi) - y^2(\pi).$
4. $v(y) = \int_0^e 2y'(xy' + y)dx + 3y^2(1) - y^2(e) - 4y(e).$
5. $v(y) = \int_0^1 e^{x+1}\left(y'^2 + 2y^2\right) dx + 2y(1)(y(0) + 1).$
6. $v(y) = \int_0^1 e^y y'^2 dx + 4e^{y(0)} + 32e^{-y(1)}.$
7. $v(y) = \int_0^{\pi/2} (y'^2 - y^2)dx + y^2(0) - y^2\left(\frac{\pi}{2}\right) + 4y\left(\frac{\pi}{2}\right).$

7.5 Euler-Poisson equation

Consider functionals dependent on higher-order derivatives

$$v[y(x)] = \int_{x_0}^{x_1} F(x, y, y', y'', \ldots, y^{(n)})dx \tag{1}$$

The function $y = y(x)$, which extremizing the functional (1) must be a solution of the equation

$$F_y - \frac{d}{dx}F_{y'} + \frac{d^2}{dx^2}F_{y''} + \cdots + (-1)^n \frac{d^n}{dx^n}F_{y^{(n)}} = 0.$$

This differential equation of order $2n$ is called the *Euler-Poisson equation* (Simon Deni Poisson (1781 – 1840) the French mathematician).

Example 1. Find the extremals of the functional

$$v[y(x)] = \int_{x_0}^{x_1} \left[2xy + (y''')^2\right] dx.$$

The Euler-Poisson equation is of the form

$$2x - \frac{d^3}{dx^3}(2y''') = 0,$$

or

$$y^{vi} - x = 0,$$

its general solution is

$$y = \frac{x^7}{7!} + C_5x^5 + C_4x^4 + C_3x^3 + C_2x^2 + C_1x + C_0.$$

Example 2. Determine the extremal of the functional

$$v[y(x)] = \int_0^{\pi/2} \left[(y'')^2 - 16y^2 + x^2\right] dx$$

that satisfies the conditions

$$y(0) = \frac{1}{2},\ y'(0) = 0,\ y\left(\frac{\pi}{4}\right) = 0,\ y'\left(\frac{\pi}{4}\right) = -1.$$

The Euler-Poisson equation is of the form

$$-32y + \frac{d^2}{dx^2}(2y'') = 0,$$

or

$$y^{iv} - 16y = 0.$$

The auxiliary equation is

$$\lambda^4 - 16 = 0$$

and the roots are $\lambda = \pm 2, \pm 2i$. The general solution is

$$y = C_1 e^{2x} + C_2 e^{-2x} + C_3 \sin 2x + C_4 \cos 2x.$$

Using the boundary conditions, we get

$$C_1 = C_2 = C_3 = 0,\ C_4 = \frac{1}{2}.$$

And the extremum can be achieved only on the curve

$$y = \frac{1}{2}\cos 2x.$$

To determine the extremum of the functional

$$v\left[y(x)\right] = \int_{x_0}^{x_1} F(x, y, y', y'', \ldots, y^{(n)})dx$$

under the conditions

$$\int_{x_0}^{x_1} G_i(x, y, y', y'', \ldots, y^{(n)})dx = l_i,\ i = 1,\ 2,\ \ldots,\ m$$

where l_i are constants, it is necessary to set up an *auxiliary functional*

$$L = \int_{x_0}^{x_1}\left(F + \sum_{i=1}^{m}\lambda_i G_i\right)dx$$

and to write the Euler-Poisson equations for it.

For the extremum of a functional dependent on any number of functions:

$$v\left[y(x)\right]$$

$$= \int_{x_0}^{x_1} F(x, y_1, y_1', y_1'', \ldots, y_1^{(n_1)}, y_2, y_2', \ldots, y_2^{(n_2)}, \ldots, y_m, y_m', \ldots, y_m^{(n_m)})dx$$

the necessary condition takes the form

$$F_{y_i} - \frac{d}{dx}F_{y_i'} + \cdots + (-1)^{n_i}\frac{d^{n_i}}{dx^{n_i}}F_{y_i}^{(n_i)} = 0,\ i = 1, 2, \ldots, n.$$

PROBLEMS

Find the extremals of the following functionals.

1. $v(y) = \int_0^1 \left(y^2 + 2y'^2 + y''^2\right)dx;$
   $y(0) = 0,\ y(1) = 0,\ y'(0) = 1,\ y'(1) = \sinh 1.$
2. $v(y) = \int_0^1 \left(y'^2 + y''^2\right)dx;$
   $y(0) = 0,\ y(1) = \sinh 1,\ y'(0) = 1,\ y'(1) = \cosh 1.$

3. $v(y) = \int_0^1 (y'')^2 dx;$
$y(0) = y'(0) = 0,\ y(1) = 0.5,\ y'(1) = 1.$

4. $v(y) = \int_0^1 y'' dx;\ y(0) = y(1) = y'(1) = 0,\ y'(0) = 1.$

5. $v(y) = \int_0^1 \left(48y - y''^2\right) dx;\ y(0) = y'(0) = 0,\ y(1) = 1,\ y'(1) = 4.$

6. $v(y) = \int_0^1 (y'' - 24xy)\, dx;\ y(0) = y'(0) = 0,\ y(1) = \frac{1}{5},\ y'(1) = 1.$

7. $v(y) = \int_0^{\pi/2} \left(y''^2 - y'^2\right) dx;$
$y(0) = y'(0) = 0,\ y\left(\frac{\pi}{2}\right) = \frac{\pi}{2},\ y'\left(\frac{\pi}{2}\right) = 0.$

8. $v(y) = \int_0^1 e^{-x} y''^2 dx;$
$y(0) = 0,\ y'(0) = 1,\ y(1) = e,\ y'(1) = 2e.$

9. $v(y) = \int_0^1 (x+1)^3 y''^{\,2} dx;$
$y(0) = 1,\ y'(0) = -1,\ y(1) = \frac{1}{2},\ y'(1) = -\frac{1}{4}.$

10. $v(y) = \int_0^1 y'''^2 dx;$
$y(0) = y'(0) = y''(0) = 0,\ y(1) = 1,\ y'(1) = 4,\ y''(1) = 12.$

11. $v(y) = \int_0^1 \left(y'''^2 - y''^2\right) dx;$
$y(0) = y''(0) = 0,\ y'(0) = 1,\ y(1) = y''(1) = \sinh 1,\ y'(1) = \cosh 1.$

12. $v(y) = \int_0^\pi \left(y'''^2 - y''^2\right) dx;$
$y(0) = y'(0) = y''(0) = 0,\ y(\pi) = \pi,\ y'(\pi) = 2,\ y''(\pi) = 0.$

13. $v(y) = \int_0^\pi \left(y'''^2 - y'^2\right) dx;$
$y(0) = y'(0) = y''(0) = 0,\ y(\pi) = y''(\pi) = \sinh\pi,\ y'(\pi) = \cosh\pi + 1.$

14. $v(y) = \int_0^\pi y''^2 dx;$
$y(0) = y(\pi) = 0,\ \int_0^\pi y'^2 dx = 1.$

15. $v(y) = \int_0^\pi y''^2 dx;$
$y(0) = y(\pi) = 0,\ \int_0^\pi y^2 dx = 1.$

7.6 Ostrogradsky equation

Consider functionals dependent on the functions of several independent variables

$$v\left[z(x,y)\right] = \int\int_D F\left(x, y, z, \frac{\partial z}{\partial x}, \frac{\partial z}{\partial y}\right) dx dy$$

the values of the function $z(x,y)$ are given on the boundary C of domain D.

The extremizing surface $z = z(x,y)$ is a solution of the *Ostrogradsky equation* after the Russian mathematician M.V.Ostrogradsky (1801 – 1861).

$$F_z - \frac{\partial}{\partial x}F_p - \frac{\partial}{\partial y}F_q = 0,$$

where

$$p = \frac{\partial z}{\partial x},$$
$$q = \frac{\partial z}{\partial y}.$$

Example 1. Write the Ostrogradsky equation for the functional

$$v\left[z(x,y)\right] = \int\int_D \left[\left(\frac{\partial z}{\partial x}\right)^2 - \left(\frac{\partial z}{\partial y}\right)^2\right] dxdy.$$

The Ostrogradsky equation is of the form

$$\frac{\partial}{\partial x}\left(z\frac{\partial z}{\partial x}\right) - \frac{\partial}{\partial y}\left(z\frac{\partial z}{\partial y}\right) = 0,$$

or

$$\frac{\partial^2 z}{\partial x^2} - \frac{\partial^2 z}{\partial y^2} = 0.$$

For the functional

$$v\left[z(x_1, x_2, \ldots, x_n)\right]$$
$$= \int\int\cdots\int_D F(x_1, x_2, \ldots, x_n, z, p_1, p_2, \ldots, p_n)dx_1dx_2\cdots dx_n,$$

where

$$p_i = \frac{\partial z}{\partial x_i}$$

the necessary condition for an extremum is the following Ostrogradsky equation

$$F_z - \sum_{i=1}^{n}\frac{\partial}{\partial x_i}F_{p_i} = 0.$$

Example 2. Write the Ostrogradsky equation for the functional

$$u[z(x,y,z)]$$
$$= \int\int\int_D \left[\left(\frac{\partial u}{\partial x}\right)^2 + \left(\frac{\partial u}{\partial y}\right)^2 + \left(\frac{\partial u}{\partial z}\right)^2 + 2uf(x,y,z)\right] dxdydz.$$

The Ostrogradsky equation is of the form

$$2f - \frac{\partial}{\partial x}\left(2\frac{\partial u}{\partial x}\right)^2 - \frac{\partial}{\partial y}\left(2\frac{\partial u}{\partial y}\right)^2 - \frac{\partial}{\partial z}\left(2\frac{\partial u}{\partial z}\right)^2 = 0,$$

or

$$\frac{\partial^2 u}{\partial x^2} + \frac{\partial^2 u}{\partial y^2} + \frac{\partial^2 u}{\partial z^2} = f.$$

The function $u(x, y)$ that extremize the functional

$$L = \int\int_D F(x, y, u, u_x, u_y, u_{xx}, u_{xy}, u_{yy}, \ldots)dxdy$$

satisfies the following Ostrogradsky equation

$$F_u - \frac{\partial}{\partial x}F_{u_x} - \frac{\partial}{\partial y}F_{u_y} + \frac{\partial^2}{\partial x^2}F_{u_{xx}} + 2\frac{\partial^2}{\partial x \partial y}F_{u_{xy}}$$

$$+\frac{\partial^2}{\partial y^2}F_{u_{yy}} - \ldots + (-1)^n \frac{\partial^n}{\partial y^n}F_{u_{yy\ldots y}} = 0.$$

PROBLEMS

Write the Ostrogradsky equation for the functionals.

1. $\int_D\int \left[\left(\frac{\partial z}{\partial x}\right)^4 + \left(\frac{\partial z}{\partial y}\right)^4 + 12zf\right] dxdy.$
2. $\int_D\int e^{z_y} \sin z_y dxdy.$
3. $\int_D\int \left(\frac{\partial^2 z}{\partial x^2} + \frac{\partial^2 z}{\partial y^2}\right)^2 dxdy.$
4. $\int_D\int \left[\left(\frac{\partial z}{\partial x}\right)^2 + \left(\frac{\partial z}{\partial y}\right)^2\right] dxdy.$
5. $\int\int\int\int \left(H_x^2 + H_y^2 + H_z^2 - E_x^2 - E_y^2 - E_z^2\right) dxdydzdt$
where
$H_x = \frac{\partial A_z}{\partial y} - \frac{\partial A_y}{\partial z}$ $H_y = \frac{\partial A_x}{\partial z} - \frac{\partial A_z}{\partial x}$ $H_z = \frac{\partial A_y}{\partial x} - \frac{\partial A_x}{\partial y}$
$E_x = -\frac{\partial A_x}{\partial t} - \frac{\partial \phi}{\partial x}$ $E_y = -\frac{\partial A_y}{\partial t} - \frac{\partial \phi}{\partial y}$ $E_z = -\frac{\partial A_z}{\partial t} - \frac{\partial \phi}{\partial z}.$
6. $\int\int G\left(\sqrt{u^2 + v^2}\right) dxdy,\ u = \frac{\partial \varphi}{\partial x}, v = \frac{\partial \varphi}{\partial y}.$
7. $\int\int\int \sqrt{1 + \left(\frac{\partial u}{\partial x}\right)^2 + \left(\frac{\partial u}{\partial y}\right)^2 + \left(\frac{\partial u}{\partial z}\right)^2}\, dxdydz.$
8. Show that a function that minimizes the integral

$$\int\int \left(U_{xx}^2 + U_{xx}U_{yy} + U_{yy}^2\right) dxdy$$

is a solution of the biharmonic equation

$$U_{xxxx} + 2U_{xxyy} + U_{yyyy} = 0.$$

Chapter 8

ANSWERS TO PROBLEMS

8.1 Separable equations

1. $3y^2 - 2x^3 = C;\ y \neq 0.$
2. $3y^2 - 2\ln|1+x^3| = C;\ 1 + x^3 \neq 0;\ y \neq 0.$
3. $\frac{1}{y} - \cos x = C,\ y \neq 0$; also $y = 0$.
4. $y = \sin(\ln|x| + C)$ if $x \neq 0$ and $|y| < 1$; also $y = \pm 1$.
5. $3y + y^3 - x^3 = C.$
6. $\ln|x| - (y^2+1)^{1/2} = C$, if $x \neq 0$; also $x = 0$.
7. $y(\ln|1-x^2| + 1) = 1.$
8. $y = (x-2)^3$; also $y = 0$.
9. $y(x+1) = 1.$
10. $y^2 - 2 = Ce^{1/x}.$
11. $\left(Ce^{-x^2} - 1\right) y = 2$; also $y = 0$.
12. $e^{-z} = 1 + Ce^t.$
13. $y^2 - 2y = x^3 + 2x^2 + 2x + 3.$
14. $y^2 = 1 + 2\ln\left(\frac{1+e^x}{2}\right).$
15. $\frac{(x-1)(y+1)}{xy} = C$; also $x = 0\ (y \neq 0)$; $y = 0\ (x \neq 0)$.
16. $x^2 + y^2 + x^4 + y^4 = C^2.$
17. $\sqrt{x} + \sqrt{y} = C.$
18. $\arcsin x + \arccos y = C.$
19. $x^2 - \sqrt{1-y^2} = C,\ y \neq \pm 1$; also $y = \pm 1$
20. $y = 1 + C(x+1);\ x \neq -1$; also $x = -1,\ y \neq 1$.
21. $e^y - e^x = C.$
22. $\sqrt{y} - \sqrt{x} = C,\ (x > 0, y > 0)$; also $y = 0\ (x > 0)$; $x = 0\ (y > 0)$.
23. $2\sqrt{y} = \ln|x| + C$; also $y = 0\ (x \neq 0)$.
24. $z = -\log|C - 10^x|.$
25. $x^2 + t^2 - 2t = C.$
26. $\cot\frac{y-x}{2} = x + C,\ y - x = 2\pi k\ , k = 0, \pm 1, \pm 2, \ldots$
27. $2x + y - 1 = Ce^x.$

28. $x + 2y + 2 = 0$.
29. $y = 2x - 1 + Ce^{-x}$.
30. $y = x - 1 + 2k\pi$; also $y = x - 1 - 2\operatorname{arccot}(C - x) + 2\pi n$.
31. $\tan(x + y) = x + C$.
32. $(2x + y + 1)^{1/2} - \ln\left[1 + (2x + y + 1)^{1/2}\right] = x + C$.
33. $x + y + 1 = \tan(x + C)$.
34. $\dfrac{1}{x} + \dfrac{1}{y} + \left|\dfrac{y}{x}\right| = C$.
35. $2e^y - 2e^{2x} + 2\arctan y + \ln(1 + y^2) = C$.
36. $y_1^2 + y_2^2 + y_3^2 = C_1$, $\dfrac{y_1^3}{3} + \dfrac{y_2^4}{4} + y_2 + 2y_3 - \dfrac{y_3^2}{2} = C_2$.
37. $y_1^3 - \dfrac{3}{4}y_2^4 + 2y_3^2 + y_5^5 = C_1$, $\sin y_1 + y_2 \arctan y_2 - \dfrac{1}{2}\ln(1 + y_2^2)$ $- \ln|\cos y_3| - \dfrac{1}{2}e^{-y_4^2} = C_2$.
38. $\displaystyle\int \frac{y_1 e^{-y_1} dy_1}{1 + y_1} - \frac{y_2^3}{2} + \ln|y_3| + y_3 = C_1$, $\dfrac{y_3^3}{3} + \dfrac{1}{2}\arctan y_2 + \ln|y_3| = C_2$.
39. $y_1 + y_2 = C_1$, $y_2 = 0$.
40. $e^{y_1 + 2y_2} - e^{y_1 - 3y_2} + y_3 + y_4 = C_1$, $y_1 + 5y_2 + 6y_3 + 7y_4 = C_2$, $y_3 y_4 = C_3$.

8.2 Homogeneous equations

1. $\ln(x^2 + y^2) = C - 2\arctan\dfrac{y}{x}$.
2. $x(y - x) = Cy$; also $y = 0$.
3. $x = \pm y\sqrt{\ln|Cx|}$; also $y = 0$.
4. $y = C\exp\dfrac{y}{x}$.
5. $y^2 - x^2 = Cy$; also $y = 0$.
6. $\sin\dfrac{y}{x} = Cx$.
7. $y = -x\ln|\ln|Cx||$.
8. $\ln\left|\dfrac{x + y}{x}\right| = Cx$.
9. $\ln|Cx| = \cot\left(\dfrac{1}{2}\ln\dfrac{y}{x}\right)$; also $y = x\exp\{2\pi k\}$, $k = 0, \pm 1, \pm 2, \ldots$
10. $x\ln|Cx| = 2\sqrt{xy}$; also $y = 0$; and $x = 0$.
11. $\arcsin\dfrac{y}{x} = \ln|Cx| \operatorname{sign} x$, $y = \pm x$.
12. $\arctan\dfrac{y}{x} = \ln\sqrt{x^2 + y^2} + C$.
13. $x + y = 0$.
14. $y\ln|y| + x = Cy$.

15. $y = x\ln|Cy|$.
16. $y = x\tan(\ln|Cx|)$.
17. $y(y-2x)^3 = C(y-x)^2$.
18. $Cy^3 = x^2 - y^2$.
19. $y = x\sqrt{\ln x^2}$.
20. $2Cy = C^2x^2 + 1$
21. $\sqrt{\frac{x}{y}} + \ln|y| = C$.
22. $y = xe^{1+Cx}$.
23. $y = 1 - x - e^{-x}$; also $y = \frac{x^2}{2}$.
24. $y(C-x) = C^2$; also $y = 4x$.
25. $y = Cx$; also $y = Cx^2$
26. $x^2 - 2Cy = C^2$.
27. $y^2 = C^2 + 2Cx$
28. $y = Cx^{-2} - \frac{1}{3}x,\ x \neq 0$.
29. $x = y\left(C + \ln|y|\right),\ y \neq 0$.
30. $y = Cx^2 - x,\ x \neq 0$; also $x = 0,\ y \neq 0$
31. $|y - x + 3| = C|x + y + 1|^3$.
32. $|y + x + 4|(y + 4x + 13)^2 = C$.
33. $-2\frac{x-2}{x+y-3} = \ln|C(x+y-3)|$.
34. $\ln|y+2| + 2\arctan\frac{y+2}{x-1} = C$.
35. $2x + y - 1 = Ce^{2y-x}$.
36. $(y - x + 2)^2 + 2x = C$.
37. $(y - x + 5)^5(x + 2y - 2) = C$.
38. $(y+2)^2 = C(x+y-1)$; also $y = 1 - x$.
39. $\ln\left|\frac{y+x}{x+3}\right| = 1 + \frac{C}{x+y}$.
40. $(x + y - 1)^3 = C\,(x - y + 3)$.
41. $(y - x - 1)\exp\left(\frac{1}{3}\frac{y-2}{x-1}\right) = C$.
42. $x^2 = (x^2 - y)\ln|Cx|$; also $y = x$.
43. $x = -y^2\ln|Cx|$; also $y = 0$.
44. $y^2\exp\left(-\frac{1}{xy}\right) = C$; also $x = 0$ and $y = 0$.
45. $(2\sqrt{y} - x)\ln|C(2\sqrt{y} - x) = x,\ 2\sqrt{y} = x$.
46. $1 - xy = Cx^3(2 + xy)$; also $xy = -2$.
47. $2\sqrt{\frac{1}{xy^2}} - 1 = -\ln|Cx|$; also $xy^2 = 1$ and $y = 0$

48. $\arcsin \frac{y^2}{|x^3|} = \ln |Cx^3|$; also $|x^3| = y^2$.
49. $x^2 y \ln |Cy| = 1$; also $y = 0$.
50. $x = Cy \exp\left(\frac{1}{xy}\right)$.
51. $x^2 e^{x^2 y^2} = Cy^2$
52. $x^2 = Cy^2 - y^4$.
53. $xy^2 \sin(\ln |Cx|) = 1$.

8.3 Exact equations

1. $x^2 - 3x^3y^2 + y^4 = C$.
2. $4y \ln x + y^4 = C$.
3. $x^2 + 3x + y^2 - 2y = C$.
4. not exact.
5. $3x^3 + xy - x - 2y^2 = C$.
6. $x^2y^2 + 2xy = C$.
7. $x + \frac{x^3}{y^2} + \frac{5}{y} = C$.
8. $x^2 + \frac{2}{3}(x^2 - y)^{3/2} = C$.
9. $x - y^2 \cos^2 x = C$.
10. $x^3 + x^3 \ln |y| - y^2 = C$.
11. $x^2 + 1 = 2(C - 2x) \sin y$.
12. $e^x \sin y + 2y \cos x = C$; also $y = 0$
13. not exact.
14. $e^{xy} \cos 2x + x^2 - 3y = C$.
15. $y \ln x + 3x^2 - 2y = C$.
16. not exact.
17. $y^2 + x^2 = C$.
18. $xy - \frac{x^2}{2} = C$.
19. $x^3 + 3x^2y^2 + y^4 = C$.
20. $x^4 + y^4 - 6x^2y^2 = C$.
21. $\sqrt{x^2 + y^2} + \frac{y}{x} = C$.
22. $x^2 + y^2 + 2 \arctan \frac{y}{x} = C$.
23. $(e^y - 1)/(1 + x^2) = C$.
24. $x + ye^{x/y} = C$.
25. $y_1y_2 + y_1y_3 + y_2y_3 = C_1,\ 2y_1y_2 + y_3^2 = C_2$.
26. $y_1y_2y_3^{-1} = C_1,\ y_1^3 + y_2^3 + y_3^3 = C_2$.
27. $y_1^4y_2y_4 + y_1y_2y_3y_4^2 = C_1,\ y_1y_2 + y_2y_3 + y_1y_3y_4 = C_2$.

28. $\ln|x| + \dfrac{x}{y} = C$ also $y = 0$.
29. $x^3y + \dfrac{1}{2}x^2y^2 = C$.
30. $x^2 + 2\ln|y| - y^{-2} = C$; also $y = 0$.
31. $e^x \sin y + 2y\cos x = C$.
32. $xy^2 - (y^2 - 2y + 2)\,e^y = C$.
33. $\ln|x| + \ln|y| + y^2 - \dfrac{x}{y} = C$.
34. $x^2 + y^2 - \ln\left|\dfrac{x-y}{x+y}\right| = C$.
35. $xy - \dfrac{x^2}{2} = C$.
36. $(x^2 + y^2)\exp\left(2\arctan\dfrac{x}{y}\right) = C$.
37. $-\dfrac{1}{x} - xy + \dfrac{y^2}{2} = C$.
38. $-\dfrac{1}{xy} + \dfrac{x^2}{2} = C$.
39. $4x - \dfrac{y}{x} = C$.
40. $x + \arctan\dfrac{x}{y} = C$.
41. $\sqrt{1+y^2} = xy + C$
42. $2x^3y^3 - 3x^2 = C$.
43. $y^2 = x^2(C - 2y)$; also $x = 0$.
44. $x^2 + \ln|y| = Cx^3$; also $x = 0$.
45. $y\sin xy = C$.
46. $\dfrac{x^2}{2} + xy + \ln|y| = C$; also $y = 0$.
47. $1 - x = xy(\arctan y + C)$; also $x = 0,\ y = 0$.
48. $x + 2\ln|x| + \dfrac{3}{2}y^2 - \dfrac{y}{x} = C$; also $x = 0$.
49. $\sin\dfrac{y}{x} = Ce^{-x^2}$.
50. $\ln y - ye^{-x} = C$; also $y = 0$.
51. $\ln\left(\dfrac{x^2}{y^2} + 1\right) = 2y + C$; also $y = 0$.
52. $x^2y\ln|Cxy| = -1$; also $x = 0,\ y = 0$.
53. $x^2 + y^2 = y + Cx$; also $x = 0$.
54. $x^2y + \ln\left|\dfrac{x}{y}\right| = C$; also $x = 0,\ y = 0$.
55. $\ln\dfrac{x+y}{y} + \dfrac{y(x+1)}{x+y} = C$; also $y = 0,\ y = -x$.
56. $y = C\ln x^2y$.

57. $\sin y = -(x^2+1)\ln C(x^2+1)$
58. $xy\,(C-x^2-y^2)=1$; also $x=0,\ y=0$.
59. $y^2 = Cx^2e^{x^2y^2}$.
60. $\dfrac{\ln(x^2+y^2)}{\cot\alpha} - 2\arctan\dfrac{y}{x} = C$.
61. $x\sqrt{1+\dfrac{y^2}{x^2}} + \ln\left|\dfrac{y}{x}+\sqrt{1+\dfrac{y^2}{x^2}}\right| = C$; also $x=0$.
62. $x^3-4y^2 = Cy\sqrt[3]{xy}$; also $x=0,\ y=0$.
63. $x^2-y^2+2xy=C$.
64. $x-\dfrac{y}{x}=C$.
65. $x^2+y-\dfrac{x}{y}+\ln y = C$.
66. $x^2y^2+2\ln\left(\dfrac{x}{y}\right)=C$.
67. $x^2+y^2=C(y-1)^2$.
68. $x^2-y=Cx$.
69. $x^2y+2x=Cy$.
70. $xy\,(x^2+y^2)-1=Cxy$.
71. $xy-\ln y=C$.
72. $x^2-y^2-1=Cx$.

8.4 Linear equations

1. $y=x$.
2. $y=\dfrac{1}{5}x^5+\dfrac{C}{x^3}$.
3. $y=Ce^{-ax}+\dfrac{e^{mx}}{m+a}$, $m\neq -a$; also $y=Ce^{mx}+xe^{mx}$ if $m=-a$.
4. $y=x\,(1+x^2)+C\,(1+x^2)$.
5. $y=(x^2+C)\,e^{-x^2}$.
6. $y=(2x+1)\,(C+\ln|2x+1|)+1$
7. $y=\sin x + C\cos x$.
8. $y=e^x\,(\ln|x|+C)$; also $x=0$.
9. $xy=C-\ln|x|$.
10. $y=x(C+\sin x)$.
11. $y=Ce^{x^2}-x^2-1$.
12. $y=C\ln^2 x-\ln x$.
13. $xy=(x^3+C)\,e^{-x}$.
14. $x=y^2+Cy$; also $y=0$.
15. $x=(C-\cos y)\sin y$.
16. $x=2\ln y-y+1+Cy^2$.
17. $x=Cy^3+y^2$; also $y=0$.

18. $(y-1)^2x = y - \ln Cy$; also $y=0,\ y=1$.
19. $y = Ce^{-x} + x^2 - 2x + 4$
20. $y = (C+x)e^{3x} - \frac{e^{-3x}}{6}$.
21. $y = Cx^2 + \frac{x^5}{3}$.
22. $y = (C+x)e^{2x} - \frac{1}{2}$.
23. $y = e^{3x} + \frac{C}{x^2}$.
24. $y = Cx^{1/2} - x + \frac{x^3}{3}$.
25. $y = C\cos x - 2\cos^2 x$.
26. $y = \frac{2ax}{3} - \frac{1}{x} + Cx^{1/2}$.
27. $x^2 + y^2 = Cy + 1$.
28. $y^2 - 2x = Cy^3$.
29. $x = y^2\left(1 + C\exp\frac{1}{y}\right)$.
30. $y = Ce^x + x^2$.
31. $y = e^x - x$.
32. $y = Ce^{-3x} + \frac{9x^2 - 6x + 11}{27}$.
33. $y = Ce^{-2x} + \frac{4x^3 - 6x^2 + 6x - 3}{8}$.
34. $y = C\sin x + x^2$.
35. $y = Ce^{-\sin x} + \sin x - 1$.
36. $y = Ce^{-x} + \sin x$.
37. $y = \ln x - \frac{1}{\ln x}$.
38. $y = e^{x^2} + \sin x$.
39. $y = Ce^{x^2} + e^{x^2}\int e^{-x^2}\,dx$.
40. $y = C\ln x + x$.
41. $y = C\sqrt{x^2+1} + e^x$
42. $y = Ce^{-\sin x} + xe^{-\sin x}$.
43. $y = C\tan x + \sin x$.
44. $y = x^2 - 2$.
45. $y = \frac{C}{x} + \frac{1}{x}\int \frac{\sin x}{x}\,dx$.
46. $y = \frac{x}{2}\sin 2x + \cos^2 x$.
47. $y = Cx^2 - \frac{1}{x}$.
48. $y = C\tan\frac{x}{2} + x\tan\frac{x}{2}$.
49. $y = Ce^{-x^2} + (\sin x - x\cos x)e^{-x^2}$.

50. $y = C + \dfrac{\cos x}{x} + \dfrac{\sin x}{x}$.
51. $y = Ce^x - 2$.
52. $y = Ce^{-x} + e^x$.
53. $y = Cx - 1$.
54. $y = Ce^x - 1$.
55. $y = Ce^{-x} + x$.
56. $y(e^x + Ce^{2x}) = 1$; also $y = 0$.
57. $y(x+1)(\ln|x+1| + C) = 1$; also $y = 0$.
58. $y^{-3} = C\cos^3 x - 3\sin x\cos^2 x$; also $y = 0$.
59. $y = x^4 \ln^2 Cx$; also $y = 0$.
60. $y^{-2} = x^4(2e^x + C)$; also $y = 0$.
61. $y^2 = x^2 - 1 + C\sqrt{|x^2 - 1|}$.
62. $x^2(C - \cos y) = y$; also $y = 0$.
63. $xy(C - \ln^2 y) = 1$.
64. $y = (Ce^{2x} - e^x)^2$.
65. $y = (Ce^{-x} + x^2 - x + 2)^{-1}$.
66. $y^{-1} = Ce^x + x + 1$.
67. $y = (Ce^{-x} + 1)^2$, $C \geq e^{-x}$; also $y = 0$.
68. $(1 + Cx + \ln x)y = 1$.
69. $y^2 + 2x^2y^2 + Cy^2e^{x^2} = 2$.
70. $a^2y^3 = Ce^{ax} - a(x+1) - 1$.
71. $ny^n = Ce^{-nx/a} + nx - a$.
72. $x = Ce^{-y} - y^2 + 2y - 2$.
73. $\left(2 - y^2 + Ce^{-y^2/2}\right)x = 1$.
74. $x = Cy^{-2} + y^4$.
75. $x = Cy + \dfrac{y^3}{2}$.
76. $x + \arctan\dfrac{x}{y} = C$.
77. $y = Ce^{2x/y} - \dfrac{2x + 3y}{4y}$.
78. $x^2y^2 + 2\ln\left|\dfrac{x}{y}\right| = C$.
79. $x^2 + y^2 = Cy^2e^{2x}$.
80. $3x(x^2 + y^2)^2 = 6x^2y + 2y^3 + 3Cx^3$.
81. $y = \dfrac{2}{x} + \dfrac{4}{Cx^5 - x}$; also $y = \dfrac{2}{x}$.
82. $y = \dfrac{1}{x} + \dfrac{1}{Cx^{2/3} + x}$; also $y = \dfrac{1}{x}$.
83. $y = x + \dfrac{x}{x + C}$; also $y = x$.
84. $y = x + 2 + \dfrac{4}{Ce^{4x} - 1}$; also $y = x + 2$.

85. $y = e^x - \dfrac{1}{x+C}$; also $y = e^x$.

86. $y = 1 + \left(Ce^{3x} - \dfrac{1}{3}\right)^{-1}$; also $y = 1$.

87. $y = \dfrac{1}{x} + \dfrac{2x}{x^2+2C}$;also $y = x^{-1}$.

88. $y = \sin x + \left(C\cos x - \dfrac{1}{2}\sin x\right)^{-1}$;also $y = \sin x$.

89. $y = \dfrac{1}{2x} + \dfrac{1}{x(C+\ln|x|)}$;also $y = \dfrac{1}{2x}$.

90. $y = x + \dfrac{1}{1+Cx}$;also $y = x$.

91. $y = -\dfrac{1}{x} + \dfrac{1}{Cx - x\ln|x|}$; also $y = -x^{-1}$.

92. $y\,(x^4 + Cx) = 4x^3 + C$; also $y = x^{-1}$.

93. $y = \dfrac{2}{x} + \dfrac{4}{x\ln|x| + Cx}$; also $y = 2x^{-1}$.

94. $y = \dfrac{2x^3 - C}{x(C+x^3)}$; also $y = 2x^{-1}$.

95. $\dfrac{1}{2\sqrt{2}}\ln\dfrac{x(xy-1)+\sqrt{2}}{x(xy-1)-\sqrt{2}} + \dfrac{1}{x} = C$.

96. $\dfrac{y\left(\sqrt[3]{x} + 3\sqrt[3]{x^2}\right)+3}{y\left(\sqrt[3]{x} - 3\sqrt[3]{x^2}\right)+3} = Ce^{6\sqrt[3]{x}}$.

97. $y = \dfrac{1}{x} + \dfrac{1}{x^2}\tan\left(\dfrac{1}{x} + C\right)$.

98. $\dfrac{x^{2/3} + yx^{5/3} - 6}{3\sqrt{2}(1+xy)x^{1/3}} = \tan\left(\dfrac{3\sqrt{2}}{x^{1/3}} + C\right)$.

99. $y = \dfrac{x^2}{u-3}$, $u = \dfrac{x^2}{v+1}$, $v = x\cot(C-x)$.

100. $y = \dfrac{t}{u-7}$, $x = t^{3/2}$, $u = \dfrac{t}{v+5}$, $v = \dfrac{t}{w-3}$, $w = \dfrac{t}{z+1}$,
$z = \sqrt{t}\cot\left(C - \sqrt{t}\right)$.

101. $y = x\sqrt{3}\dfrac{Ce^{2x\sqrt{3}}+1}{Ce^{2x\sqrt{3}}-1}$.

102. $y = \dfrac{x^2}{x\tanh(C+x)-1} - 3$.

8.5 Nonlinear equations

1. $y = Ce^x$; $\qquad y = Ce^{-x}$.
2. $y = C + x^2$; $\qquad y = C - x^2$.

3. $y=\left[1+(C-x)^2\right]^{-1}; \qquad y=1; \qquad y=0.$
4. $y=\pm(C+x)^2; \qquad y=\pm(C-x)^2; \qquad y=0.$
5. $y=\frac{1}{4}(C+x)^2; \qquad y=Ce^x.$
6. $y=(C+x)^{3/2}; \qquad y=-(C+x)^{3/2}; \qquad y=0.$
7. $y=C; \qquad y=C-\sqrt{x}; \qquad y=C+\sqrt{x}.$
8. $y=\left[C+(x-1)^{4/3}\right]^{3/4}.$
9. $y=C; \qquad y=C+\frac{1}{2}x^2.$
10. $y=\pm\left[1-(C+x)^2\right]^{1/2}; \qquad y=\pm 1.$
11. $y=\ln|x+\sqrt{x^2-1}|+C, \quad x>1; \qquad y=-\ln|x+\sqrt{x^2-1}|+C,$ $x<-1.$
12. $y=C\exp\left(x+2e^{x/2}\right); \qquad y=C\exp\left(x-2e^{x/2}\right); \qquad y=0.$
13. $y=C+\frac{1}{2}x^2; \qquad x=-C+\frac{1}{2}y^2.$
14. $2y^2+2(x-C)^2=C^2; \qquad y=\pm x.$
15. $y^2=\frac{1}{2}C^2-(x+C)^2; \qquad y=\pm x.$
16. $y=\cos(x-C); \qquad y=\pm 1.$
17. $y=C-x^2; \qquad y=C+2x^2.$
18. $y=\sec^2\left(\frac{x+C}{2}\right); \qquad y=0; \qquad y=1.$
19. $y=\frac{C}{x^3}\pm 2\sqrt{\frac{x}{7}}.$
20. $y=C+\frac{1}{2}x^2; \qquad y=(C\pm x)^2.$
21. $y=\pm(x+C)^{3/2}; \qquad y=0.$
22. $x=3p^2+2p+C, \quad y=p^2(1+2p).$
23. $y=\frac{1}{2}x^2+Cx+C^2; \qquad y=\frac{1}{4}x^2.$
24. $x=(p+1)e^p+C, \quad y=p^2e^p; \qquad y=0.$
25. $x=2\arctan p+c, \quad y=\ln(1+p^2); \qquad y=0.$
26. $x=\pm\frac{1}{2p\sqrt{\ln(Cp)}}, \quad y=\mp\left(\sqrt{2\ln(Cp)}-\frac{1}{\sqrt{2\ln(Cp)}}\right).$
27. $x=\sin p+C, \quad y=p\sin p+\cos p; \qquad y=1.$
28. $y=\frac{1}{2}Cx^2+C^2; \qquad y=-\frac{1}{16}x^4.$
29. $y=C(x+1)+e^x; \qquad y=-\frac{1}{4}(x+1)^2.$
30. $x=\left(C\sqrt{p}-1\right)p^{-2}, \quad y=xp-x^2p^3; \qquad y=0.$
31. $y=\left[C+(x-1)^{4/3}\right]^{3/4}.$

32. $x = p\left(1 + p^2\right), \quad y = \left(\frac{3}{4}p^2 + \frac{1}{2}\right)p^2 + C.$

33. $x = \sin p + \ln p, \quad y = p + \cos p + p \sin p + C.$

34. $y = C\left(x - C\right)^2; \qquad y = \frac{4}{27}x^3; \qquad y = 0.$

35. $x = Cp^2e^p, \quad y = C\left(p+1\right)e^p; \quad y = 0.$

36. $x = \frac{2p}{p^2-1}, \quad y = \frac{2}{p^2-1} - \ln\left|p^2 - 1\right| + c.$

37. $y = \frac{1}{C}\ln x - C^2; \qquad x = -3\left(\ln\sqrt{y}\right)^{2/3}.$

38. $y = Cx + C^{-2}; \qquad 4y^3 = 27x^2.$

39. $x = p + \sin p, \quad y = \frac{1}{2}p^2 + p\sin p + \cos p + C.$

40. $y = \ln\frac{2C}{1 + C^2x^2}; \qquad y = -\ln|x|.$

41. $y = Cx - C - 2.$

42. $y = x\left(1+p\right) + p^2, \quad x = Ce^{-p} - 2p + 2.$

43. $y = Cx + 2\sqrt{1 + C^2}; \qquad y^2 - x^2 = 2, \ y > 0.$

44. $y = Cx + \frac{1}{4}C^2; \qquad y = -x^2.$

45. $y = Cx + C^2; \qquad y = \frac{1}{4}x^2.$

46. $y = Cx + C^2; \qquad y = -\frac{1}{4}x^2.$

47. $y = Cx - \frac{1}{3}C^3; \qquad y = \pm\frac{2}{3}x^{3/2}.$

48. $y = C\frac{2}{p} + \frac{1}{3}p^2, \quad x = C\frac{1}{p^2} + \frac{2}{3}p; \qquad y = 0.$

49. $x = 3p^2 + \frac{C}{p^2}, \quad y = 2p^3 + 2\frac{C}{p}; \qquad y = 0.$

50. $x = \frac{C}{\sqrt{p}} + \frac{2}{3}p, \quad y = -xp + p^2.$

51. $x = \frac{C}{\left(p-1\right)^2} + 2p + 1, \quad y = \frac{cp^2}{\left(p-1\right)^2} + p^2; \quad y = x - 2;$
$y = 0.$

52. $y = Cx^2 + \frac{1}{C}; \qquad y = \pm 2x.$

53. $y = \pm 2.$

54. $y = 4x.$

55. $y = 0; \qquad y = \frac{2}{9}x^2.$

56. $y = \frac{x^2}{4}.$

57. $x = e^p - p^2, \quad y = pe^p - e^p - \frac{2}{3}x^3 + C.$

58. $y = \left(x - C\right)^{3/2} + C; \qquad y = x - \frac{4}{27}.$

59. $x = \dfrac{C}{\sqrt{p}} - \dfrac{1}{p}, \quad y = \ln p - C\sqrt{p} + 1.$

60. $x = 2p - \ln p, \quad y = p^2 - p + C.$

61. $(y-x)^2 = 2C(x+y) - C^2; \quad y^{2/3} - x^{2/3} = C; \qquad y = 0.$

62. $y = \dfrac{1}{16}Cx^3 - \dfrac{1}{Cx}.$

63. $x = 2\sqrt{1+p^2} - \ln\left(1+\sqrt{1+p^2}\right) + \ln(Cp), \quad y = p\sqrt{1+p^2}.$

64. $x = 4p^3 - \ln(Cp), \quad y = 3p^4 - p; \qquad y = 0.$

65. $x = \dfrac{C}{p^{2/3}} - \dfrac{2}{5}p, \quad y = -2Cp^{1/3} - \dfrac{1}{5}p^2.$

66. $x = \dfrac{\sin p}{p} + \dfrac{\cos p}{p^2} + \dfrac{C}{p^2}, \quad y = \sin p + 2\dfrac{\cos p}{p} + 2\dfrac{C}{p}; \qquad y = 0.$

8.6 Applications in physics

1. The falling time T is a root of the equation:

$$\text{a)} \quad z(T) = h - \frac{mg}{k}\left(T - \frac{m}{k}\left(1 - e^{-kT/m}\right)\right) = 0,$$

if $k^2h/m^2g >> 1$, we obtain $T \approx kh/mg$;

$$\text{b)} \quad z(T) = h - \frac{m}{k}\ln\cosh\left(\sqrt{\frac{gk}{m}}T\right) = 0,$$

then

$$T + \sqrt{\frac{m}{kg}}\ln\left(\exp\frac{kh}{m} + \sqrt{\left(\exp\frac{2kh}{m} - 1\right)}\right),$$

if $h >> m/k$, we obtain $T \approx h(k/mg)^{1/2}$.

2.

$$v(t) = \frac{k_1}{2k_2}\left(1 - a\tanh\left(\frac{k_1 at}{2m} + C\right)\right) < 0,$$

$$z(t) = h + \frac{k_2}{2k_1}\left[t - \frac{2m}{k_1}\left(\ln\cosh\left(C + \frac{tk_1a}{2m}\right) - \ln\cosh C\right)\right],$$

where $a = (1 + 4mgk_2/k_1^2) > 1$ and $C = \operatorname{arctanh}(1/a)$.

If $t >> (2m/k_1a)$ we obtain $v_{max} = v(t\to\infty) = k_1(a-1)/2k_2$.

$$\lim_{k_1\to 0} v_{max} = \sqrt{\frac{mg}{k_2}}, \qquad \lim_{k_2\to 0} v_{max} = \frac{mg}{k_1}.$$

The falling time T is a root of the equation $z(T) = 0$.

3. a) $T = \frac{m}{k} \ln\left(1 + \frac{kv_0}{mg}\right), \quad h = \left(\frac{m}{k}\right)^2 g\left[\frac{kv_0}{mg} - \ln\left(1 + \frac{kv_0}{mg}\right)\right]$

b) $T = \sqrt{\frac{m}{gk}} \arctan\left(\sqrt{\frac{k}{mg}} v_0\right),$

$h = -\frac{m}{k} \ln \cos \arctan\left(\sqrt{\frac{k}{mg}} v_0\right) = \frac{m}{2k} \ln\left(1 + \frac{kv_0}{mg}\right).$

4. Using the results of the problems 3 and 1 we obtain:

a) $T = T_u + T_d$, where $T_u = (m/k)\ln(1 + kv_0/mg)$ and T_d is a root of the equation

$$T_d - \frac{m}{k}\left(1 - \exp\left(-\frac{k}{m}T_d\right)\right) = \frac{m}{k}\left[\frac{kv_0}{mg} - \ln\left(1 + \frac{kv_0}{mg}\right)\right].$$

If $kv_0/mg >> 1$ we find $T \approx v_0/g + m/k$; if $k \to 0$ we find $T = 2v_0/g$.

b) $T = T_u + T_d$, where

$$T_u = \sqrt{\frac{m}{gk}} \arctan\left(v_0\sqrt{\frac{k}{mg}}\right), \quad T_d = \sqrt{\frac{m}{kg}} \ln\left(\sqrt{1 + \frac{kv_0^2}{mg}} + \sqrt{\frac{kv_0^2}{mg}}\right).$$

If $kv_0^2/mg >> 1$ we find

$$T \approx \sqrt{\frac{m}{gk}}\left(\frac{\pi}{2} + \ln 2 + \frac{1}{2}\ln \frac{kv_0^2}{mg}\right);$$

if $k \to 0$ we obtain $T = 2v_0/g$.

5. $$v(t) = v_0\left(1 + \alpha v_0^\alpha \frac{kt}{m}\right)^{-1/\alpha},$$

$$x(t) = \frac{m}{(\alpha - 1)\, kv^{\alpha-1}}\left[\left(1 + \alpha v_0^\alpha \frac{kt}{m}\right)^{1-1/\alpha} - 1\right].$$

If $\alpha \to -1/2$, then

$$v(t) = v_0\left(1 - \frac{kt}{2m\sqrt{v_0}}\right)^2, \quad x(t) = \frac{2k}{3v_0^{3/2}}\left[1 - \left(1 - \frac{kt}{2m\sqrt{v_0}}\right)^3\right],$$

where $0 \le t \le 2m\sqrt{v_0}/k$;

if $\alpha \to 0$, then

$$v(t) = v_0 e^{-kt/m}, \quad x(t) = \frac{v_0 m}{k}\left(1 - e^{-kt/m}\right) \quad t \ge 0;$$

if $\alpha \to 1$, then

$$v(t) = \frac{v_0}{1 + v_0 kt/m}, \quad x(t) = \frac{k}{m}\ln\left(1 + \frac{v_0 kt}{m}\right) \quad t \ge 0;$$

if $\alpha \to 2$, then

$$v(t) = v_0\left(1 + 2v_0^2\frac{k}{m}t\right)^{-1/2}, \quad x(t) = \frac{kv_0}{m}\left[\left(1 + 2v_0^2 t\frac{k}{m}\right)^{1/2} - 1\right],$$

$t \le 0$.

6. $v(t) = -x_0\omega\sin(\omega t)$, $\quad x(t) = x_0\cos(\omega t)$, $\quad \omega = \sqrt{\dfrac{k}{m}}$.

7. $x(t) = h\left(1 - \dfrac{\alpha t^2}{mh^4}\right)^{1/2}$, $\quad 0 \le t \le h^2\sqrt{\dfrac{m}{\alpha}} \equiv T$, where T is the falling time on the center.

8. a) $z(t) = H\left(1 - 5k\dfrac{r^2}{R^2}\sqrt{\dfrac{g}{2H}}t\right)$, $\quad T = \dfrac{1}{5k}\sqrt{\dfrac{2H}{g}}\dfrac{R^2}{r^2}$;

b) $t = \dfrac{1}{k}\sqrt{\dfrac{2H}{g}}\dfrac{R^2}{r^2}\left[\dfrac{8}{15} - \sqrt{\dfrac{z}{H}}\left(1 - \dfrac{2z}{3H} + \dfrac{z^2}{5H^2}\right)\right]$,

$T = \dfrac{8}{15k}\sqrt{\dfrac{2H}{g}}\dfrac{R^2}{r^2}$;

c) $t = \dfrac{8}{15k}\sqrt{\dfrac{R}{g}}\dfrac{R^2}{r^2}\left[2 - \sqrt{\dfrac{z}{2R}}\left(\dfrac{2z}{2R} - \dfrac{3z^2}{4R^2}\right)\right]$, $\quad T = \dfrac{16}{15k}\sqrt{\dfrac{R}{g}}\dfrac{R^2}{r^2}$;

d) $z(t) = H\left(1 - 3k\dfrac{r^2}{H^2}\sqrt{\dfrac{g}{2H}}t\right)^{2/3}$, $\quad T = \dfrac{1}{3k}\sqrt{\dfrac{2H}{g}}\dfrac{H^2}{r^2}$;

e) $t = \dfrac{1}{k}\sqrt{\dfrac{2H}{g}}\dfrac{H^2}{r^2}\left[\dfrac{2}{3} - \sqrt{\dfrac{z}{H}}\left(1 - \dfrac{z}{3H}\right)\right]$, $\quad T = \dfrac{2}{3k}\sqrt{\dfrac{2H}{g}}\dfrac{H^2}{r^2}$.

9. $z(t) = 2R\left[1 - \left(\dfrac{3k\pi r^2}{8RH}\sqrt{\dfrac{g}{R}}t\right)^{2/3}\right]$, $\quad T = \dfrac{8RH}{3k\pi r^2}\sqrt{\dfrac{R}{g}}$.

10. $I(t) = \dfrac{\mathcal{E}}{R}\left(1 - e^{-tR/L}\right)$.

11. $I(t) = -\dfrac{Q}{RC}e^{-t/RC}$.

12. $I(t) = I_0 e^{-tR/L}$.

13. $I(t) = -Q_0\omega\sin(\omega t)$, where $\omega^2 = 1/LC$.

14. $m_0 - m(t) + m_0\dfrac{V/v_0 - 1}{1-b}\ln\left|1 - \dfrac{(1-b)(m_0 - m(t))}{b\rho(V - v_0)}\right|$

$= -ksA(1-b)t$,

where $b = \dfrac{p_s - p_0}{A\rho}$, $s = \pi r^2$. If $V >> v_0$ and $p_0 \to p_s$ we obtain

$$m(t) = m_0 - \frac{V}{A}(p_s - p_0)\left[1 - \exp\left(-\frac{ksAt}{V}\right)\right].$$

15. $-\dfrac{3}{2}\ln\left|\dfrac{M-\mu}{M-\mu_0}\right| + \dfrac{1}{2}\ln\left|\dfrac{M^3 - m}{M^3 - m_0}\right|$

$$-\arctan \frac{\sqrt{3}M\left(\mu_0-\mu\right)}{M\left(\mu_0+\mu\right)+2\left(M^2+\mu\mu_0\right)}=k\pi\left(\frac{3R}{\pi\rho H}\right)^{2/3}\frac{A}{V}M^2t,$$

where $\mu=m^{1/3}$, $\mu_0=m_0^{1/3}$, $M=\left(m_0-\frac{V}{A}\left(p_s-p_0\right)\right)^{1/3}$, ρ is the density of the liquid.

16. $p\left(h\right)=p_0\exp\left(-\frac{\mu h}{RT\left(1+h/r\right)}\right)$, where r is the radius of earth.

17. $p'\left(r\right)=-kp\left(r\right)\int_0^r\frac{x^2}{r^2}p\left(x\right)dx$, where $k=4\pi\gamma\left(\frac{\mu}{RT}\right)^2$.

18. $T\left(t\right)=20+80\cdot 2^{-t/10}$, $T\left(t_0\right)=25^o$ at $t_0=$40 min.

19. $$\frac{1}{2}\ln\left|\frac{M^3}{M^3-\mu_0}\right|-\frac{3}{2}\ln\left|\frac{M}{M-m_0}\right|-\arctan\frac{\sqrt{3}\mu_0}{\mu_0+2M}$$
$$=4k\pi\left(\frac{3}{4\pi\rho}\right)^{2/3}\frac{A}{V}M^2T,$$

where $\mu_0=m_0^{1/3}$, $m_0=4\pi r_0^3\rho/3$,

$M=\left(m_0-\frac{V}{A}\left(p_s-p_0\right)\right)^{1/3}$, ρ is the density of the liquid.

20. $T\left(t\right)=\left(T_0+k\int_0^t T_{ext}\left(s\right)e^{ks}\,ds\right)e^{-kt}$.

21. $m\left(t\right)=m_0e^{-kt}$, $\quad k=\frac{\ln 2}{T_{1/2}}$, $\quad t\approx 970\cdot 10^6$ year.

22. $T\left(x\right)=\frac{kI_0t}{c_V}\frac{\cosh k\left(d-x\right)}{\sinh kd}$, where c_V is the heat capacity of the liquid.

23. $I\left(t\right)=I_0\cdot 2^{-t/35}$, $I\left(200\right)\approx 0.02\cdot I_0$.

8.7 Miscellaneous problems

1. $(2x+y-3)^2=C(6x+2y-5)$.
2. $x^2e^{x^2y^2}=Cy^2$.
3. $\ln(y-x-1)=C+\frac{x-2}{x-y+1}$.
4. $(x-y)(x+7y-4)^3=C$.
5. $y^2=x\ln Cy^2$.
6. $x^6+y^4=Cy^2$.
7. $3x^2+3y^2-3xy+3x-3y+1=C$.
8. $(y-4x-2)^3=C(x+2y+5)$.
9. $\ln(4x+8y+5)+8y-4x=C$.
10. $\sqrt{y}-x^2\sqrt{y^3}=C$.
11. $x^2-y^2+2xy-4x+8y=C$.
12. $x^2\left(\sqrt{x^4y^2+1}-x^2y\right)=C$.
13. $x^2\sin y+y^2\cos x+y=C$.

14. $\dfrac{x^4}{4} + xy(\ln y - x - 1) + \sin y = C.$
15. $(x+y)(x^2 - 7xy + y^2) = C.$
16. $3xy^2 + x^2y + 3y + x^2 = C.$
17. $x^2 \cos y + y^2 \cos x = C.$
18. $\sqrt{x^2+y^2+1} - \arctan \dfrac{y}{x} = C.$
19. $3x^2y + x^3y^3 = C.$
20. $xy^2 + 2x^2y - 2 = Cx.$
21. $e^x(x^2+y^2) = C.$
22. $xy + x + y = C(x+y)(x+y+2).$
23. $x^3 + xy + y^3 = C(x+y).$
24. $y = Ce^{xy}.$
25. $4xy + 2x^2y^2 \ln x + 1 = Cx^2y^2.$
26. $x^2 + y^2 + 2\arctan \dfrac{y}{x} = C.$
27. $x^2 + y^2 = C(y-1)^2.$
28. $x^2y^2 = \ln \dfrac{Cy^2}{x^2}.$
29. $x^6 - 3x^2y^4 - 2y^6 = C.$
30. $\tan y = Ce^{-x^2/2} + 1.$
31. $2e^{-y} = Ce^{-x} - e^x.$
32. $2\sqrt{1+y} = 2x(x + 5\ln x) + Cx.$
33. $\sqrt{1+y^2} = Ce^{-x} + x^2 - 2x + 3.$
34. $e^{-3y} = xe^x + Ce^x.$
35. $y^3(Ce^{\cos x} + 3) = 1.$
36. $\sin y = Ce^{\cos x} + 1.$
37. $y^{-3} = C\cos^3 x + 2\sin^3 x - 3\sin x.$
38. $xy(C + \ln^2 x) + 2 = 0.$
39. $y^{-3} = x^3\left(3\ln x^3 - \dfrac{1}{2}\right) + \dfrac{3}{2}x.$
40. $\dfrac{x}{x+y} = C\exp\dfrac{x(a-y)}{a(x+y)}.$
41. $2y = (y^2 - x^2)\ln C\dfrac{x+y}{x-y}.$
42. $x^4 - y^3 = Cxy^2.$
43. $xy + x + y = C(x+y+2)(x+y)..$
44. $y = \dfrac{Cx^2+3}{Cx^3+x}.$
45. $y\left(Ce^{2/x} - 1\right) = x\left(Ce^{2/x} + 1\right)$, $y_1(x) = x.$
46. $xy\left(C + \int e^{-2x^2/3}dx\right) = e^{-2x^2/3}$, $y_1(x) = x.$
47. $y = x + e^{x^2}\left(C - \int e^{x^2}dx\right)^{-1}$, $y_1(x) = x.$

48. $y = \dfrac{Cx^2 - x}{C - x^2}$.
49. $y = \dfrac{1}{Cx - x\ln x} - \dfrac{1}{x}$.
50. $y = \dfrac{1 - Cx^2}{C - x}$.
51. $2 - \left(1+\sqrt{5}\right)xy = \left[2 - \left(1-\sqrt{5}\right)xy\right]Cx^{-\sqrt{5}}$.
52. $y = \dfrac{1}{\cos x} + \dfrac{3\cos^2 x}{C - \cos^3 x}$, $y_1(x) = \dfrac{1}{\cos x}$.
53. $y = -\dfrac{1}{x^2} - \dfrac{Ce^{-2x}+1}{x\left(Ce^{-2x}-1\right)}$, $y_1(x) = \dfrac{1}{x} - \dfrac{1}{x^2}$.
54. $C(y-x)^2 + x^2 - 1 = 0$.
55. $\left(x^2 - y\right)^2 = C\left(x^2 + y^2\right)$.
56. $x^3 + xy^2 + 2y = C$.
57. $\displaystyle\int \frac{dz}{\phi(z)} = -\frac{1}{x} + C,\ z = \frac{y}{x}$.
58. $y = 2x\tan(2x + C) + 2,\ -\dfrac{\pi}{4} - \dfrac{C}{2} < x < \dfrac{\pi}{4} - \dfrac{C}{2}$.
59. $y = x + 1 + e^{(2x+x^2)/2}\left(C - \int e^{(2x+x^2)/2}dx\right)^{-1}$.
60. $y = x + \dfrac{1}{1 + Cx}$.
61. $y = \dfrac{1}{4}\left(e^{2x} - 1\right) + \dfrac{1}{2}x$.
62. $y = \left(Cx^{8/3} - 2x^2\right)^{3/2}$.
63. $y = 2e^x - 1$.

8.8 Reduction of order

1. $sp(s)\,p'^2(s) = 1;\qquad y'(x) = \left(C_1 \pm 3\sqrt{y}\right)^{2/3};$
 $x = C_2 - \dfrac{2}{3}C_1\left(C_1 \pm 3\sqrt{y}\right)^{1/3} + \dfrac{1}{6}\left(C_1 \pm 3\sqrt{y}\right)^{4/3}$.
2. $sp'(s) = p(s);\qquad y = C_1 e^{C_2 x}$.
3. $spp'' + 2p^2p' = 2sp'^2$.
4. $sp^2\left(p''p - p'^2\right) = 1$.
5. $sp^3p'\left(p''p + p'^2\right) = 1$.
6. $sp\left(p''p + p'^2 - 3pp'^2\right) = p'^5$.
7. $2sp' + p + p^3\left(p''p + p'^2\right)^4 = 0$.
8. $3sp\left(p''p + p'^2\right) + 4p^2\left(p'''p^2 + 4p''p'p + p'^3\right) = 1$.
9. $p' = 2(s+1);\quad y = -1 + C_1\tan\left(C_1x + C_2\right),\qquad y = C$.
10. $8pp' - p^2 + e^s = 0;\quad x + C_2 = \displaystyle\int\left(C_1e^{y/4} + \frac{1}{3}e^y 3\right)^{-1/2} dy$.
11. $s\left(p''p + p'^2\right) + p'p^2 = pp'^2$.

15. $s^2 + p^2 p'^2 = 1; \quad y'(x) = \pm\left(C \pm \left(y\sqrt{1-y^2} + \arcsin y\right)\right)^{1/2}.$

16. $s^2 - p^2 p'^2 = 1;$

$x + C_2 = \pm \int \left(y\sqrt{y^2-1} - \ln\left(y + \sqrt{y^2-1}\right) + C_1\right)^{-1/2} dy.$

17. $s - 2p^2\left(1 + p'^2\right) = 0.$

18. $z' - z + \cos x = 0; \quad y = C_1 + C_2 x + C_3 e^x + \frac{1}{2}(\sin x - \cos x).$

21. $y = C_1 + C_2 x + C_3 \int dx \int \exp\left(x^2\right) dx.$

22. $y = C_1 + C_2 e^x + \int e^x\, dx \int e^{-x} \sin x^2\, dx.$

25. $y = C_1 \pm \int \left(C_2 + \frac{2}{3}x^3\right)^{1/2} dx.$

26. $y = C_1 \ln x + C_2$.

27. $xz' + z\exp\left(\frac{z}{x}\right) = 0; \quad x = C_1 \exp\left(-\int \frac{du}{u + e^u}\right),$

$y = C_2 + \int xu(x)\, dx.$

28. $y = C_1 - \int \frac{dx}{\ln\left(C_2 x\right)}.$

32. $y'' = C\exp\left(3(x+1)e^{-x}\right).$

34. $y = \pm C_1\left(x + \sqrt{1+x^2}\right)^{C_2} \exp\left(C_2 x\left(x + \sqrt{1+x^2}\right)\right).$

35. $y = C_1\left(x + \sqrt{1+x^2}\right)^{C_2}$.

36. $y = C_2 x e^{C_1/x}.$

37. $y = C_2 e^{C_1 x^2}.$

38. $y = C_2 \frac{e^{C_1 x^3}}{x^{1/3}}$.

39. $y = C_1 x^{C_2}$.

40. $y = C_2 \exp\left(\frac{C_1 x}{2} + \frac{1}{2C_1 x}\right).$

41. $y = C_2 \exp\left[\cos\left(\ln C_1 x\right)\right].$

42. $y = C_2 \exp\left(\int z\, dx\right), \quad z' + z^2 = C_1 + \frac{x^2}{2}.$

44. $y = C_3 \exp\left(\frac{x^2}{2} - x + C_1 e^x + C_2 e^{2x}\right).$

45. $y = C_1 \exp\left(C_2 x + \int dx \int f(x)\, dx\right).$

46. $z' + g(x)\, z = f(x).$

47. $y = C_1 \left|1 - \frac{1}{1 + C_2 x}\right|^{C_2}; \qquad y = Ce^{1/x}; \qquad y = C.$

48. $|y| = C_2\left(\left|x + C_1\right|^{C_1^2} \left|x - \frac{1}{C_1}\right|\right)^{\frac{1}{1+C_1^2}}; \qquad y = C.$

49. $y = C_1 \cos\ln x + C_2 \sin\ln x + \frac{x^3}{10}.$

51. $y'^2 + 2\arctan\sinh y = C.$

52. $y'^2 y'' + \frac{e^y}{x} = C.$

53. $y'' + y^2 + \ln y = C.$
54. $y = C_1 x + C_2; \quad y = C_3 + C_2 \ln|x + C_1|.$
55. $\arctan y + C_1 x = C_2.$
56. $ye^{C_1 x} = C_2.$
57. $x \pm \int (C_1 + 2\int f(y)\,dy)^{-1/2}\,dy = C_2.$
58. $y = C_1 C_3^x + C_2 C_3^{-x}$.
59. $y = \pm\sqrt{C_1 + C_2 x}$.
60. $y' + y\dfrac{x}{1+x^2} = C_1.$
61. $y = C.$
62. $y = C_1 \pm C_2 e^{\pm x}.$
63. $y = C_1 \pm \sin(x \pm C_2).$
64. $y = C_1 \pm \cosh(x \pm C_2).$
65. $C_1\sqrt{x} - C_1^2 y = \ln|1 + C_1\sqrt{x}| + C_2; \qquad 2y - x = C; \quad y = C.$
66. $y = C_1 (C_2 + x)^2; \qquad y = C.$
67. $y = \dfrac{C_3}{x} - \left(\dfrac{C_1}{x} + 1\right)\ln|C_2(x + C_1)|; \qquad y = C_1 + \dfrac{C_2}{x}.$
68. $y = C_1 + C_2\left(\ln x - \dfrac{1}{x}\right).$
69. $y = C_3 + C_2 x + C_1 x^2 - \left(C_1 - \dfrac{1}{2}\right)\sin^2 x.$
70. $y = C_1(1 \pm \cosh(x + C_2))^2; \qquad y = Ce^{\pm 2x}$.
71. $y = C_2\left(C_1 + \dfrac{3}{2}x^2\right)^{-1/3}$.
72. $x = C_1 + C_2 y - y\ln y.$
73. $y = C_1 x + C_2 x\ln|x| + x^2$.
74. $y = \ln|C_1 x + \dfrac{C_2}{x}|.$
75. $y = \sqrt{C_1 + C_2 x^2}$.
76. $y = C_2 \sec\left(C + \dfrac{x^2}{2}\right).$
77. $(x - C_1)^2 + (y + C_2)^2 = 1.$
78. $y\sin^2(C_1 + C_2 x) = 2C_2^2; \quad y\sinh^2(C_1 + C_2 x) = 2C_2^2;$
$y(x + C)^2 = 2$.
79. $2\sqrt{C_1 y - 1} = C_1 \ln|C_2 x|.$
80. $y = C_2 \pm \arcsin(C_1 x).$
81. $y = 3\tanh^2\dfrac{x\sqrt{3}}{2} - 2.$
82. $y = x.$
83. $y = 1.$
84. $y = C\,[x - 2 + (2 + x)\,e^{-x}].$
85. $y = \ln(x^2 + \pi e x + e).$
86. $y = y_0 + y_0' x + y_0'' \int_0^x (x - t)\,e^{-t^2}\,dt.$

87. $y = \dfrac{x^3}{3}$.
88. $x = -\sqrt{5}\int_1^y \left(1+4s^{5/2}\right)^{-1/2} ds.$
89. $x(x+1)u' - 2u = 0,$
$y = C_1(1+x^{-1}) + C_2(x+1-x^{-1}-2x^{-1}(x+1)\ln|x+1|).$
90. $u' + 2u = 0, \quad y = x^{-1}(C_1 e^x + C_2 e^{-x}).$
91. $u' + 2u\tanh^{-1}x = 0, \quad y = C_1(e^x - 1) + C_2(e^x+1)^{-1}.$
92. $xu'' + (3x-1)u' + 2(x-1)u = 0,$
$x(1-x)v' - (x^2-4x+5)v = 0; \quad y = C_1 e^x + C_2 x + C_3 e^{-x}$.
93. $x^3(2x-1)u'' + x^2(10x-6)u' + 6x(x-1)u = 0,$
$(2x-1)v' - 2v = 0; \; y = C_1 x + C_2 x^{-1} + C_3(1 + x\ln|x|).$
94. $u' + 4u\sin^{-1}2x = 0, \quad y = C_1\tan x + C_2(1 + x\tan x).$

8.9 Linear homogeneous equations

1. $y'' + y = 0.$
2. $y'' - y = 0.$
3. $y^{iv} - y = 0.$
4. $y''' - y'' + 2y = 0.$
5. $y''' + \tan(x)\,y'' - y' - \tan(x)\,y = 0.$
6. $y'''(x^3\ln x - x^3 - x^2 + x) - y''(x^3\ln x - x^3 + 2x - 2)$
$+y'(x^2+2x) - y(x+2) = 0.$
7. $y''\sinh^2 2x + y'\sinh 4x - 4y = 0.$
8. $y'''\left(e^{-x} - 1 + x - \dfrac{x^2}{2}\right) + y''\dfrac{x^2}{2} - y'x + y = 0.$
9. $(x^2-2x+2)y''' - x^2y'' + 2xy' - 2y = 0.$
10. $xy''' + y'' - xy' - y = 0.$
11. Yes. 12. No. 13. Yes. 14. Yes. 15. No. 16. No. 17. Yes.
18. No. 19. No. 20. Yes.
21. $y = C_1 x + C_2 e^x$.
22. $y = C_1(x^2+2x-1) + C_2 x^{-1}.$
23. $y = e^{2x}(C_1 x(x-1) + C_2).$
24. $y = C_1(x^2+1) + C_2 e^x.$
25. $y = C_1(x-2) + C_2 x^2.$
26. $y = C_1(x-1) + C_2 e^{-x}.$
27. $y = C_1 x + C_2 x^2 + C_3 x^{-1}.$
28. $y = C_1(x+1) + C_2 x(x+2) + C_3 e^x.$
29. $y = C_1 x^2 + C_2\left(\ln x + \dfrac{1}{2}\right).$
30. $y = C_1 x + C_2 x_2 + C_3\cos x.$
31. $y = C_1 x + C_2 e^{x^2 - x}.$

32. $y = C_1 x^{-1} + C_2 (x - 4)$.
33. $y = C_1 x + C_2 e^{x^3}$.
34. $y = C_1 (x^2 + 2) + C_2 x^3$.
35. $y = C_1 (x + 1) + C_2 (x - 1)^{-1}$.
36. $y = C_1 e^{2x} + C_2 (3x - 2) e^{-x}$.
37. $y = C_1 (x - 2) + C_2 \sqrt{x}, \quad x > 0$.
38. $y = C_1 + (1 - x)(C_1 \ln(1 - x) + C_2), \quad x < 1$.
39. $y = C_1 x + C_2 e^x + C_3 e^{x^2}$.
40. $y = C_1 (x^2 + 1) + C_2 (x + (x^2 + 1) \arctan x)$.
41. $z''' + \frac{13}{3} z' + \frac{46}{27} z = 0$.
47. $z'' - \frac{3}{4x} z = 0$.
49. $y''(t) - y(t) \frac{1}{4t^2} = 0$.
54. $(t^2 - 1) y''(t) - 2y(t) = 0$.
56. $y'''(t) + \left(\frac{3}{2t} - 1\right) y''(t) - \left(\frac{2}{t} - 1\right) y'(t) + y(t) = 0$.
59. $y^{iv}(t) - 6y'''(t) + 10y''(t) - 5y'(t) + y(t) = 0$.
64. $(1 - x^2) z'' - 3xz' = 0$.
65. $y = (1 + x)(\ln 4 - 1) + 2\left(1 + \frac{1}{2}x - (1 + x) \ln\left|\frac{x + 1}{x}\right|\right)$.
66. The point $x = 0$ is singular and $y(x)$ does not exist.
67. $y = 1 + x \arctan x$.
68. $y = x$.
69. $y = (6 - e^{-1}) x + (1 - 3e)(1 + \ln x)$.
70. $y = 3x + 1 + \frac{x}{2} \ln \frac{1 - x}{1 + x}$.
71. $y = 2x \left(1 - \sqrt{-\ln x}\right)$.
72. The point $x = 0$ is singular and $y = (1 - C) x + C \sinh x$, where C is a constant.
73. $y = x$.
74. $y = \frac{1}{3}(e^x - 1 - x + x^2)$.
75. $\varphi_1 = x + e^x, \quad \varphi_2 = x$.
76. $\varphi_1 = \frac{1}{2}(x + e^{1-x}), \quad \varphi_2 = \frac{1}{2}(x - e^{1-x})$.
77. The point $x = 0$ is singular and φ_i do not exist.
78. $\varphi_1 = \frac{1}{4}(x^2 - 2x + 2 + 2e^x), \quad \varphi_2 = \frac{1}{4}(-x^2 + 2x - 2 + 2e^x)$.

8.10 Linear nonhomogeneous equations

1. $y = C_1 x + C_2 x \int \frac{e^{-x^2/2}}{x^2} dx + 2e^x + 1$.

2. $y = C_1 x + C_2 x \int \frac{e^{-x}}{x}\,dx + x^2.$
3. $y = C_1 x + C_2 \frac{1}{x} - \sin x - \frac{\cos x}{x}.$
4. $y = C_1 x + C_2 x e^{1/x} - \frac{x}{2}\left(\cos \frac{1}{x} + \sin \frac{1}{x}\right).$
5. $y = C_1 (x+2) + \frac{C_2}{x} + \frac{3}{2} + \left(1 + \frac{x}{2}\right) \ln |x|.$
6. $y = C_1 (x-2) + C_2 e^{x/2} + (x^2 - 2x + 5).$
7. $y = C_1 x + \frac{3}{2} + \frac{1+2x}{x}\left(C_2 - \frac{1}{2} \ln |x|\right).$
8. $y = C_1 x + C_2 \ln x - x \ln (x) \left(1 - \frac{1}{2} \ln x\right).$
9. $y = C_1 \frac{\sin x}{x} + C_2 \frac{\cos x}{x} + \sin x + \frac{\cos x}{x} \ln |\cos x|.$
10. $y = C_1 \sqrt{1+x} + C_2 \sqrt{1-x} + \frac{1}{15}(2x^2 + 14).$
11. $y = C_1 \cos x + C_2 e^x + \frac{1}{2} e^{2x} \sin x.$
12. $y = C_1 \sin x + C_2 e^x - \frac{1}{2}(\cos x + \sin x + 2x \sin x).$
13. $y = C_1 (1 + x^2) + C_2 e^x - (x^2 + 3)(x + 1).$
14. $y = C_1 x^3 + C_2 e^x - 1 + \frac{x}{2}.$
15. $y = C_1 x + C_2 \sqrt{1 - x^2} + \frac{1}{3}(2 - x^2).$
16. $y = C_1 (x+1) + \frac{C_2}{x-1} + x.$
17. $y = \frac{C_1}{x} + C_2 e^{1/x} + 1 - \frac{\ln |x|}{x}.$
18. $y = C_1 e^{x^2} + C_2 + (x^2 - 1) e^{x^2}.$
19. $y = C_1 e^x + C_2 x^2 + x^2 e^x - \frac{x^3 e^x}{3}.$
20. $y = \frac{C_1 + C_2 x^2}{1 - x} + \frac{1}{2(x-1)}\left(\frac{2}{3}x^3 + \frac{1}{2}x^2 - x^2 \ln |x| - 2x\right).$
21. $y = C_1 (1 + x^2) + \frac{C_2}{x} + 2x.$
22. $y = \frac{C_1}{x} + \frac{C_2}{x^2} + 12(x^2 + 4x - 6).$
23. $y = C_1 (2x - 1) + C_2 e^{-x} + \frac{1}{2}(x^2 + 1).$
24. $y = C_1 x + C_2 x^2 + C_3 x^3 + \frac{8}{15} x^{7/2} - \frac{5}{3} x^{5/2}$
$+ x^{3/2} - (x^3 - 2x^2 + x) \ln |\sqrt{x} + 1|.$
25. $y = C_1 (x+1) + C_2 (x + 2x) + C_3 e^x$
$-\frac{1}{6}x^4 - \frac{2}{3}x^3 - 3x^2 - 6x - 6.$

26. $y = C_1 x + C_2 x^2 + C_3 e^x + (x^2 - x - 1) - x(x+2)\ln x$.
27. $y = C_1 x + C_2 x^2 + C_3 e^x + C_4 e^{-x} + x \ln x$.
28. $y = C_1 + C_2 \ln x + \dfrac{C_3}{x} + \arctan x$.
29. $y = C_1(x+1) + C_2 x^2 + C_3 e^x + C_4 e^{-x} + x^5$.
30. $y = C_1 x + C_2 e^x + C_3 e^{-x} - x \ln x$.
31. $y = C_1 x + C_2 e^x + \dfrac{C_3}{x} - \dfrac{1}{2} e^{-x}$.
32. $y = C_1 x + C_2 e^x + C_3 \exp(-x^2) + x^2$.
33. $y_p = x^{-1}$.
34. $y_p = x \ln x$.
35. $y_p = x e^x$.
36. $y_p = 1 + x + \dfrac{1}{2} x^2$.
37. $y_p = -x$.
38. $y_p = 1$.
39. $y_p = x^4$.
40. $y_p = e^x \left(\dfrac{x^2}{2} - x \right)$.
41. $y_p = e^{-x}$.
42. $y_p = x$.
43. $y_p = x^3$.
44. $y_p = e^x - 1 - x - \dfrac{1}{2} x^2$.
45. $y_p = x^3$.
46. $y_p = \sin x$.
47. $y_p = (x-2)^4 \ln |x-2|$.
48. $y_p = (x^2+1)^{-1}$.
49. $y_p = x^3 + 4x^2 - 2x + 3$.
50. $y_p = \dfrac{8}{15} x^{7/2}$.
51. $y_p = x^3$.
52. $y_p = e^{x^2/2}$.
53. $K(x,s) = x e^{2s-2x} - s e^{2s-2x}$.
54. $K(x,s) = \sin \omega (x - s)$.
55. $K(x,s) = e^{-s} \sin(e^x - e^s)$.
56. $K(x,s) = \dfrac{1}{3} s \left(1 - \dfrac{s^3}{x^3} \right) \exp\left(\dfrac{1}{x^3} - \dfrac{1}{s^3} \right)$.
57. $K(x,s) = \dfrac{x^2}{s} \ln \dfrac{x}{s}$.
58. $K(x,s) = s \sin(\ln xs)$.
59. $K(x,s) = (x-s)\sqrt{\dfrac{1+s^2}{1+x^2}}$.
60. $K(x,s) = \exp(-s - e^{-s} + e^{-x}) - \exp(-s - 2e^{-s} + 2e^{-x})$.

61. $K(x,s) = e^{s-x} + x - s - 1.$

62. $K(x,s) = \dfrac{sx}{6} - \dfrac{s^3}{2x} + \dfrac{s^4}{3x^2}.$

63. $y = e^{1-x} + x\ln x.$

64. $y = \dfrac{3}{2}x + \dfrac{x^2}{3} - \dfrac{7}{6x}.$

65. $y = -\dfrac{1}{6}\left(x + 1 + \dfrac{2}{x} + \dfrac{4}{x+1}\right).$

66. $y = \dfrac{1}{4}x^2 + \dfrac{5}{2}e^{2x^2-2} - \dfrac{7}{4}e^{3x^2-3}.$

67. $y = \dfrac{1}{12}\exp\dfrac{1}{x} + \dfrac{1}{4}\exp\left(4 - \dfrac{3}{x}\right) - \dfrac{1}{3}\exp\left(3 - \dfrac{2}{x}\right).$

68. $y = \dfrac{3}{4} + \dfrac{1}{2}e^{2x}.$

8.11 Linear equation with constant coefficients

1. $y = C_1 + C_2e^{2x}.$
2. $y = C_1e^{2x} + C_2e^{x/2}.$
3. $y = e^{-x}(C_1\cos 3x + C_2\sin 3x).$
4. $y = C_1\cos 2x + C_2\sin 2x.$
5. $y = C_1e^{2x} + e^{-x}\left(C_2\cos\sqrt{3}x + C_3\sin\sqrt{3}x\right).$
6. $y = C_1e^{x} + C_2e^{-x} + C_3\cos x + C_4\sin x.$
7. $y = e^{x}(C_1\cos x + C_2\sin x) + e^{-x}(C_3\cos x + C_4\sin x).$
8. $y = e^{\sqrt{3}x}(C_1\cos x + C_2\sin x) + C_3\cos 2x.$
   $+C_4\sin 2x + e^{-\sqrt{3}x}(C_5\cos x + C_6\sin x).$
9. $y = e^{x}(C_1 + C_2x).$
10. $y = e^{-x/2}(C_1 + C_2x).$
11. $y = C_1 + C_2e^{x} + C_3e^{-x} + C_4e^{3x} + C_5e^{-3x}.$
12. $y = e^{x/\sqrt{2}}\left(C_1\cos\dfrac{x}{\sqrt{2}} + C_2\sin\dfrac{x}{\sqrt{2}}\right).$
    $+e^{-x/\sqrt{2}}\left(C_3\cos\dfrac{x}{\sqrt{2}} + C_4\sin\dfrac{x}{\sqrt{2}}\right).$
13. $y = C_1\cos x + C_2\sin x + C_3x\cos x + C_4x\sin x$
14. $y = C_1e^{x} + C_2xe^{x} + C_3e^{-x}.$
15. $y = C_1e^{x} + C_2xe^{x} + C_3x^2e^{x}.$
16. $y = Ce^{x} + C_2e^{2x} + C_3e^{-x}.$
17. $y = C_1 + C_2x + C_3e^{2x} + C_4xe^{2x}.$
18. $y = C_1\cos x + C_2\sin x + e^{\sqrt{3}x/2}\left(C_3\cos\dfrac{x}{2} + C_4\sin\dfrac{x}{2}\right).$
    $+e^{-\sqrt{3}x/2}\left(C_5\cos\dfrac{x}{2} + C_6\sin\dfrac{x}{2}\right).$
19. $y = C_1e^{x} + C_2e^{-x} + C_3e^{2x} + C_4e^{-2x}.$
20. $y = C_1e^{x} + C_2xe^{x} + C_3x^2e^{x} + C_4e^{-x} + C_5xe^{-x} + C_6x^2e^{-x}.$

21. $y = C_1 + C_2x + C_3e^x + C_4e^{-x} + C_5\cos x + C_6\sin x$.
22. $y = C_1 + C_2e^x + C_3e^{2x} + C_4\cos x + +C_5\sin x$.
23. $y = C_1 + C_2e^{2x} + e^{-x}\left(C_3\cos\sqrt{3}x + C_4\sin\sqrt{3}x\right)$.
24. $y = e^x\left[(C_1 + C_2x)\cos x + (C_3 + C_4x)\sin x\right]$.
$+e^{-x}\left[(C_5 + C_6x)\cos x + (C_7 + C_8x)\sin x\right]$.
25. $y = (C_1 + C_2x)\cos x + (C_3 + C_4x)\sin x$.
26. $y = C_1 + (C_2 + C_3x)\cos 2x + (C_5 + C_6x)\sin 2x$.
27. $y = e^x(C_1 + C_2x) + C_3e^{-2x}$.
28. $y = C_1\cos x + C_2\sin x + C_3\cos\sqrt{3}x + C_4\sin\sqrt{3}x$
29. $y = C_1e^{-x} + C_2e^{-4x}$.
30. $y = C_1e^{5x} + C_2e^{3x}$.
31. $y = e^x(C_1\cos 3x + C_2\sin 3x)$.
32. $y = C_1 + C_2e^{-3x}$.
33. $y = C_1e^{ax} + C_2e^{-ax}$.
34. $y = C_1\sin 3x + C_2\cos 3x$.
35. $y = e^{2x} - 1$.
36. $y = e^{3x}(C_1\cos 2x + C_2\sin 2x)$
37. $y = e^{2x}(C_1 + C_2x)$.
38. $y = C_1e^{-x} + C_2e^{-2x/3}$.
39. $y = 5e^{-x} - 3e^{-3x}$
40. $y = e^{2x}(\sin 5x + \cos 5x)$.
41. $y = 2e^{x/2} - 2xe^{x/2}$.
42. $y = e^{-3x}(1 + 2x)$.
43. $y = \exp\left\{\left(\sqrt{3} - 1\right)x\right\} - \exp\left\{-\left(\sqrt{3} - 1\right)x\right\}$.
44. $y = 2e^{x/2} + e^{-x}$
45. $y = C_1e^{-x} + C_2e^{2x} + C_3e^{-3x}$.
46. $y = C_1e^{-3x} + C_2e2x + C_3e^{-2x}$.
47. $y = C_1e^{-x} + C_2e^{3x} + C_3xe^{3x}$.
48. $y = e^{-2x}(C_1 + C_2x + C_3x^2)$.
49. $y = C_1 + C_2x + C_3e^{2x}$.
50. $y = C_1 + C_2e^{2x} + C_3e^{-2x}$.
51. $y = C_1 + C_2e^{2x} + e^{-x}\left(C_3\cos\sqrt{3}x + C_4\sin\sqrt{3}x\right)$.
52. $y = C_1 + C_2x + C_3\cos\sqrt{3}x + C_4\sin\sqrt{3}x$.
53. $y = C_1 + C_2x + C_3x^2 + C_4e^{3x}$.
54. $y = e^{-2x}\left(C_1 + C_2x + C_3x^2 + C_4x^3\right)$.
55. $y = e^x\left(C_1 + C_2x + C_3x^2 + C_4x^3\right)$.
56. $y = C_1\cos\sqrt{2}x + C_2\sin\sqrt{2}x + C_3\cos\sqrt{3}x + C_4\sin\sqrt{3}x$.
57. $y = C_e^x + e^{x\cos(2\pi/5)}\left[C_2\cos\left(x\sin\dfrac{2\pi}{5}\right) + C_3\sin\left(x\sin\dfrac{2\pi}{5}\right)\right]$.
$+e^{x\cos(2\pi/5)}\left[C_4\cos\left(x\sin\dfrac{4\pi}{5}\right) + C_5\sin\left(x\sin\dfrac{4\pi}{5}\right)\right]$.
58. $y = C_1 + C_2x + C_3x^2 + C_4e^{2x} + C_5e^x$.

59. $y = C_1 + C_2x + C_3e^{-2x} + e^{-x}\left(C_4 \cos\sqrt{3}x + C_5 \sin\sqrt{3}x\right)$.

60. $y = C_1 + C_2x + C_3x^2 + C_4x^3 + e^{2x}\left(C_5 \cos 3x + C_6 \sin 3x\right)$.

61. $y = \sin x - 2\cos x + 2$.

62. $y = C_1e^x + C_2e^{2x} + C_3e^{3x}$.

63. $y = C_1e^x + e^{2x}(C_2 \cos 3x + C_3 \sin 3x)$.

64. $y = C_1 \cos x + C_2 \sin x + C_3 \sin 2x + C_4 \cos 2x$.

65. $y = e^{2x}\left(C_1 + C_2x + C_3x^2\right)$.

66. $y = e^{-x/2}(C_1 + C_2x)\cos x\dfrac{\sqrt{3}}{2} + e^{-x/2}(C_3 + C_4x)\sin x\,\dfrac{\sqrt{3}}{2}$.

67. $y = e^{-x}\left(C_1 + C_2x + C_3x^2 + c_4x^3\right)$.

68. $y = e^{3x/2}\left(C_1 \cos x\dfrac{\sqrt{7}}{2} + C_2 \sin x\,\dfrac{\sqrt{7}}{2}\right)$.
$+e^{-x}(C_3 \cos x\sqrt{2} + C_4 \sin x\sqrt{2})$.

69. $y = C_1 + C_2x + C_3x^2 + C_4x^3 + e^{-x}\left(C_5 + C_6x + C_7x^2\right)$.

70. $y = C_1 \cos 2x + C_2 \sin 2x + e^{-x}(C_3 + C_4x)$.

71. $y = C_1e^{-x} + C_2e^{3x} + \dfrac{1}{5}e^{4x}$

72. $y = C_1 \cos x + C_2 \sin x + (2x - 2)e^x$.

73. $y = C_1e^x + C_2e^{-x} + xe^x + x^2 + 2$.

74. $y = C_1e^x + C_2e^{-2x} + \left(\dfrac{x^2}{2} - \dfrac{x}{3}\right)e^x$.

75. $y = C_1e^x + C_2e^{2x} + \dfrac{1}{10}\sin x + \dfrac{3}{10}\cos x$.

76. $y = C_1 \cos x + C_2 \sin x - 2x\cos x$.

77. $y = C_1e^x + C_2e^{4x} - (2x^2 - 2x + 3)e^{2x}$.

78. $y = C_1e^x + C_2e^{2x} + (0.1x - 0.12)\cos x - (0.3x + 0.34)\sin x$.

79. $y = C_1e^x + C_2e^{-3x} + \left(\dfrac{x^3}{12} - \dfrac{x^2}{16} + \dfrac{x}{32}\right)e^x$.

80. $y = e^{2x}(C_1 \cos 2x + C_2 \sin 2x) + \dfrac{1}{4}e^{2x} + \dfrac{1}{10}\cos 2x + \dfrac{1}{20}\sin 2x$.

81. $y = \dfrac{3}{4}(1 - \cos x) + \dfrac{1}{8}x^2$.

82. $y = C_1e^x + C_2xe^x + C_3e^{-x} + \dfrac{1}{2}xe^{-x} + 3$

83. $y = C_1e^{-x} + C_2 \cos x + C_3 \sin x + \dfrac{1}{2}xe^{-x} + 4(x - 1)$.

84. $y = C_1 + C_2e^x + C_3e^{-x}\cos x$.

85. $y = (x - 4)\cos x - \left(\dfrac{3}{2}x + 4\right)\sin x + 3x + 4$.

86. $y = C_1 + C_2x + C_3e^{-2x} + C_4e^{2x} - \dfrac{1}{3}e^x - \dfrac{1}{48}x^4 - \dfrac{1}{16}x^2$.

87. $y = C_1 \cos x + C_2 \sin x + C_3x\cos x + C_4x\sin x + 3 + \dfrac{1}{25}\cos 2x$.

88. $y = C_1 + C_2e^x + C_3e^{2x} - xe^x + \dfrac{1}{4}\left(x^2 + 3x\right)$.

89. $y = C_1 + C_2x + C_3x^2 + C_4e^{-x} + e^{x/2}\left(C_5 \cos \frac{\sqrt{3}}{2}x + C_6 \sin \frac{\sqrt{3}}{2}x\right)$
$+\frac{x^4}{24}$.

90. $y = C_1 + C_2x + C_3x^2 + C_4e^{-x} + \frac{1}{20} \sin 2x + \frac{1}{40} \cos 2x$.

91. $y = C_1e^x + C_2e^{-x} + C_3 \cos x + C_4 \sin x - 3x + \frac{x \sin x}{4}$.

92. $y_p = -\frac{x^2}{8} - \frac{3}{5} \sin x + \frac{1}{8}xe^{-2x}$.

93. $y = C_1e^{3x} + C_2e^{-3x} + e^{3x}\left(\frac{6}{37} \sin x - \frac{1}{37} \cos x\right)$.

94. $y = (C_1 + C_2x + x^3)\, e^x$.

95. $y = \left(C_1 - \frac{x^2}{4}\right) \cos x + \left(C_2 + \frac{x}{4}\right) \sin x$.

96. $y = (C_1 + C_2x)e^{-2x} + \left(\frac{x}{16} - \frac{1}{32}\right) e^{2x}$.

97. $y = C_1 + C_2e^{5x} - \frac{1}{5}x^3 - 0.12x^2 - 0.048x + 0.02(\cos 5x - \sin 5x)$

98. $y = C_1 \cos ax + C_2 \sin ax + \frac{e^x}{a^2+1}$.

99. $y = \frac{1}{2}e^{3x}\,(x^2 - 2x + 2) + C_1e^x + C_2e^{2x}$

100. $y = \frac{1}{3}e^{2x} + C_1e^{-x} + e^x\left(\frac{x^2 - x}{4} + C_2\right)$.

101. $y = \frac{e^{2x}}{32}\,(2x^2 - 3x) + \frac{1}{5} - \frac{x^3}{12} + \frac{x}{8} + C_1 + C_2e^{2x} + C_3e^{-2x}$.

102. $y = -e^{-x} \cos x + \frac{x^3}{6}e^{-x} + e^{-x}(C_1 + C_2x)$.

103. $y = \frac{x}{2}e^{-x} \sin x + e^{-x}x + e^{-x}(C_1 \cos x + C_2 \sin x)$.

104. $y = \frac{x^2e^x}{2} + \frac{1}{4}e^{-x} + \frac{1}{2} \cos x + e^x(C_1 + C_2x)$.

105. $y = \frac{1}{2}x \sin x + \frac{1}{8} \cos 3x + C_1 \cos x + C_2 \sin x$.

106. $y = C_1 \cos ax + C_2 \sin ax + \frac{\sin bx}{b^2 - a^2},\ b \neq a;$
$y = C_1 \cos ax + C_2 \sin ax - \frac{x \cos ax}{2a},\ b = a.$

141. $y_p = -\ln \cos x - (\sin x) \ln(\sec x + \tan x)$.

142. $y_p = -\frac{1}{2}x^2$.

143. $y_p = e^{4x}/30$.

144. $y_p = (e^{-x} + e^{-2x}) \ln(e^x + 1)$.

145. $y_p = \sin 2x \ln|\cos x| - x \cos 2x$.

146. $y_p = \frac{4}{5}(x+1)^{5/2}e^{-x}$.
147. $y_p = -\frac{\cos 2x}{\cos x}$.
148. $y_p = 1 + xe^{-x} + (e^x + e^{-x})\ln(1-e^{-x})$
149. $y_p = -\sqrt{\cos 2x}$.
150. $y_p = x^{-1}$.
151. $y_p = \frac{1}{2}\int^x \left[e^{x-t} - \sin(x-t) - \cos(x-t)\right] g(t)dt$.
152. $y_p = \frac{1}{2}\int^x [\sinh(x-t) - \sin(x-t)]\, g(t)dt$.
153. $y_p = \frac{1}{2}\int \left(\frac{x}{t^2} - \frac{2x^2}{t^3} + \frac{x^3}{t^4}\right) g(t)dt$.
154. $y_p = -\frac{1}{2}(1+x)^2 e^x$.
155. $y_p = x^4 - 12x^2 - 22x + 25$.
156. $y_p = x^4 - 4x^3 + 2x + 23$.
157. $y_p = x^3 + 7x^2 + 30x + 55$.
158. $y_p = 8x^2 - 2x^3 + \frac{1}{2}x^4$.
159. $y_p = -\frac{x^2}{8}$.
160. $y_p = \frac{1}{2}x^2 - 2x$.
161. $y_p = e^{2x}\left(9x^2 - 24x + 26\right)/27$.
162. $y_p = \frac{e^x}{5}$.
163. $y_p = \frac{1}{41}\cos 3x$.
164. $y_p = \frac{5}{8}\sin x$.
165. $y_p = \frac{1}{12}e^{2x}x^4$.
166. $y_p = \frac{1}{6}x^3 e^x$.
167. $y_p = \frac{\cos x - \sin x}{2}$.
168. $y_p = \frac{x \sin s}{2}$.
169. $y_p = \frac{xe^x}{4}$
170. $y_p = x^2 - x$.
171. $y_p = \frac{x^2}{4}\sin x + \frac{x}{4}\cos x$.
172. $y_p = e^{-x}\left(x^2 - 2\right)$.
173. $y = \frac{1}{5}\left(e^{3t} + 4e^{-2t}\right)$.

174. $y = 2e^{-t} - e^{-2t}$.
175. $y = e^t \sin t$.
176. $y = e^{2t} - te^{2t}$.
177. $y = te^t - t^2e^t + \frac{2}{3}t^3e^t$.
178. $y = \cosh t$.
179. $y = 2e^t \cosh t - \frac{2}{\sqrt{3}}e^t \sinh t$.
180. $y = \frac{1}{\omega^2 - 4}\left[(\omega^2 - 5)\cos \omega t + \cos 2t\right]$.
181. $y = \frac{1}{5}(\cos t - 2\sin t + 4e^t \cos t - 2e^t \sin t)$.
182. $y = \frac{1}{5}(e^{-t} - e^{-t}\cos t + 7e^t \sin t)$.
183. $y = \frac{5}{3}\sin t - \frac{1}{3}\sin 2t$.
184. $y = \frac{1}{3}e^{2t} + \frac{2}{3}e^{-t}$.
185. $y = 2 + 2e^t - e^{-t} - \sin t$.
186. $y = \frac{1}{2}\left(e^t - e^{-t}\right) - \sin t$.
187. $y = 2e^t - e^{2t} + e^{3t}$.
188. $y = 2$.
189. $y = t^2 - 2t + 3 - e^{-2t}$.
190. $y = 4e^{-2t} + t - 1$.
191. $y = t^3 - 3t^2 + 1 - e^{-t}$.
192. $y = e^{-2t}(2\cos t + 7\sin t)$.
193. $y = e^t(\sin 2t - 2\cos 2t) + 2$.
194. $y = e^t(t^2 - 2t + 3)$.
195. $y = e^t$.
196. $y = e^t + t + 1$.
197. $y = 3 - 2^{t/2}\cos\frac{\sqrt{3}}{2}t - e^{-t}$.
198. $y = -t\cos t$.
199. $y = (2 - \sqrt{2})\cos x\sqrt[4]{2} + (2 + \sqrt{2})\cosh x\sqrt[4]{2}$.
200. $y = \left(e^{2x} + 2e^{-x}\right)/3$.
201. $y = (\cos x + \sin x + e^x)/2$.
202. $y = x$.
203. $y = x\sin x$.
204. $y = 6\sinh 2x$.
205. $y = 2x\sinh x + 3\cosh x + \cos x\sqrt{3}$.
206. $y = 2x - x^3/3$.

8.12 Equations with polynomial coefficients

1. $y = C_1 x + C_2 x^2 + C_3 x^3.$
2. $y = C_1 + C_2 + C_3 x \ln|x|.$
3. $y = C_1 x^2 + C_2 x^2 \ln|x| + C_3 x^2 \ln^2|x|.$
4. $y = C_1 + C_2 x + C_3 x^2.$
5. $y = \dfrac{C_1}{x} + C_2 x^2 \cos 3\ln|x| + C_3 x^2 \sin 3\ln|x|.$
6. $y = C_1 x + C_2 \cos 2\ln|x| + C_3 \sin 2\ln|x|.$
7. $y = C_1 + C_2 x + \dfrac{C_3}{x}.$
8. $y = C_1 \cos\ln|x| + C_2 \sin\ln|x| + \ln|x| (C_3 \cos\ln|x| + C_4 \sin\ln|x|).$
9. $y = \left(C_1 x + \dfrac{C_2}{x}\right) \cos\ln|x| + \left(C_3 x + \dfrac{C_4}{x}\right) \sin\ln|x|.$
10. $y = C_1 x + C_2 x^2 + \dfrac{C_3}{x} + \dfrac{C_4}{x^2}.$
11. $y = C_1 x + \dfrac{C_2}{x} + C_3 x \cos 2\ln|x| + C_4 x \sin 2\ln|x|.$
12. $y = C_1 x + \dfrac{1}{x} (C_2 \cos\ln|x| + C_3 \sin\ln|x|)$
 $+ x^2 (C_4 \cos 3\ln|x| + C_5 \sin 3\ln|x|).$
13. $y = C_1 (6x-3)^{-3} + \sqrt{6x-3}$
 $\times \left(C_3 \cos\ln|\sqrt{6x-3}| + C_4 \sin\ln|\sqrt{6x-3}|\right).$
14. $y = C_1 (x+8) + C_2 (x+8)^2 + (x+8)^{1/3}$
 $\times (C_3 \cos 2\ln|x| + C_4 \sin 2\ln|x|).$
15. $y = C_1 (4x+1)^8 + \sqrt{4x+1} \left(C_2 \cos\dfrac{\sqrt{3}}{2} \ln|x| + C_3 \sin\dfrac{\sqrt{3}}{2} \ln|x|\right).$
16. $y = C_1 \cos\sqrt{2} \arcsin x + C_2 \sin\sqrt{2} \arcsin x.$
17. $y = C_1 \cosh\dfrac{\sqrt{5}}{2} \ln|x + \sqrt{x^2-1}| + C_2 \sinh\dfrac{\sqrt{5}}{2} \ln|x + \sqrt{x^2-1}|.$
18. $y = C_1 x + C_2 \sqrt{1-x^2} + C_3 \exp(\arcsin x) + C_4 \exp(-\arcsin x).$
19. $y = C_1 x + C_2 \sqrt{1-x^2} + C_3 \cos\dfrac{1}{\sqrt{2}} \arcsin x + C_4 \sin\dfrac{1}{\sqrt{2}} \arcsin x.$
20. $y = C_1 \exp\left(2\ln|x + \sqrt{x^2+1}|\right) + C_2 \exp\left(-2\ln|x + \sqrt{x^2+1}|\right).$
21. $y = C_1 \cos\sqrt{5} \ln|x + \sqrt{x^2+1}| + C_2 \sin\sqrt{5} \ln|x + \sqrt{x^2+1}|.$
22. $y = C_1 \left(x + \sqrt{x^2+1}\right) + C_2 \left(x + \sqrt{x^2+1}\right)^{-1}$
 $+ C_3 \cos\ln|x + \sqrt{x^2+1}| + C_4 \sin\ln|x + \sqrt{x^2+1}|.$
23. $y = x (C_1 e^x + C_2 e^{2x}).$
24. $y = e^x (C_1 x + C_2) + x e^{-x} (C_3 x + C_4).$
25. $y = x (C_1 + C_2 e^x + C_3 e^{-x}).$
26. $y = \exp\left(-\dfrac{x^2}{2}\right) (C_1 x^2 + C_2 x + C_3).$
27. $y = x (C_1 e^{-x} + C_2 e^{2x} + C_3 e^{3x}).$

28. $y = e^{-1/x}(C_1 \cos x + C_2 \sin x)$.

29. $y = \exp\left(\frac{1}{x^2}\right)(C_1 e^x + C_2 e^{2x})$.

30. $y = x^3 e^x (C_1 + C_2 x)$.

31. $y = e^{x^2}(C_1 + C_2 e^x + C_3 e^{-x})$.

32. $y = x e^{2x}(C_1 + C_2 x)$.

33. $y = \frac{e^x}{x}\left(C_1 e^x + C_2 \cos\sqrt{3}x + C_3 \sin\sqrt{3}x\right)$.

34. $y = e^{-x-1/x}\left(C_1 e^x + e^{-x/2}\left(C_2 \cos\ \frac{\sqrt{3}}{2}x + C_3 \sin\frac{\sqrt{3}}{2}x\right)\right)$.

35. $y = x^{-4}(C_1 e^x + C_2 e^{-x} + C_3 e^{2x} + C_4 e^{-2x})$.

36. $y = e^{-x}x^{-2}(C_1 \cos x + C_2 \sin x)$.

37. $y = x^{-1}(C_1 + C_2 x + C_3 x^{-1})$.

38. $y = e^{-x}(C_1 + C_2 x + C_3 x^{-1})$.

39. $y = e^{-x}(C_1 x + C_2 x^{-1} + C_3 x^2 + C_4 x^{-2})$.

40. $y = e^x\left(C_1 \cos\sqrt{7}\ln|x + \sqrt{x^2+1}| + C_2 \sin\sqrt{7}\ln|x + \sqrt{x^2+1}|\right)$.

41. $y = \exp\left(-\frac{x^2}{2}\right)(C_1 x + C_2 e^x + C_3 e^{-x})$.

42. $y = 1 + \frac{1}{6}x^3 - \frac{1}{24}x^4 - \frac{1}{120}x^5 + 0 + \ldots$.

43. $y = \frac{1}{2}x^2 + \frac{3}{2\cdot 5!}x^3 + \ldots$.

44. $y = 1 - \frac{1}{24}x^4 - \frac{1}{120}x^6 + \ldots$.

45. $y = x^2 - x^3 + \frac{5}{4}x^4 - \frac{13}{10}x^5 + \frac{17}{15}x^6 + \ldots$.

46. $y = 1 + \frac{1}{2}x^2 + \frac{1}{16}x^4 + \frac{1}{160}x^6 \ldots$

47. $y = x + \frac{1}{6}x^3 + \frac{1}{48}x^5 + \ldots$.

48. $y = \frac{1}{3!}x^3 + \frac{1}{4!}x^4 + \frac{1}{5!}x^5 + \frac{1}{6!}x^6 + \frac{1}{7!}x^7 + \frac{1}{8!}x^8$
$+ \frac{1}{9!}x^9 + \left(\frac{1}{10!} - \frac{1}{6\cdot 7\cdot 8\cdot 9\cdot 10}\right)x^{10} + \ldots$.

49. $y = 2e^{x^2-x} + e^{x^2-2x}$.

50. $y = \frac{x^2+5}{2x}$.

51. $y = e^{x-1}\frac{1}{2}\left(\frac{3}{x} - x\right)$.

52. $y = e^{-x}\frac{1}{\sqrt{3}}\sin\sqrt{3}\arcsin x$.

53. $y = x^2(2e^{x-1} + 4e^{1-x} - 5)$.

54. $y = \frac{1}{\sqrt{7}}\sinh\sqrt{7}\ln|x + \sqrt{x^2+1}|$.

8.13 Series solutions

1. $\rho = 1, \quad |x-2| < \rho.$
2. $\rho \to \infty.$
3. $\rho = 1, \quad |x-1| < \rho.$
4. $\rho = 1, \quad |x-\zeta| < \rho.$
5. $\rho = 1, \quad |x^2-2x+3| < \rho.$
6. $\rho = 1, \quad |2x+3| < \rho.$
7. $y_1 = \sum_{n=0}^{\infty} \frac{x^{2n}}{2^n n!}, \quad y_2 = \sum_{n=0}^{\infty} \frac{2^n n! x^{2n+1}}{(2n+1)!}.$
8. $a_{n+2} = \frac{a_{n+1}+a_n}{n+2}, \quad n \geq 0$

$$y_1 = 1 + \frac{1}{2}(x-1)^2 + \frac{1}{6}(x-1)^3 + \frac{1}{6}(x-1)^4 + \frac{1}{15}(x-1)^5 + \ldots,$$
$$y_2 = x - 1 + \frac{1}{2}(x-1)^2 + \frac{1}{2}(x-1)^3 + \frac{1}{4}(x-1)^4 + \frac{3}{20}(x-1)^5 + \ldots.$$

9. $a_{n+4} = -\frac{a_n}{(n+4)(n+3)},$

$$y_1 = \Gamma(\tfrac{3}{4}) \sum_{n=0}^{\infty} \frac{(-1)^n}{n!\Gamma(n+\frac{3}{4})} \left(\frac{x^4}{16}\right)^n,$$
$$y_2 = x\Gamma(\tfrac{5}{4}) \sum_{n=0}^{\infty} \frac{(-1)^n}{n!\Gamma(n+\frac{5}{4})} \left(\frac{x^4}{16}\right)^n.$$

10. $y_1 = 1 - \frac{1}{2}x^2 + \frac{1}{12}x^4 - \frac{1}{48}x^6 + \ldots,$

$$y_2 = x - \frac{1}{6}x^3 + \frac{1}{30}x^5 - \frac{13}{1680}x^7 + \ldots.$$

11. $a_n = \frac{1}{n(n-1)}\left(ba_{n-2} + \frac{1}{4}a_{n-4}\right), \quad n \geq 2,$

$$y_1 = 1 + \frac{b}{2}x^2 + \frac{1}{48}(2b^2+7)x^4 + \frac{b}{1440}(2b^2+7)x^6 + \cdots,$$
$$y_2 = x + \frac{b}{6}x^3 + \frac{1}{240}(2b^2+3)x^5 + \frac{b}{10080}(2b^2+13)x^7 + \cdots.$$

12. $a_{n+2} = \frac{na_{n+1}}{n+2} - \frac{a_n}{(n+1)(n+2)}, \quad n \geq 0$

$$y_1 = 1 - \frac{1}{2}x^2 - \frac{1}{6}x^3 - \frac{1}{24}x^4 - \frac{1}{60}x^5 + \ldots,$$
$$y_2 = x - \frac{1}{6}x^3 - \frac{1}{12}x^4 - \frac{1}{24}x^5 + \ldots.$$

13. $y_1 = 1 - \frac{1}{3}(x-1)^3 - \frac{1}{12}(x-1)^4 - \frac{1}{90}(x-1)^6 + \ldots,$

$$y_2 = x - 1 - \frac{1}{12}(x-1)^4 - \frac{1}{20}(x-1)^5 - \frac{1}{252}(x-1)^7 + \ldots.$$

14. $y_1 = \Gamma\left(\tfrac{2}{3}\right) \sum_{n=o}^{\infty} \frac{1}{n!\Gamma\left(n+\frac{2}{3}\right)} \left(\frac{x^3}{9}\right)^n,$

$$y_2 = x\Gamma\left(\tfrac{4}{3}\right)\sum_{n=0}^{\infty}\frac{1}{n!\Gamma\left(n+\frac{4}{3}\right)}\left(\frac{x^3}{9}\right)^n.$$

15. $a_{n+3} = \dfrac{a_{n+1}+a_n}{(n+3)(n+2)} \quad n \geq 0$

$$y_1 = 1 + \frac{1}{2}(x-1)^2 + \frac{1}{6}(x-1)^3 + \frac{1}{24}(x-1)^4 + \frac{1}{30}(x-1)^5 + \ldots,$$

$$y_2 = x - 1 + \frac{1}{6}(x-1)^3 + \frac{1}{12}(x-1)^4 + \frac{1}{120}(x-1)^5 + \ldots.$$

16. $y_1 = 1 + \dfrac{1}{2}x^2 + \dfrac{1+2^2}{4!}x^4 + \dfrac{(1+2^2)(1+4^2)}{6!}x^6 + \ldots,$

$$y_2 = x + \frac{2}{3!}x^3 + \frac{2(1+3^2)}{5!}x^5 + \frac{2(1+3^2)(1+5^2)}{7!}x^7 + \ldots.$$

17. $y_1 = 1 - \dfrac{1}{3!}x^3 + \dfrac{1}{5!}x^5 + \dfrac{1}{180}x^6 + \ldots,$

$$y_2 = x - \frac{1}{12}x^4 + \frac{1}{180}x^6 + \ldots.$$

18. $y_1 = 1 - \dfrac{1}{6}x^3 + \dfrac{1}{12}x^4 - \dfrac{1}{40}x^5 + \ldots,$

$$y_2 = x - \frac{1}{12}x^4 + \frac{1}{20}x^5 + \ldots.$$

19. $y_1 = \sqrt{x}\Gamma\left(\tfrac{5}{4}\right)\sum_{n=0}^{\infty}\dfrac{(-1)^n}{n!\Gamma\left(n+\frac{5}{4}\right)}\left(\dfrac{x^2}{8}\right)^n,$

$$y_2 = \Gamma\left(\tfrac{3}{4}\right)\sum_{n=0}^{\infty}\frac{(-1)^n}{n!\Gamma\left(n+\frac{3}{4}\right)}\left(\frac{x^2}{8}\right)^n.$$

20. $y_1 = x^{1/3}\Gamma\left(\tfrac{4}{3}\right)\sum_{n=0}^{\infty}\dfrac{(-1)^n}{n!\Gamma\left(n+\frac{4}{3}\right)}\left(\dfrac{x^2}{4}\right)^n,$

$$y_2 = x^{-1/3}\Gamma\left(\tfrac{2}{3}\right)\sum_{n=0}^{\infty}\frac{(-1)^n}{n!\Gamma\left(n+\frac{2}{3}\right)}\left(\frac{x^2}{4}\right)^n.$$

21. $y_1 = x^{1/3}\Gamma\left(\tfrac{7}{6}\right)\sum_{n=0}^{\infty}\dfrac{(-1)^n}{n!\Gamma\left(n+\frac{7}{6}\right)}\left(\dfrac{x^2}{12}\right)^n,$

$$y_2 = \Gamma\left(\tfrac{5}{6}\right)\sum_{n=0}^{\infty}\frac{(-1)^n}{n!\Gamma\left(n+\frac{5}{6}\right)}\left(\frac{x^2}{12}\right)^n.$$

22. $y_1 = x^{\sqrt{2}}\Gamma\left(1+2^{3/2}\right)\sum_{n=0}^{\infty}\dfrac{(-1)^n}{n!\Gamma\left(n+1+2^{3/2}\right)}x^n,$

$$y_2 = x^{-\sqrt{2}}\Gamma\left(1-2^{3/2}\right)\sum_{n=0}^{\infty}\frac{(-1)^n}{n!\Gamma\left(n+1-2^{3/2}\right)}x^n.$$

23. $y_1 = \sum_{n=0}^{\infty}\dfrac{x^n}{(n!)^2},$

$$y_2 = y_1\ln|x| - 2\sum_{n=1}^{\infty}\frac{S_0(n)}{(n!)^2}x^n.$$

24. $y_1 = \sum_{n=0}^{\infty} \frac{(-1)^n}{(n!)^2} (x)^n$,

$y_2 = y_1 \ln|x| - 2\sum_{n=1}^{\infty} \frac{(-1)^n S_0(n)}{(n!)^2} (2x)^n$.

25. $y_1 = \sum_{n=0}^{\infty} \frac{(-1)^n}{n!(n+1)!} x^n$,

$y_2 = -y_1 \ln|x| + \frac{1}{x}\left(1 - \sum_{n=1}^{\infty} \frac{S_0(n) + S_0(n-1)}{n!(n-1)!} (-x)^n\right)$.

26. $y_1 = \frac{1}{x}\sum_{n=0}^{\infty} \frac{(-x)^n}{(n!)^2}$,

$y_2 = y_1 \ln|x| - \frac{2}{x}\sum_{n=1}^{\infty} \frac{(-1)^n S_0(n)}{(n!)^2} x^n$.

27. $y_1 = x\sum_{n=0}^{\infty} \frac{(-x)^n}{n!(n+1)!} x^n$,

$y_2 = -y_1 \ln|x| + \left(1 + \sum_{n=1}^{\infty} \frac{(-1)^n}{(n!)^2} (1 - 2nS_0(n))\, x^n\right)$.

28. $y_1 = \sum_{n=0}^{\infty} \frac{x^n}{n!} = e^x$,

$y_2 = \sqrt{x}\left(\sum_{n=0}^{\infty} \frac{(2x)^n}{(2n+1)!!}\right) = \frac{1}{2} e^x \int_0^x \frac{e^{-x}}{\sqrt{x}}$.

29. $y_1 = \sum_{n=0}^{\infty} \frac{(-1)^n}{n!(n+1)!} x^{n-1}$,

$y_2 = -y_1 \ln|x| + \frac{1}{x^2}\left(1 - \sum_{n=1}^{\infty} \frac{S_0(n) + S_0(n-1)}{n!(n-1)!} (-x)^n\right)$.

30. $y_1 = x^{3/2}\Gamma\left(\frac{5}{2}\right) \sum_{n=0}^{\infty} \frac{(-1)^n}{n!\Gamma\left(n+\frac{5}{2}\right)} \left(\frac{x^2}{4}\right)^n$,

$y_2 = x^{-3/2}\left(1 + \Gamma\left(-\frac{1}{2}\right) \sum_{n=1}^{\infty} \frac{(-1)^n}{n!\Gamma\left(n-\frac{1}{2}\right)} \left(\frac{x^2}{4}\right)^n\right)$.

31. $y_1 = x\sum_{n=0}^{\infty} \frac{(-1)^n}{n!(n+1)!} \left(\frac{x^2}{4}\right)^n$,

$y_2 = -y_1 \frac{1}{2} \ln|x| + \frac{1}{x}\left(1 - \sum_{n=1}^{\infty} \frac{S_0(n) + S_0(n-1)}{n!(n-1)!} (-1)^n \left(\frac{x^2}{4}\right)^n\right)$.

32. $y_1 = 1 + \sum_{n=1}^{\infty} \left[\frac{(2n-1)!!}{(2n)!!}\right]^2 x^n$,

$y_2 = y_1 \ln|x| + 4\sum_{n=1}^{\infty} \left[\frac{(2n-1)!!}{(2n)!!}\right]^2 (\psi(2n+1) - \psi(n+1))\, x^n$,

where $\psi(x) = \Gamma'(x)$. By replacing $x \to 1 - x$ in these solutions, the solutions about the singular point $\zeta = 1$ are obtained.

33. $y_1 = \sum_{n=0}^{\infty} \frac{(-1)^n}{(n!)^2} \left(\frac{x^2}{4}\right)^n,$

$y_2 = y_1 \ln|x| - \sum_{n=1}^{\infty} \frac{(-1)^n S_0(n)}{(n!)^2} \left(\frac{x^2}{4}\right)^n.$

34. $y_1 = x^{\nu} \Gamma(\nu) \sum_{n=0}^{\infty} \frac{(-1)^n}{n!\Gamma(n+\nu)} \left(\frac{x^2}{4}\right)^n,$

$y_2 = x^{-\nu} \Gamma(-\nu) \sum_{n=0}^{\infty} \frac{(-1)^n}{n!\Gamma(n-\nu)} \left(\frac{x^2}{4}\right)^n.$

35. $y_1 = \cos(\ln|x|) \left(1 + \frac{1}{96}x^2 - \frac{19}{2464000}x^4 + \dots\right)$

$\quad - \sin(\ln|x|) \left(\frac{1}{96}x^2 + \frac{13}{2464000}x^4 + \dots\right),$

$y_2 = \sin(\ln|x|) \left(1 + \frac{1}{96}x^2 - \frac{19}{2464000}x^4 + \dots\right)$

$\quad + \cos(\ln|x|) \left(\frac{1}{96}x^2 + \frac{29}{2464000}x^4 + \dots\right).$

36. $y_1 = \cos(\ln|x|) \left(1 - \frac{2}{5}x + \frac{1}{10}x^2 + \frac{17}{780}x^3 + \dots\right)$

$\quad - \sin(\ln|x|) \left(-\frac{1}{5}x + \frac{1}{20}x^2 + \frac{1}{130}x^3 + \dots\right),$

$y_2 = \sin(\ln|x|) \left(1 - \frac{2}{5}x + \frac{1}{10}x^2 + \frac{17}{780}x^3 + \dots\right)$

$\quad + \cos(\ln|x|) \left(-\frac{1}{5}x + \frac{1}{20}x^2 + \frac{1}{130}x^3 + \dots\right).$

37. $y_1 = \Gamma\left(\frac{2}{3}\right) \sum_{n=0}^{\infty} \frac{1}{n!\Gamma\left(n+\frac{2}{3}\right)} \frac{1}{(9x^3)^n},$

$y_2 = x^{-1} \Gamma\left(\frac{4}{3}\right) \sum_{n=0}^{\infty} \frac{1}{n!\Gamma\left(n+\frac{4}{3}\right)} \frac{1}{(9x^3)^n}.$

38. $y_1 = x^{-1/2} \Gamma\left(\frac{5}{4}\right) \sum_{n=0}^{\infty} \frac{(-1)^n}{n!\Gamma\left(n+\frac{5}{4}\right)} \frac{1}{(8x^2)^n},$

$y_2 = \Gamma\left(\frac{3}{4}\right) \sum_{n=0}^{\infty} \frac{(-1)^n}{n!\Gamma\left(n+\frac{3}{4}\right)} \frac{1}{(8x^2)^n}.$

39. $y_1 = x^{-1/2} \left(1 + \sum_{n=1}^{\infty} \left[\frac{(2n-1)!!}{(2n)!!}\right]^2 \frac{1}{x^n}\right),$

$y_2 = -y_1 \ln|x|$

$\quad + 4x^{-1/2} \sum_{n=1}^{\infty} \left[\frac{(2n-1)!!}{(2n)!!}\right]^2 \left(\psi(2n+1) - \psi(n+1)\right) \frac{1}{x^n},$

where $\psi(x) = \Gamma'(x)$.

40. $y_1 = \sum_{n=0}^{\infty} \frac{(-1)^n}{(2n+1)!} x^{-2n} = x \sin x^{-1},$

$y_2 = x\sum_{n=1}^{\infty} \frac{(-1)^n}{(2n)!} x^{-2n} = x\cos x^{-1}.$

41. $y = \frac{\sqrt{3}e^{ix^3/3}}{2\pi x}\sum_{n=0}^{\infty} \frac{\Gamma\left(n+\frac{2}{3}\right)\Gamma\left(n+\frac{1}{3}\right)}{n!}\left(\frac{3i}{2x^3}\right)^n,$
$y_1 = Rey, \quad y_2 = Imy, \quad \rho = 0.$

42. $y_1 = \frac{e^{-x^2/2}}{\pi\sqrt{2x}}\sum_{n=0}^{\infty} \frac{\Gamma\left(n+\frac{3}{4}\right)\Gamma\left(n+\frac{1}{4}\right)}{n!} x^{2n},$

$y_2 = \frac{e^{-x^2/2}}{\pi\sqrt{2x}}\sum_{n=0}^{\infty} \frac{\Gamma\left(n+\frac{3}{4}\right)\Gamma\left(n+\frac{1}{4}\right)}{n!}\frac{(-1)^n}{x^{2n}}, \quad \rho = 0.$

43. $y = e^{-x/2}\exp\left(-ie^x\right)\frac{1}{\pi}\sum_{n=0}^{\infty} \frac{\left(\Gamma\left(n+\frac{1}{2}\right)\right)^2}{n!}\left(\frac{ie^{-x}}{2}\right)^n,$
$y_1 = Rey, \quad y_2 = Imy, \quad \rho = 0.$

44. $y = \frac{e^{-ix^2/2}}{\pi\sqrt{2x}}\sum_{n=0}^{\infty} \frac{\Gamma\left(n+\frac{3}{4}\right)\Gamma\left(n+\frac{1}{4}\right)}{n!}\left(\frac{i}{x^2}\right)^n,$
$xy_1 = Rey, \quad y_2 = Imy, \quad \rho = 0.$

45. $y = \frac{e^{ix^3/3}}{x} 1 - \frac{i}{2x} - \frac{1}{8x^2} - \left(\frac{1}{6} - \frac{i}{48}\right)\frac{1}{x^3}$
$- \left(\frac{143}{384} - \frac{i}{48}\right)\frac{1}{x^4} + \ldots,$
$y_1 = Rey, \quad y_2 = Imy \quad \rho = 0.$

46. $y = \sqrt{\frac{x}{\ln|x|}}\exp\left(\frac{i}{2}\ln^2 x - \frac{i}{8}\ln|\ln|x||\right)$
$\sum_{n=0}^{\infty} \frac{\Gamma\left(n+\frac{3}{4}+\frac{i}{16}\right)\Gamma\left(n+\frac{1}{4}+\frac{i}{16}\right)}{n!\Gamma\left(\frac{3}{4}+\frac{i}{16}\right)\Gamma\left(\frac{1}{4}+\frac{i}{16}\right)}\left(\frac{-i}{\ln^2|x|}\right)^n,$
$y_1 = Rey, \quad y_2 = Imy \quad \rho = 0.$

47. $y_{1,2} = \frac{2\exp\left(\pm\frac{2}{3}x^{3/2}\right)}{\pi x^{1/4}}\sum_{n=0}^{\infty} \frac{\Gamma\left(n+\frac{1}{6}\right)\Gamma\left(n+\frac{5}{6}\right)}{n!}\left(\frac{\pm 3}{4x^{3/2}}\right)^n,$
$\rho = 0.$

48. $y = \frac{2\exp\left(-\frac{2}{3}ix^{3/2}\right)}{\pi x^{1/4}}\sum_{n=0}^{\infty} \frac{\Gamma\left(n+\frac{1}{6}\right)\Gamma\left(n+\frac{5}{6}\right)}{n!}\left(\frac{3i}{4x^{3/2}}\right)^n,$
$y_1 = Rey, \quad y_2 = Imy, \quad \rho = 0.$

8.14 Linear boundary value problems

1. $y = 0.$
2. $y = 0.$
3. $y = C(x+1).$
4. $y = C_1 x + C_2.$
5. $y = C\left(\cos x - \frac{1-\sin 1}{\cos 1}\sin x\right).$

6. $y = 0$.
7. $y = 0$.
8. $y = C\,(x - e^x)$.
9. $y = C\left(x + 1 + \dfrac{1}{x-1}\right)$.
10. $y = C\,(x - 4)$.
11. $y = \dfrac{1}{6} cx\,(x^2 - 1)$.
12. $y = \dfrac{1}{6}(x^3 - 4x - 4)$.
13. y does not exist.
14. y does not exist.
15. $y = (1 - x)$.
16. $y = 1 + \dfrac{1}{6} x\,(x^2 - 7)$.
17. y does not exist.
18. $y = 1 + \dfrac{x^3}{6} + c\,(x + 1)$.
19. $y = 1 - \sin x - \cos x$.
20. y does not exist.
21. $y = 1$.
22. $y = \dfrac{\sinh x}{\sinh 1} - 2x$.
23. $y = \dfrac{e^{-x} - e^{-1}}{1 - e^{-1}} + x - 1$.
24. $y = e^x - 2$.
25. $y = 2x - \pi + \pi \cos x + c \sin x$.
26. $y = 1 - x$.
27. $y = \dfrac{2\pi}{2 + \pi}\,(x + (1 + x^2) \arctan x)$.
28. $y = 5e^{-1}\left(1 - e^{x/2}\right) + x^2 + \dfrac{e - 5}{2e} x$.
29. $y = 2x^3$.
30. $y = Cx^{-2}$.
31. $y = e^{-x} - 1$.
32. $y = -e^{-x-ix}$.
33. $y = 2e^{-x}$.
34. $y = 3x^2$.
35. y does not exist.
36. $y = C \sin x$.
37. $y = C_1 \sin x + c - 2 \cos x$.
38. $y = (x - 1)\, e^x$.
39. $y = \dfrac{1}{2} \arctan^2 x$.

40. $y = \left(1 + x - \frac{x^2}{2}\right) e^x$.

41. $y = x^{-1/2}\left(1 - \ln|x|\right)$.

42. 1) $a \neq n\pi$ then $y = 0$, where n is an integer number; 2) $a = n\pi$ $y = C\sin(n\pi x)$, where n is an integer number and C is an arbitrary constant ; 3) such parameter a does not exist.

43. $G(x,s) = \begin{cases} \cos s \sin x, & 0 \le s \le x \\ \sin s \cos x, & x \le s \le 1 \end{cases}$.

44. $G(x,s) = \frac{1}{2}\sin|x - s|$.

45. $G(x,s) = \begin{cases} -e^{-x}\cosh s, & 0 \le s \le x \\ -e^{-s}\cosh x, & x \le s \le 2 \end{cases}$.

46. $G(x,s)$ does not exist.

47. $G(x,s) = \begin{cases} (1 - s^3)/3s^3 x, & 1 \le s \le x \\ (1 - x^3)/3s^3 x, & x \le s \le 2 \end{cases}$.

48. $G(x,s) = \begin{cases} 1/s - 1, & 1 \le s \le x \\ 1/x - 1, & x \le s \le 3 \end{cases}$.

49. $G(x,s) = \begin{cases} (\sin s/\sin 1)\sin(1 - x), & 0 \le s \le x \\ (\sin x/\sin 1)\sin(1 - s), & x \le s \le 1 \end{cases}$.

50. $G(x,s) = \begin{cases} s(x - 1), & 0 \le s \le x \\ x(s - 1), & x \le s \le 1 \end{cases}$.

51. $G(x,s) = \begin{cases} x - 1, & 0 \le s \le x \\ s - 1, & x \le s \le 1 \end{cases}$.

52. $G(x,s) = \begin{cases} s(x - 2)/2, & 0 \le s \le x \\ x(s - 2)/2, & x \le s \le 1 \end{cases}$.

53. $G(x,s) = \begin{cases} -s, & 0 \le s \le x \\ -x, & x \le s \le 1 \end{cases}$.

54. $G(x,s) = \begin{cases} -s, & 0 \le s \le x \\ -x, & x \le s \le \infty \end{cases}$.

55. $G(x,s) = \begin{cases} -e^{s-x}, & 0 \le s \le x \\ -1, & x \le s \le \infty \end{cases}$.

56. $G(x,s) = -\frac{1}{2}e^{-|x-s|}$.

57. $G(x,s) = \begin{cases} -\frac{1}{3x}, & 0 \le s \le x \\ -\frac{x^2}{3s^3}, & x \le s \le 1 \end{cases}$.

8.15 Eigenvalues and eigenfunctions

1. $\hat{L}^+ y \equiv -x^2 y' - (1 + 2x)\, y$.
2. $\hat{L}^+ y \equiv x^2 y'' + 2xy' + (n^2 - x^2)\, y = 0$.
3. $\hat{L}^+ y \equiv x^2 y'' + 6xy' - (n^2 - 4 + x^2)\, y$.
4. $\hat{L}^+ y \equiv (1 - x^2)\, y'' - 2xy' + 4y$.
5. $\hat{L}^+ y \equiv (1 + x^2)\, y'' + 2xy' - y$.
6. $\hat{L}^+ y \equiv x^3 y'' - 2x(1 + 3x)\, y' + 6(1 + x)\, y$.

7. No. 8. No. 9. Yes. 10. No.

11. $\lambda_n = n^2\pi^2$, $\varphi_n = \sin(n\pi x)$, where $n = 1, 2, \ldots$.

12. $\lambda_n = n^2\pi^2$, $\varphi_n = \cos(n\pi x)$, where $n = 0, 1, 2, \ldots$.

13. $\lambda_n = (n + 1/2)^2 \pi^2$, $\varphi_n = \sin((n + 1/2)\pi x)$, where $n = 0, 1, 2, \ldots$

14. $\lambda_n = 4n^2\pi^2$, $\varphi_n = \cos(2n\pi x)$, where $n = 0, 1, 2, \ldots$; $\varphi_n = \sin(2n\pi x)$, where $n = 1, 2, \ldots$.

15. $\lambda^* = 1$, $\varphi^* = \sin x + \cos(x - 1)$, $\lambda_n = n^2\pi^2$, $\varphi_n = (-1)^n n\pi \cos(n\pi x) + \sin(n\pi x)$, where n is an integer number, $n \neq 0$.

16. $\lambda^* = 0$, $\varphi^* = 1 - x$; $\lambda_n = z_n^2$, $\varphi_n = \sin z_n x - z_n \cos z_n x$, where z_n are roots of equation $\tan z = z$.

17. $\lambda_n = z_n^2$, $\varphi_n = \sin z_n x$, where z_n are roots of equation $\tan \pi z = -z$.

18. $\lambda_n = z_n^2$, $\varphi_n = \cos z_n x$, where z_n are roots of equation $\cot z = z$.

19. $\lambda_n = z_n^2$, $\varphi_n = \sin z_n x + z_n \cos z_n x$, where z_n are roots of the equation $(z^2 - 1)\tan z = 2z$.

20. $\lambda_n = n^2\pi^2 + \dfrac{1}{4}$, $\varphi_n = e^{-x/2}\sin(n\pi x)$, where n is an integer number, $n \neq 0$.

21. $\lambda_n = z_n^2$, $\varphi_n = \cos\left(z_n \ln|x + \sqrt{x^2+1}|\right)$, where z_n are roots of the equation $\tan z = \dfrac{\ln|1+\sqrt{2}|}{\sqrt{2}z}$.

22. $\lambda_n = \left(\dfrac{n\pi}{\ln a}\right)^2 + \dfrac{1}{4}$, $\varphi_n = \sqrt{x}\sin\dfrac{n\pi \ln x}{\ln a}$, where $n = 1, 2, \ldots$.

23. $\lambda_n = 1 + i\dfrac{2n\pi}{\ln 2}$, $\varphi_n = x\left(\cos\left(2n\pi\dfrac{\ln x}{\ln 2}\right) - 1 + i\sin\left(2n\pi\dfrac{\ln x}{\ln 2}\right)\right)$, where $n = 1, 2, 3 \ldots$.

24. $\lambda_n = n^2\pi^2$, $\varphi_n = e^x \sin(n\pi x)$, where $n = 1, 2, \ldots$.

25. $\lambda^* = -z^2$, $\varphi^* = e^{-2x}\sinh\sqrt{3}zx$, where z is a root of the equation $3z = 2\tanh 3z$; $\lambda_n = z_n^2$, $\varphi_n = e^{-2x}\sin\left(\sqrt{3}z_n x\right)$, where z_n are roots of the equation $2\tan 3z = 3z$.

26. $\lambda_n = -\dfrac{z_n + n\pi}{\sin z_n}(\cos z_n + i\sin z_n)$, $\varphi_n = -\lambda_n e^{-x} + e^{-\lambda_n x}$, where z_n are roots of the equation $(-1)^n \sin z \exp((z + \pi n)\cot z) = -e^{-1}(z + \pi n)$, and $n = 0, 1, 2, \ldots$.

27. $y = -\dfrac{2}{\pi}\sum_{n=0}^{\infty} \dfrac{\sin(\pi(2n+1)x)}{(2n+1)\left(-1 + \pi^2(2n+1)^2\right)}$.

28. $y = -\sum_{n=1}^{\infty} \dfrac{2\pi(1 - (-1)^n) + (-1)^n}{n(n^2\pi^2 - 1)}\sin(n\pi x)$.

29. $y = \dfrac{3}{\pi}\sum_{n=1}^{\infty} \dfrac{(-1)^n \sin(\pi n x)}{n(-1 + \pi^2 n^2)} - x$.

30. $y = x - \sum_{n=1}^{\infty} \dfrac{\sin(\pi n x)}{\pi n(-1 + \pi^2 n^2)}$.

8.16 Systems with constant coefficients

1. $\begin{cases} x = C_1 \cos 2t + C_2 \sin 2t, \\ y = (2C_2 - C_1)\cos 2t - (2C_1 + C_2)\sin 2t. \end{cases}$

2. $\begin{cases} x = C_1 e^{-t} + C_2 e^{3t}, \\ y = 2C_1 e^{-t} - 2C_2 e^{3t}. \end{cases}$

3. $\begin{cases} x = C_1 e^{3t} + C_2 e^{-3t}, \\ y = 2C_1 e^{3t} - 4C_2 e^{-3t}. \end{cases}$

4. $\begin{cases} x = e^{2t}\left(C_1 \cos t + C_2 \sin t\right), \\ y = e^{2t}\left[(C_1 + C_2)\cos t + (C_2 - C_1)\sin t\right]. \end{cases}$

5. $\begin{cases} x = C_1 e^{t} + C_2 e^{3t}, \\ y = -2C_1 e^{t} - C_2 e^{3t}. \end{cases}$

6. $\begin{cases} x = e^{2t}\left(C_1 + C_2 + C_2 t\right), \\ y = -e^{2t}\left(C_1 + C_2 t\right). \end{cases}$

7. $\begin{cases} x = 2e^{2t}\left(C_1 \cos t - C_2 \sin t\right), \\ y = e^{2t}\left[-(C_1 + C_2)\cos t + (C_1 - C_2)\sin t\right]. \end{cases}$

8. $\begin{cases} x = e^{t}\left(C_1 \cos 3t + C_2 \sin 3t\right), \\ y = e^{t}\left(C_1 \cos 3t - C_2 \sin 3t\right). \end{cases}$

9. $\begin{cases} x = e^{3t}\left(C_1 + C_2 t\right), \\ y = e^{3t}\left(C_1 + C_2 + C_2 t\right). \end{cases}$

10. $\begin{cases} x = e^{t}\left(C_1 + C_2 t\right), \\ y = e^{t}\left(2C_1 - C_2 + 2C_2 t\right). \end{cases}$

11. $\begin{cases} x = e^{-t}\left(C_1 + 2C_2 t\right), \\ y = e^{-t}\left(C_1 + 3C_2 t\right). \end{cases}$

12. $\begin{cases} x = e^{2t}\left(C_1 + 3C_2 t\right), \\ y = e^{2t}\left(C_2 - C_1 - 3C_2 t\right). \end{cases}$

13. $\begin{cases} x = C_2 e^{2t} + C_3 e^{3t}, \\ y = C_1 e^{t} + C_2 e^{2t}, \\ z = C_1 e^{t} + C_2 e^{2t} + C_3 e^{3t}. \end{cases}$

14. $\begin{cases} x = C_1 e^{2t} + e^{3t}\left(C_2 \cos t + C_3 \sin t\right), \\ y = e^{3t}\left[(C_1 + C_2)\cos t + (C_3 - C_2)\sin t\right], \\ z = C_1 e^{2t} + e^{3t}\left[(2C_2 - C_3)\cos t + (2C_3 + C_2)\sin t\right]. \end{cases}$

15. $\begin{cases} x = C_1 e^{t} + C_2 e^{2t} + C_3 e^{5t}, \\ y = C_1 e^{t} - 2C_2 e^{2t} + C_3 e^{5t}, \\ z = -C_1 e^{t} - 3C_2 e^{2t} + 3C_3 e^{5t}. \end{cases}$

16. $\begin{cases} x = -C_2 e^{t} + C_3 e^{2t}, \\ y = e^{t}\left(C_1 + C_2 t\right) - 2C_3 e^{2t}, \\ z = e^{t}\left(C_2 - C_1 - C_2 t\right) + 2C_3 e^{2t}. \end{cases}$

17. $\begin{cases} x = C_1 e^{t} + C_2 e^{2t} + C_3 e^{-t}, \\ y = C_1 e^{t} - 3C_3 e^{-t}, \\ z = C_1 e^{t} + C_2 e^{2t} - 5C_3 e^{-t}. \end{cases}$

18. $\begin{cases} x = C_1 + C_2 t + 4C_3 e^{3t}, \\ y = C_2 - 2C_1 - 2C_2 t + 4C_3 e^{3t}, \\ z = C_1 - C_2 + C_2 t + C_3 e^{3t}. \end{cases}$

19. $\begin{cases} x = C_1 + C_c e^{t}, \\ y = 3C_1 + C_3 e^{t}, \\ z = -C_1 + (C_2 - C_3)\, e^{t}. \end{cases}$

20. $\begin{cases} x = e^{-t}(C_2 + C_3 t), \\ y = 2C_1 e^{t} - e^{-t}(2C_2 + C_3 + 2C_3 t), \\ z = C_1 e^{t} - e^{-t}(C_2 + C_3 + C_3 t). \end{cases}$

21. $\begin{cases} x = C_1 e^{2t} + C_3 e^{-5t}, \\ y = C_2 e^{2t} + 3C_3 e^{-5t}, \\ z = e^{2t}(C_1 - 2C_2) + 2C_3 e^{-5t}. \end{cases}$

22. $\begin{cases} x = e^{t}(C_1 + C_3 t), \\ y = e^{t}(C_2 + 2C_3 t), \\ z = e^{t}(C_1 - C_2 - C_3 - C_3 t). \end{cases}$

23. $\begin{cases} x = C_1 - 3C_2 e^{2t}, \\ y = -2C_2 e^{2t} + C_3 e^{-t}, \\ z = C_1 + C_2 e^{2t} - 2C_3 e^{-t}. \end{cases}$

24. $\exp(At) = \begin{pmatrix} e^{3t} & 0 \\ 0 & e^{-2t} \end{pmatrix}.$

25. $\exp(At) = \begin{pmatrix} \cosh t & \sinh t \\ \sinh t & \cosh t \end{pmatrix}.$

26. $\exp(At) = \begin{pmatrix} 1 - 2t & -4t \\ t & 1 + 2t \end{pmatrix}.$

27. $\exp(At) = \begin{pmatrix} e^{t} & 0 \\ 0 & e^{t} \end{pmatrix}.$

28. $\exp(At) = \begin{pmatrix} \cosh t & i \sinh t \\ -i \sinh t & \cosh t \end{pmatrix}.$

29. $\exp(At) = \begin{pmatrix} \cos t & \sin t \\ -\sin t & \cos t \end{pmatrix}.$

30. $\exp(At) = E \cosh t + A \sinh t.$

31. $\exp(At) = e^{\alpha t} \begin{pmatrix} \cos\beta & \sin\beta \\ -\sin\beta & \cos\beta \end{pmatrix}.$

32. $\vec{x} = e^{3t} E \vec{C}.$

33. $\vec{x} = \begin{pmatrix} \cos t & \sin t \\ -\sin t & \cos t \end{pmatrix} \vec{C}.$

34. $\vec{x} = e^{t} E \begin{pmatrix} \cos t & \sin t \\ -\sin t & \cos t \end{pmatrix} \vec{C}.$

35. $\vec{x} = e^{t} E \begin{pmatrix} 1 & 2t \\ 0 & 1 \end{pmatrix} \vec{C}.$

36. $\vec{x} = e^{t} E \begin{pmatrix} \cos\sqrt{12}t & \frac{3}{\sqrt{12}} \sin\sqrt{12}t \\ -\frac{4}{\sqrt{12}} \sin\sqrt{12}t & \cos\sqrt{12}t \end{pmatrix} \vec{C}.$

37. $\vec{x} = e^{2t}E\begin{pmatrix} \cosh\sqrt{3}t & \sqrt{3}\sinh\sqrt{3}t \\ \frac{1}{\sqrt{3}}\sinh\sqrt{3}t & \cosh\sqrt{3}t \end{pmatrix}\vec{C}.$

38. $\vec{x} = e^{-t}E\begin{pmatrix} 1+2t & -2t \\ 1+2t & -2t \end{pmatrix}\vec{C}.$

39. $\vec{x} = e^{t/2}\left(E\cosh\frac{3}{2}t + \frac{1}{3}\begin{pmatrix} -5 & -8 \\ 2 & 5 \end{pmatrix}\sinh\frac{3}{2}t\right)\vec{C}.$

40. $\vec{x} = e^{t}\left(E\cos 2t + \begin{pmatrix} 1 & -1 \\ 2 & -1 \end{pmatrix}\sin 2t\right)\vec{C}.$

41. $\vec{x} = \left(E\cosh\sqrt{17}t + \begin{pmatrix} -3 & 2 \\ 4 & 3 \end{pmatrix}\frac{1}{\sqrt{17}}\sinh\sqrt{17}t\right)\vec{C}.$

42. $\vec{x} = (E\cosh t + A\sinh t)\vec{C}.$

43. $\vec{x} = e^{t}(E\cosh t + A\sinh t)\vec{C}.$

44. $\vec{x} = e^{-t}\left[E + \left(A - \frac{1}{3}E\mathrm{Sp}A\right)t\right]\vec{C}$, where $\mathrm{Sp}A = \sum a_{ii}$ is the spur of A.

45. $\vec{x} = \left(E + At + \frac{1}{2}At^2\right)\vec{C}.$

46. $\vec{x} = ((E - A^2)(E - At) + A^2\cosh t + A^3\sinh t)\vec{C}.$

47. $\vec{x} = \begin{pmatrix} 0 & e^{t} & e^{-t} \\ e^{2t} & -e^{t} & -e^{-t} \\ e^{2t} & -e^{t} & -2e^{-t} \end{pmatrix}\vec{C}.$

48. $\vec{x} = \begin{pmatrix} 1 & e^{t} & e^{t} \\ -1 & e^{t} & e^{t}(1+t) \\ 1 & -e^{t} & -te^{t} \end{pmatrix}\vec{C}.$

49. $\vec{x} = \begin{pmatrix} e^{-t} & e^{-t}\cos 2t & e^{-t}\sin 2t \\ e^{-t} & -e^{-t}\sin 2t & e^{-t}\cos 2t \\ -e^{-t} & e^{-t}\cos 2t & e^{-t}\sin 2t \end{pmatrix}\vec{C}.$

50. $\begin{cases} x = e^{t}(C_1\cos t + C_2\sin t) - \sin t, \\ y = e^{t}(-C_1\sin t + C_2\cos t) - \cos t. \end{cases}$

51. $\begin{cases} x = C_1e^{t} + C_2e^{-t} + te^{t} - t^2 - 2, \\ y = C_1e^{t} - C_2e^{-t} + (t-1)e^{t} - 2t. \end{cases}$

52. $\begin{cases} x = C_1e^{t} + 2C_2e^{2t} + e^{t}\ln|1 + e^{2t}| + 2e^{t}\arctan e^{t}, \\ y = C_1e^{t} + 3C_2e^{2t} - e^{t}\ln|1 + e^{2t}| + 3e^{2t}\arctan e^{t}. \end{cases}$

53. $\begin{cases} x = 2C_1e^{8t} - 2C_2 - 6t + 1, \\ y = 3C_1e^{8t} + C_2 + 3t. \end{cases}$

54. $\begin{cases} x = 3C_1e^{t} + C_2e^{-t} + 3\sin t, \\ y = C_1e^{t} + C_2e^{-t} - \cos t + 2\sin t. \end{cases}$

55. $\begin{cases} x = C_1\cos t + C_2\sin t + t\sin t - t\cos t, \\ y = C_1(\sin t + \cos t) + C_2(\sin t - \cos t) - 2t\cos t + \sin t + \cos t. \end{cases}$

56. $\begin{cases} x = C_1e^{t}\cos t + C_2e^{t}\sin t + e^{t} + t + 1, \\ y = -C_1e^{t}(\cos t + \sin t) + C_2e^{t}(\cos t - \sin t) - 2e^{t} - 2t - 1. \end{cases}$

57. $\begin{cases} x = C_1\cos t + C_2\sin t + t(\cos t + \sin t) + (\cos t - \sin t)\ln|\cos t|, \\ y = (C_1 - C_2)\cos t + (C_1 + C_2)\sin t + 2\cos(t)\ln|\cos t| + 2t\sin t. \end{cases}$

58. $\begin{cases} x = C_1 \cos t + C_2 \sin t + \tan t, \\ y = -C_1 \sin t + C_2 \cos t + 2. \end{cases}$

59. $\begin{cases} x = C_1 \cos t + C_2 \sin t + \frac{t}{2} \cos t + 1, \\ y = -C_1 \sin t + C_2 \cos t - \frac{t}{2} \sin t - \frac{1}{2} \cos t. \end{cases}$

60. $\begin{cases} x = e^{-4t} \left(C_1 + C_2 t\right) + \frac{4}{25} e^t - \frac{1}{36} e^{2t}, \\ y = -e^{-4t} \left(C_1 + C_2 + C_2 t\right) + \frac{1}{25} e^t + \frac{7}{36} e^{2t}. \end{cases}$

61. $\begin{cases} x = C_1 e^t + C_2 e^{-t} + \sin t, \\ y = -C_1 e^t + C_2 e^{-t}. \end{cases}$

62. $\begin{cases} x = C_1 \left(\cos 2t - \sin 2t\right) + C_2 \left(\cos 2t + \sin 2t\right), \\ y = C_1 \cos 2t + C_2 \sin 2t + e^{-t}. \end{cases}$

63. $\begin{cases} x = C_1 e^t + C_2 \sin t + C_3 \cos t, \\ y = -C_1 e^t + C_2 \cos t - C_3 \sin t + t, \\ z = C_2 \sin t + C_3 \cos t + 1. \end{cases}$

64. $\begin{cases} x = e^{-t} \left(C_1 \left(1 + t\right) + C_2 t - C_3 t\right) + t - 1, \\ y = e^{-t} \left(C_1 t + C_2 \left(1 + t\right) - C_3 t\right) + 1, \\ z = e^{-t} \left(2C_1 t + 2C_2 t + C_3 \left(1 - 2t\right)\right) + t. \end{cases}$

65. $\vec{y} = -\frac{1}{5} \begin{pmatrix} 1 \\ -4 \end{pmatrix} e^{-3x} + \frac{1}{5} \begin{pmatrix} 1 \\ 1 \end{pmatrix} e^{2x}.$

66. $\vec{y} = \begin{pmatrix} \cos 2x + \sin 2x \\ 2 \sin x \end{pmatrix}.$

67. $\vec{y} = \frac{1}{6} \begin{pmatrix} 1 \\ 2 \end{pmatrix} e^{-x} + \frac{2}{3} \begin{pmatrix} 2 \\ 1 \end{pmatrix} e^{2x}$

$\quad -\frac{1}{2} \begin{pmatrix} 1 \\ 2 \end{pmatrix} e^{x} + 2 \begin{pmatrix} 1 \\ 1 \end{pmatrix} x e^{x}.$

68. $\vec{y} = \begin{pmatrix} \cos \sqrt{2}x \\ \sqrt{2} \sin \sqrt{2}x \end{pmatrix}.$

69. $\begin{cases} x = e^t \left(C_1 \cos t + C_2 \sin t\right) + e^{-t} \left(C_3 \cos t + C_4 \sin t\right), \\ y = e^t \left(C_1 \sin t - C_2 \cos t\right) + e^{-t} \left(C_4 \cos t - C_3 \sin t\right). \end{cases}$

70. $\begin{cases} x = 3C_1 e^t + C_2 e^{-t}, \\ y = C_1 e^t + C_2 e^{-t}. \end{cases}$

71. $\begin{cases} x = -2C_2 e^{3t} + C_3 e^t, \\ y = C_1 e^{-t} + C_2 e^{3t}. \end{cases}$

72. $\begin{cases} x = e^t \left(C_1 + C_2 t\right) + C_3 e^{-t}, \\ y = e^t \left(-2C_1 - C_2 - 2C_2 t\right) - 4C_3 e^{-t}. \end{cases}$

73. $\begin{cases} x = 2C_1 e^t + e^{-t/2} 2 \left(C_2 \cos \frac{\sqrt{7}}{2} t + C_3 \sin \frac{\sqrt{7}}{2} t\right) - t - \frac{1}{2}, \\ y = C_1 e^t - e^{-t/2} \left(C_2 \left(3 + \sqrt{7}\right) \cos \frac{\sqrt{7}}{2} t + \left(3 - \sqrt{7}\right) C_3 \sin \frac{\sqrt{7}}{2} t\right) \\ \qquad -\frac{1}{2}. \end{cases}$

74. $\begin{cases} x = C_1 + C_2 e^t + 5C_3 \cos t + 5C_4 \sin t, \\ y = -C_1 - C_2 e^t + \left(3C_4 - 4C_3\right) \cos t - \left(3C_3 + 4C_4\right) \sin t. \end{cases}$

8.17 Linear systems

1. $\begin{cases} y = C_1 + C_2 x, \\ z = 2C_2 + \frac{C_1}{x}. \end{cases}$

2. $\begin{cases} y = e^{x}\left(C_1 \cos x + C_2 \sin x\right), \\ z = e^{-x}\left(C_1 \sin x - C_2 \cos x\right). \end{cases}$

3. $\begin{cases} y = C_1 e^{-1/x} + 2C_2 e^{-2/x}, \\ z = -C_1 e^{-1/x} - 3C_2 e^{-2/x}. \end{cases}$

4. $\begin{cases} x = C_1 + C_2 t + \frac{C_3}{t}, \\ y = C_1 + \frac{C_3}{t}, \\ z = -C_1 - C_2 t - 2\frac{C_3}{t}. \end{cases}$

5. $\begin{cases} x - y = C_1 t, \\ x + z = C_2 e^{t}, \\ x + y - z = C_3. \end{cases}$

6. $\begin{cases} y = C_1 (x - 2), \\ z = C_2 (x - 2). \end{cases}$

7. $\begin{cases} y = C_1 (1 + x^2) + C_2 x e^{x}, \\ z = C_1 x + C_2 e^{x}. \end{cases}$

8. $\begin{cases} y = \frac{C_1}{x} + C_2 x, \\ z = \frac{C_1}{x} - C_2 x. \end{cases}$

9. $\begin{cases} y = C_1 \cos x, \\ z = C_1 x + C_2 = 0. \end{cases}$

10. $\begin{cases} y = C_1 \left(\cos x - \frac{\sin x}{x}\right) - C_2 \left(\sin x + \frac{\cos x}{x}\right), \\ z = \frac{1}{x}\left(C_1 \sin x + C_2 \cos x\right). \end{cases}$

11. $\begin{cases} y = \frac{C_1}{x} + C_2 x^3, \\ z = -C_1 + 3C_2 x^4. \end{cases}$

12. $\begin{cases} y = C_1 e^{x} \cos x - C_2 \sin x, \\ z = C_1 e^{x} \sin x + C_2 \cos x. \end{cases}$

13. $\begin{cases} y = x - C_1 x^{\frac{1+\sqrt{5}}{2}} - C_2 x^{\frac{1-\sqrt{5}}{2}}, \\ z = C_1 x^{\frac{1+\sqrt{5}}{2}} + C_2 x^{\frac{1-\sqrt{5}}{2}}. \end{cases}$

14. $\begin{cases} y = \frac{2C_1}{x^3} - \frac{C_2}{x^4} + \frac{3}{10}x + \frac{1}{15}x^2, \\ z = -\frac{C_1}{x^3} + \frac{C_2}{x^4} - \frac{1}{20}x + \frac{2}{15}x^2. \end{cases}$

15. $\begin{cases} y - \frac{1}{3}z = \frac{C_1}{\sin x} + \frac{x}{2\sin x} - \frac{1}{2}\cos x + \frac{1}{9}\cos^2 x \cot x, \\ y - z = C_2 \sin x + x \sin x - \cos x \sin x - \sin x \ln\left|\tan \frac{x}{2}\right|. \end{cases}$

16. $\begin{cases} y + z = C_1 \cos^2 x + \cos^2 x \ln\left|\tan\left(\frac{\pi}{4} + \frac{x}{2}\right)\right| + \cos x, \\ y + 2z = C_2 \cos^3 x + \cos^3 x \ln\left|\tan x\right| + \cos x. \end{cases}$

17. $\begin{cases} y = 2C_1 e^{-2x^3} - C_2 e^{-3x^3} + \frac{5}{9}, \\ z = C_2 e^{-3x^3} - C_1 e^{-2x^3} - \frac{1}{18}. \end{cases}$

18. $\begin{cases} y = \frac{C_1}{x^2} + \frac{x}{3}, \\ z = C_2 e^{x} - \frac{C_1}{x^2} - \frac{x}{3}. \end{cases}$

19. $\begin{cases} x = C_1 t + C_2 t^2 + \frac{C_3}{t} + 1, \\ y = C_1 t - C_2 t^2 + \frac{2C_3}{t} + 2, \\ z = 2C_1 t + 3C_2 t^2 + \frac{C_3}{t} + 3. \end{cases}$

20. $\begin{cases} x = C_1 t^2 + t, \\ y = C_1 + C_2 t, \\ z = \frac{C_1}{t} + C_2 + C_3 t. \end{cases}$

21. $\begin{cases} x + y - z = \frac{C_1}{t} + \frac{1}{2}t + \frac{1}{4}t^3 - 4, \\ x + y + 2z = C_2 t^2 + \frac{1}{2}t + \frac{1}{4}t^3 + 8, \\ x - y = \frac{C_3}{t^2} + \frac{1}{4}t - \frac{1}{4}t^3. \end{cases}$

22. $y = -e^{-\sin x}, \quad z = 2e^{\sin x}$.

23. $y = x^2 + \dfrac{1}{x}, \quad z = 2x - \dfrac{1}{x^2}$.

24. $y = 1 + x, \quad z = 1 - x$.

25. $y = (1 + x)^2, \quad z = x + 2$.

26. $y = (1 + x^2)^2, \quad z = x^2 + 2$.

27. $y = x + \dfrac{1}{1 - x}, \quad z = 1 + x$.

28. $y = \dfrac{1}{2 - x}, \quad z = x$.

29. $x = \dfrac{t}{3}, \quad y = -\dfrac{t}{3}$.

8.18 Nonlinear systems

1. $y = C_1 e^{C_2 x}, \quad z = C_1 C_2 e^{C_2 x}$.

2. $y = \left(C_1 e^x + C_2 e^{-x} - \dfrac{1}{2}\sin x\right)^{1/2}, \quad z = C_1 e^x - C_2 e^{-x} - \dfrac{1}{2}\cos x$.

3. $y = \dfrac{\sin(x + C_2)}{\sqrt{1 + C_1 e^{-2x}}}, \quad z = \dfrac{\cos(x + C_2)}{\sqrt{1 + C_1 e^{-2x}}}$.

4. $y = C_2 e^{C_1 x^2}, \quad z = \dfrac{e^{-C_1 x^2}}{2C_1 C_2}$.

5. $y = C_2 e^{C_1 x}, \quad z = x + \dfrac{C_2}{C_1} e^{C_1 x}; \quad y = 0, \quad z = C + x$.

6. $x = e^t (2C_1 C_2 t + C_3) - C_2 e^t + C_2^2 e^{2t} - C_1^2$,
$y = C_1 + C_2 e^t, \quad z = e^t (2C_1 C_2 t + C_3) + C_2^2 e^{2t}$.

7. $x = \dfrac{C_1}{C_2} t e^{-C_1 t^2}, \quad y = 2C_2 e^{C_1 t^2}$.

8. $x = C_2 \exp\left(\dfrac{e^{2t}}{2C_1^2}\right), \quad y = C_1 e^{-t}$.

9. $y = C_2 e^{C_1 x^2}, \quad z = 2\dfrac{C_1}{C_2} x e^{-C_1 x^2}; \quad y = 0, \quad z = Cx$.

10. $x = C_1 e^{-t} + C_2 \exp\left(\dfrac{e^{2t}}{8C_1^2}\right), \quad y = C_1 e^{-t} - C_2 \exp\left(\dfrac{e^{2t}}{8C_1^2}\right)$.

11. $y = -\dfrac{1}{C_1} + \dfrac{C_1}{2}(x + C_2) - \dfrac{C_1}{4}(x + C_2)^2,$
$z = \dfrac{C_1}{4}(x + C_2)^2 + \dfrac{1}{C_1}.$

12. $x = e^{2t}(C_1 - 4C_2 t), \quad y = (2C_2 e^{2t} + C_3 e^t)^{1/2}, \quad z = C_3 e^{2t}.$

13. $y = \dfrac{x + C_1}{x + C_2}, \quad z = \dfrac{x(C_2 - C_1)}{(x + C_2)^2}.$

14. $y = \ln|C_1 + e^x|, \quad x + z\left(C_2 + \dfrac{1}{2}z\right) = 0.$

15. $y = C_1 x^2 + \dfrac{1}{4C_1}, \quad \dfrac{1}{3}z^3 - xz - \dfrac{C_1}{3}x^3 - \dfrac{1}{4C_1}x = C_2;$
$y = x, \quad \dfrac{1}{3}z^3 - xz - \dfrac{1}{2}x^2 = C; \quad y = -x, \quad \dfrac{1}{3}z^3 - xz + \dfrac{1}{2}x^2 = C.$

16. $ze^{-x} + y = C_1, \quad ze^{-y} + x = C_2.$

17. $\dfrac{x}{y} = C_1, \quad xy + z^2 = C_2.$

18. $x - y + z = C_1, \quad \ln|x| + \dfrac{y}{z} = C_2.$

19. $\sqrt{y} - \sqrt{x} = C_1, \quad z - \sqrt{x} = C_2.$

20. $x^2 + y^2 + z^2 = C_1, \quad lx + my + nz = C_2.$

21. $y - z^2 = C_1, \quad (y - z)^2 + 2x = C_2.$

22. $(y - t)z = C_1, \quad (y - t)\exp\left(\dfrac{t}{z(y - t)}\right) = C_2.$

23. $\dfrac{x}{y} = C_1, \quad xy - 2\sqrt{1 + z^2} = C_2.$

24. $y^2 + z^2 = C_1, \quad x(y - z) = C_2.$

25. $x - y + z = C_1, \quad \ln|x| + \dfrac{z}{y} = C_2.$

26. $x^2 - z^2 = C_1, \quad y^2 - u^2 = C_2, \quad \dfrac{x + z}{y + u} = C_3.$

27. $x + z = C_1, \quad y + u = C_2, \quad (x - z)^2 + (y - u)^2 = C_3.$

28. $x^2 + y^2 + z^2 = C_1, \quad \dfrac{yz}{x} = C_2.$

29. $\sin x - \sin y = C_1, \quad \sin x - z = C_2.$

30. $\dfrac{x}{y} = C_1, \quad xy - z^2 = C_2.$

31. $ux = C_1 e^t + C_2 e^{-t}, \quad uy = C_1 e^t - C_2 e^{-t},$
$uz = C_1 e^t - C_2 e^{-t} + C_3, \quad u = 1 - C_1 e^t - C_2 e^{-t} - C_3 t.$

32. $ux = C_1 e^t, \quad uy = C_2 e^t + C_3 e^{-t}, \quad uz = C_2 e^t - C_3 e^{-t},$
$u = 1 - (C_1 + C_2)e^t + C_3 e^{-t}.$

33. $ux = (C_1 + C_2 t)e^t + C_3 e^{2t}, \quad uy = (C_1 - 2C_2 + C_2 t)e^t,$
$uz = (C_1 - C_2 + C_2 t)e^t + C_3 e^{2t}, \quad u = \exp\left(-\dfrac{C_2}{C_1 - 2C_2 + C_2 t}\right).$

34. $ux = C_1 e^t + C_2 e^{5t}, \quad uy = -C_1 e^t + 3C_2 e^{5t},$
$u = 1 - 5C_1 e^t + \dfrac{7}{5}C_2 e^{5t}.$

35. $ux = C_1 e^{2t}, \quad uy = C_2 e^t,$

$$u = \left[1 - 2\left(\frac{C_2}{C_1}\right)^2 (C_2^2 e^{2t} - 2C_1 C_2 e^t + C_1^2) \exp\left(\frac{C_1}{C_2} e^t\right)\right]^{1/2}.$$

36. $ux = C_1 + C_2 e^{2t}, \quad uy = C_1 - C_2 e^{2t},$

$$u^{-1} = 1 - \frac{1}{2C_2^2}\sqrt{C_1^2 - C_2^2 e^{4t}}.$$

37. $x = e^{pt}(C_1 \cos qt + C_2 \sin qt), \quad y = e^{pt}(C_2 \cos qt - C_1 \sin qt).$

38. $x = -\dfrac{C_1 + t}{(C_1 + t)^2 + C_2^2}, \quad y = \dfrac{C_2}{(C_1 + t)^2 + C_2^2}.$

39. $ux = C_2 \cos t - C_2^2 - C_1 \sin t - C_1^2,$

$uy = C_2^2 - C_2 \sin t + C_1^2 - C_1 \cos t,$

$u = (C_1 + C_2 - \cos t)^2 + (C_1 - C_2 + \sin t)^2.$

40. $2x = \ln|C_2^2 + (C_1 + t)^2|, \quad y = \arctan\dfrac{C_2}{C_1 + t}.$

41. $x = (C_1 + t)^2 - C_2^2, \quad y = 2C_2(C_1 + t).$

42. $ux = C_2^2 + (2t + C_1 + 1)(2t + C_1 - 1), \quad uy = 2C_2,$

$u = C_2^2 + (2t + C_1 + 1)^2.$

43. $y = C_3\left[\tan\left(C_2 + \dfrac{t}{2}\right)\right]^{C_1}, \quad z = 1 - \dfrac{1}{C_1}\cos(t + 2C_2);$

$y = 0,$ z is an arbitrary function.

44. $x = C_1 e^t + C_2 e^{-t} - C_3 - t, \quad y = \sqrt{C_3 + t}.$

45. $x = C_2 e^{t/2}\cos\left(C_1 + \dfrac{\sqrt{3}}{2}t\right), \quad y = C_3 e^{-t/2}\sin\left(C_1 + \dfrac{\sqrt{3}}{2}t - \dfrac{\pi}{6}\right);$

$x = 0$ or $y = 0$, y or x are arbitrary functions, respectively.

46. $y = C_1 x^2 + C_2 x + C_3, \quad yz = -C_1.$

47. $x = y = C_3(C_2 + C_c t)^{2/C_1}, \quad z = C_1 x + C_2, \quad C_1 \neq 0;$

$x = y = C_2 e^{2t/C_1}, \quad z = C_1, \quad C_1 \neq 0.$

48. $t + C_3 = \pm\displaystyle\int \frac{\sin\theta d\theta}{\sqrt{2(C_2 - \cos\theta)\sin^2\theta - C_1^2}},$

$\varphi + C_4 = C_1 \int \dfrac{dt}{\sin^2\theta(t)}.$

49. $s = \dfrac{g \sin\alpha}{6 + \cos^2\alpha} t^2 + C_1 t + C_2, \quad \dfrac{1}{2} s(t)\cos\alpha + C_3 t + C_4.$

50. $t + C_4 = \pm\displaystyle\int \sqrt{\frac{1 + \sin^2\varphi}{2(C_3 - \cos\varphi - C_1^2/4)}}\, d\varphi,$

$x = \dfrac{1}{2}(C_1 t + \sin\varphi) + C_2.$

51. $t + C_4 = \pm\displaystyle\int \frac{\sin\theta d\theta}{\sqrt{2(C_3 - \cos\theta)\sin^2\theta - C_1^2 - C_2^2 + 2C_1 C_2 \cos\theta}},$

$\varphi + C_5 = \displaystyle\int \frac{C_1 - C_2 \cos\theta(t)}{\sin^2\theta(t)} dt, \quad \psi + C_6 = \int \frac{C_2 - C_1 \cos\theta(t)}{\sin^2\theta(t)} dt.$

8.19 Linear partial equations

1. $u = f\left(\dfrac{y}{x}, \dfrac{z}{x}\right)$.
2. $z = f(xy + y^2)$
3. $u = \left(\dfrac{x-y}{z}, \dfrac{(x+y+2z)^2}{z}\right)$.
4. $z^2 = 2x + f\left(\dfrac{y}{z}\right)$.
5. $3xyz = y^3 + f(xy)$.
6. $z = xy + xf\left(\dfrac{y}{x}\right)$.
7. $z = a\sin\left[xy + f\left(\dfrac{y}{x}\right)\right]$.
8. $z[y - xf(y)] = y^2 f(y) + xy$.
9. $z + x^2 + y^2 = xf\left(\dfrac{y}{x}\right)$.
10. $z = rf(\ln r + \phi);\ x\tan\phi = y,\ r^2 = x^2 + y^2$.
11. $(x+y+2z)(x+y+z)^2 = f[(x-y)(x+y+2z)]$.
12. $y + ze^{-x} = f(x + ze^{-y})$.
13. $y + z + u = x^2 f[x(y-z), x(y-u)]$.
14. $12u = x^4 - 2x^3(y+z) + 6x^2yz + f(y-x, z-x)$.
15. $u = \dfrac{xy}{z}\ln x + xf\left(\dfrac{y}{x}, \dfrac{z}{x}\right)$.
16. $u = x^2\ln x + x^2 f\left(\dfrac{y}{x}, \dfrac{z}{x}\right)$.
17. $z^2 = xy + f\left(\dfrac{y}{x}\right)$.
18. $z = e^{mx}\cos(py + \alpha) + \phi(y + ax);\ m\tan\alpha = ap$.
19. $z^2 + xy = a^2 + h^2$.
20. $2xyz - x^2 = 2$.
21. $z^2 = x^2 - y^2$.
22. $z^2 = 2xy - 2(x+y-1) + (x+y-1)^4$.
23. $(x^2+y^2+z^2)^2 = 2a^2(x^2+y^2+xy)$.
24. $z = x^2y$.
25. $2x^2 - y^2 - z^2 = a^2$.
26. $\left[(y^2z-2)^2 - x^2 + z\right]y^2z = 1$.
27. $x^2 + z^2 = 5(xz - y)$.
28. $3(x+y+z)^2 = x^2 + y^2 + z^2$.
29. $xz = (xz - y - x + 2z)^2$.
30. $(1+yz)^3 = 3yz(1+yz-x) + y^3$.

8.20 Pfaffian equation

1. $x^2yz^3 = C.$
2. $x^2z^2 + yx + z^3 = C.$
3. $z = x + C(xy - 1).$
4. $x^2yz = C - x^3$; also $x = 0$.
5. $z = Ce^{x^2} - y^2 - x.$
6. $e^{x^2}(x + y + z^2) = C$
7. $z - y = Cxz.$
8. $\dfrac{1}{x-y} + \dfrac{1}{y-z} + \dfrac{1}{z-x} = C.$
9. $(1 - z^2)x + y = Cz.$
10. not integrable.
11. $x^2 - y^2 + z^2 = \dfrac{1}{2}(x + y).$
12. not integrable.
13. $z = x + C(1 + xy).$
14. $xy + yz + xz = C(x + y + z).$
15. $z = (y + C)(x - y^2).$
16. $z = Ce^{y/x}.$
17. $z = e^{xy}(y + C).$
18. $z = x^2(y + \ln x - x + C).$
19. $z = C(x + y)^2.$
20. $x - \dfrac{z}{y+a} = C$
21. $y(z + x) = C(y + z).$
22. $\dfrac{y+z}{x} + \dfrac{z+x}{y} = C.$
23. $x^2 + xy^2 + x^2z - u = C.$
24. $\dfrac{x}{y} - \ln z = C.$

8.21 Nonlinear partial equations

1. $z = (3x + ay + b)^{2/3} - a^2/9.$
2. $\dfrac{3}{2}z = (x - a)^{3/2} - (a - y)^{3/2} + b.$
3. $z = ax + by + f(a, b).$
4. $2az = ak + 4 + (a + ay + b)^2.$
5. $az^2 = (x + b)^2 + 4y^2.$
6. $4z = (x + a)^2 + (y + a)^2 + (x - y)\left[\dfrac{1}{2}(x - y)^2 - a^2\right]^{1/2}.$
$-\sqrt{2}a^2 \ln\left\{x - y + [(x - y)^2 - 2a^2]^{1/2}\right\} + b.$

7. $z = ax^3 + \dfrac{y^3}{9a} + b.$
8. $z = ax + by + a^3b^3.$
9. $z = x\sin a + ay + b.$
10. $z^3a^2 = y + ax + b.$
11. $z^2a^2 = x + ay + b.$
12. $z = ax + by + a^2.$
13. $\sqrt{R^2 - z^2} + \dfrac{ax}{\sqrt{a^2+1}} + \dfrac{y}{\sqrt{a^2+1}} + b = 0.$
14. $z = ax + by - a^2b.$
15. $z = (2b^2+1)x + by + a.$
16. $z = ax + y\ln a + b.$
17. $z = a^3x + a^{-2}y + b.$
18. $az = (x + ay + b)^2.$
19. $z = \pm\cosh\left(\dfrac{x + ay + b}{\sqrt{1+a^2}}\right).$
20. $z^2 - a^2 = (x + ay + b)^2$, or $z = b.$
21. $z^2(1 + a^3) = 8(x + ay + b)^3.$
22. $z = be^{ax + a^2y}.$
23. $3z = (2x + a)^{3/2} + 3ay + 3b.$
24. $2az = a^2x^2 + y^2 + 2ab.$
25. $az = ax^2 + a^2x + e^{ay} + ab.$
26. $(2z - ay^2 - 2b)^2 = 16ax.$
27. $z = a(e^x + e^y) + b.$
28. $az = a^2x + a\sin x + \sin y + ab.$
29. $z = ax + by - a^2b$, singular solution $z^2 = x^2y.$
30. $zx = ax + by - a^2b$, singular solution $z^2 = y.$
31. $z = 3x^3 - 3ax^2 + a^2x + 2y^2 - 4ay^3 + 3a^2y^2 - a^3y + b.$
32. $3a(x + ay + b) = (1 + a^3)\ln z$, singular solution $z = b.$
33. $z(1 + a^2 + b^2) = (x_1 + ax_2 + bx_3 + c)^2.$
34. $z = ax - \left(2 + 3a + \dfrac{1}{2}a^2\right)y + b.$
35. $z = a\tan(x + ay + b)$, singular solution $z = b.$
36. $y^2[(x-a)^2 + y^2 + 2z] = b.$
37. $z^2 = 2ax + a^2y^2 + b.$
38. $z = ax + be^y(y + a)^{-2}.$
39. $z^2 = 2(a^2+1)x^2 + 2ay + b.$
40. $z = ax + 3a^2y + b.$
41. $(z^2 + a^2)^3 = 9(x + ay + b).$
42. $z = x^3 + ax + \dfrac{2}{3}(y + a)^{3/2} + b.$
43. $z = ax + by + a^2 + b^2.$
44. $z = -2 - \ln xy.$

45. $3z = xy - x^2 - y^2$.
46. $8z^3 = -27x^2y^2$.
47. $xz = -y$.
48. $z = 0$.
49. $z^2 = 1$.
50. $z = 0$.

8.22 Phase plane. Linear systems

1. Unstable spiral point.
2. Saddle point.
3. Stable center.
4. Asymptotically stable spiral point.
5. Unstable improper node.
6. Unstable proper node.
7. Unstable improper node.
8. Unstable improper node.
9. Asymptotically stable spiral point.
10. Saddle point.
11. Center.
12. Unstable improper node.
13. Critical points lie on the curve $y = \frac{3}{2}x$.
14. Critical points lie on the curve $y = 2x$.
15. Unstable spiral point.
16. Saddle point.
17. Unstable spiral point.
18. Asymptotically stable improper node.
19. Asymptotically stable improper node.
20. Asymptotically stable improper node.
21. Asymptotically stable if $a < 0$, unstable if $a \geq 0$.
22. Unstable.
23. Stable if $a \leq -\frac{1}{2}$.
24. Unstable.
25. Stable.
26. $0 < |a| < 1$ spiral point; $a = 0$ center; $|a| \geq 1$ improper node.
27. $a > 0$ saddle point; $a < 0$ improper node; $a = 0$ rest line.
28. $a = 0$ center; $|a| \geq \sqrt{2}$ improper node, $|a| < \sqrt{2}$ spiral point..
29. $a < 1$ saddle point; $a > 1$ improper node; $a = 1$ rest line.
30. $a > 0$ saddle point; $a < 0$ improper node; $a = 0$ rest line.

8.23 Almost linear systems

1. $(-1,\ 2k\pi)$ stable, $(-1,\ (2k+1)\pi)$ unstable.
2. $(2k\pi\ ,\ 0)$ unstable $((2k+1)\pi\ ,0)$ stable.
3. $(3,\ 2)$ unstable, $(0,\ -1)$ stable.
4. $(2,\ 1)$ stable $(-2,\ 1)$ unstable.
5. The zero solution is stable.
6. The zero solution is stable.
7. The zero solution is stable.
8. The zero solution is stable.
9. The zero solution is unstable.
10. The zero solution is stable.
11. The zero solution is unstable.
12. The zero solution is asymptotically stable.
13. The zero solution is asymptotically stable.
14. The zero solution is stable.
15. The zero solution is stable.
16. The zero solution is unstable.
17. The zero solution is unstable.
18. The zero solution is stable.
19. The zero solution is stable
20. The zero solution is unstable.
21. The zero solution is stable.
22. The zero solution is unstable.
23. The zero solution is stable.
24. The zero solution is stable.
25. The zero solution is stable.
26. The zero solution is unstable.
27. The zero solution is stable.
28. $a > 0,\ ab > 1 + a^2$.
29. $a > 2/3,\ b > 0,\ 9b - 6a + 4 < 0$.
30. $b > 0,\ 3a - b > 0$.
31. Unstable for all a.
32. $b > 0,\ a > b + 1$.
33. $0 < a < 2$.
34. $a > 0,\ b > 0,\ 8a - a^2 b > 4$.
35. The zero solution is stable.
36. The zero solution is stable.
37. The zero solution is stable.
38. The zero solution is stable.
39. The zero solution is stable.
40. The zero solution is stable.

41. The zero solution is stable.
42. The zero solution is unstable.
43. The zero solution is unstable.

8.24 Liapunov's second method

1. $v = 2x^2 + y^2$.
2. $v = 2x^2 + y^2$.
3. $v = 2x^2 + y^2$.
4. $v = 2x^2 + y^2$.
5. $v = 2x^2 + y^2$.
6. $v = 2x^2 + y^2$.
7. $v = 2x^2 + y^2$.
8. $v = x^4 + y^4$.
9. $v = 2x^2 - y^2$.
10. $v = x^2 + y^2$.
11. $v = x^2 + y^2$.
12. $v = 2x^2 + y^2$.
13. $v = x^2 + y^2$.
14. $v = x^2 - y^2$.
15. $v = 2x^2 + y^2$.
16. $v = x^2 + 3y^2$.
17. $v = x^4 - y^4$.
18. $v = x^4 + 2y^2$.
19. $v = x^2 + y^2$.
20. $v = x^2 + y^2$.
21. $v = -x^2 + 2y^2$.
22. $v = x^2 + y^2$.
23. $v = x^2 - y^2$.
24. $v = x^2 + 2y^2 + 3z^2$.

8.25 Euler's equation

1. $y = \dfrac{x^3 - x}{12}$.
2. $y = \dfrac{\mathrm{e}}{x} - \ln x$.
3. $y = x^3 - x$.
4. There is no solution.
5. $y = \dfrac{\sinh x}{\sinh 1} - \dfrac{x}{2}$

6. $y = \frac{1}{2}xe^x - \sinh x$.
7. $y = e^x - e^{-3x}$.
8. $y = [e^{-x} + (1+e)xe^{-x} - 1]$.
9. $y = 0$, if $\alpha = 0$; there is no solution if $\alpha \neq 0$.
10. $y = x + 1$.
11. $y = \frac{\sinh x}{\sinh 1}$.
12. $y = e^{2(1-x)}$.
13. There is no solution.
14. $y = C_1 + C_2 x - \frac{x^2}{4}$.
15. There is no solution.
16. $y = -x$.
17. $y = x$.
18. $y = \frac{1}{2}(1+x)$.
19. $y = \left(\frac{2x}{3}\right)^{3/2}$.
20. $y = -\frac{x^2}{4} - \frac{x}{2} + \frac{3}{4}$.
21. $y = \frac{1}{4}(x - x^2)$.
22. $y = \sin 2x,\ z = -\frac{x^2}{2} + \frac{32+\pi^2}{8\pi}x$.
23. $y = -\frac{1}{6}(x^3 + 5x - 6),\ z = x$.
24. $y = \sin x,\ z = \sin x$.
25. $y = \frac{x^2}{2} + 1,\ z = 1$.
26. $y_1 = \sin x,\ y_2 = -\sin x$.
27. $y_1 = \sinh x,\ y_2 = -\sinh x$.
28. $y_1 = \sin x,\ y_2 = -\sin x$.
29. $y_1 = e^x,\ y_2 = e^{-x}$.
30. $y_1 = x^4,\ y_2 = x^3$.
31. $y_1 = \ln x + 1,\ y_2 \equiv 0$.
32. $y_1 = -\frac{x}{2}\sin x - \cos x - C\sin x + x,\ y_2 = -\frac{x}{2}\sin x + C\sin x$.
33. $y_1 = \frac{x^2}{2} + 1,\ y_2 \equiv 1$.

8.26 Conditional extremum

1. $y_1 = \cos x + \sin x,\ y_2 = \cos x - \sin x$.
2. $y_1 = (x-2)e^{-x},\ y_2 = (1-x)e^{-x}$.

3. $y_1 = 2 - x,\ y_2 = x + 1.$
4. $y_1 = x^2 + x,\ y_2 = x^2 - x.$
5. There is no solution.
6. $y_1 = \frac{x}{2}\sin x,\ y_2 = \frac{1}{2}(\sin x - x\cos x).$
7. $y_1 = x^2 + \cos x + \sin x,\ y_2 = x^2 - \cos x - \sin x.$
8. $y_1 = 2x - 3,\ y_2 = 1 - x.$

8.27 Isoperimetric problem

1. $y = \frac{5}{2}x^3 - \frac{3}{2}x.$
2. $y = 60x^3 - 96x^2 + 36x.$
3. $y = \pi^{-1}(x - 2\sin x).$
4. $y = x \pm \sin x.$
5. $y = xe^{-x}.$
6. $y = 3x^2 + 2x + 1.$
7. $y = \frac{2x - x^2}{4}.$
8. $y = \pm\sqrt{\frac{2}{\pi}}\sin nx.$
9. $y_1 = -6x^2 + 6x,\ y_2 = 3x^2 - 2x.$
10. $y_1 = 3x^2 - 2x,\ y_2 = 3x^2 - 6x$
    and $y_1 = -3x^2 + 4x,\ y_2 = -3x^2.$
11. $y_1 = 2\sin k\pi x,\ y_2 = -2\sin k\pi x.$
12. $y_1 = 3x - x^3,\ y_2 = x^3 - x$
    and $y_1 = x^3 + x,\ y_2 = -x^3 + x.$
13. $y = -\frac{5}{2}x^2 + \frac{7}{2}x,\ z = x.$

8.28 Movable end points

1. $y = -2x,\ x_1 = 1.$
2. $y = x/\sqrt{2},\ x_1 = \sqrt[6]{2}.$
3. $y = \frac{3}{4} - x,\ x_0 = \frac{1}{2},\ x_1 = 23/8.$
4. $y = \frac{1}{4}(1 - x)x.$
5. $y = \frac{1}{4}(x^2 - 1).$
6. $y = C\sin x.$
7. There is no solution.
8. $y = \sqrt{2 - (x - 1)^2},\ x_1 = 2.$
9. $y = (x - 1 - \pi/4)\sin x.$

10. $y = \dfrac{x^2 - x - 1}{2}$.
11. $\sqrt{c^2 + \dfrac{a^2}{2}}$.
12. $\dfrac{4}{\sqrt{5}}$.
13. $\sqrt{20}$.
14. $2\sqrt{2} - 1$.
15. $\dfrac{1}{\sqrt{10}}$.
16. $\dfrac{\sqrt{11}}{2}$.
17. $\sqrt{17 + 4\sqrt{6}}\left(\dfrac{5}{2} - \sqrt{6}\right)$.
18. 1.
19. $\sqrt{3} - 1$.

8.29 Bolza problem

1. $y \equiv 0$.
2. $y = \sinh x$.
3. $y = e^x + \sin x$.
4. $y = \ln x + 1$
5. $y = -e^x(e^3 + 1)^{-1}$.
6. $y = 2\ln(x + 1)$.
7. $y = \sin x + \cos x$.

8.30 Euler-Poisson equation

1. $y = (1 - x)\sinh x$.
2. $y = \sinh x$.
3. $y = \dfrac{x^2}{2}$
4. $y = x(x - 1)^2$.
5. $y = x^4$.
6. $y = \dfrac{1}{10}\left(x^5 + 3x^3 - 2x^2\right)$.
7. $y = x + \cos x$.
8. $y = xe^x$.
9. $y = (x + 1)^{-1}$.
10. $y = x^4$.
11. $y = \sinh x$.

12. $y = x - \sin x.$
13. $y = \sinh x - \sin x.$
14. $y = \pm\sqrt{\dfrac{2}{\pi}}\sin x.$
15. $y = \pm\sqrt{\dfrac{2}{\pi}}\sin x.$

8.31 Ostrogradsky equation

1. $\left(\dfrac{\partial z}{\partial x}\right)^2 \dfrac{\partial^2 z}{\partial x^2} + \left(\dfrac{\partial z}{\partial y}\right)^2 \dfrac{\partial^2 z}{\partial y^2} = f.$
2. $z = y.$
3. $\Delta\Delta z = 0.$
4. $\dfrac{\partial^2 z}{\partial x^2} + \dfrac{\partial^2 z}{\partial y^2} = 0.$
5. $\dfrac{\partial H_z}{\partial y} - \dfrac{\partial H_y}{\partial z} = \dfrac{\partial E_x}{\partial t}$

$$\frac{\partial H_x}{\partial z} - \frac{\partial H_z}{\partial x} = \frac{\partial E_y}{\partial t}$$
$$\frac{\partial H_y}{\partial x} - \frac{\partial H_x}{\partial y} = \frac{\partial E_z}{\partial t}$$
$$\frac{\partial E_x}{\partial x} + \frac{\partial E_y}{\partial y} + \frac{\partial E_z}{\partial z} = 0.$$

6. $(c^2 - u^2)\dfrac{\partial^2 \varphi}{\partial x^2} - 2uv\dfrac{\partial^2 \varphi}{\partial x \partial y} + (c^2 - v^2)\dfrac{\partial^2 \varphi}{\partial y^2} = 0,$

$$c^2 = \frac{-q^2 G'(q)}{qG''(q) - G'(q)}$$

7. $\left(1 + u_y^2 + u_z^2\right) u_{xx} + (1 + u_z^2 + u_x^2)\, u_{yy} + \left(1 + u_x^2 + u_y^2\right) u_{zz}$
$-2\left(u_x u_y u_{xy} + u_y u_z u_{yz} + u_x u_z u_{xz}\right) = 0.$

Bibliography

[1] *R P Agnew, Differential Equations. New York: McGraw-Hill, 1962.*

[2] *L A Al'sevich and L P Cherenkova, Practice on Ordinary Differential Equations Minsk 1990 (in Russian).*

[3]. *V I Arnold, Ordinary Differential Equations. Cambridge: M.I.T. Press, 1973.*

[4] *R Bellman, Stability Theory of Differential Equations. New York, 1953.*

[5] *G Birkhoff and G C Rotta, Ordinary Differential Equations. New York: Wiley, 1989.*

[6] *G A Bliss, Calculus of Variation. Open Court. La Salle, III 1925.*

[7] *W E Boyce and R C Di Prima, Elementary Differential Equations and Boundary Value Problems. New York: Wiley, 1969.*

[8] *F Brauer, Ordinary Differential Equations. New York: Benjamin, 1967.*

[9] *M Braun, Differential Equations and Their Applications. New York: Springer-Verlag, 1975.*

[10] *L Cesari, Asymptotic Behavior and Stability Problems in Ordinary Differential Equations. New York: Academic Press, 1963.*

[11] *E A Coddington, An Introduction to Ordinary Differential Equation. Englewood Cliffs, N.J.: Prentice-Hall, 1961.*

[12] *E A Coddington and N Levinson, Theory of Ordinary Differential Equation. New York: McGraw-Hill, 1955.*

[13] *R Courant and D Hilbert, Methods of Mathematical Physics. Interscience I,1953, II 1962.*

[14] *A B Dickenson, Differential Equations. Reading, Mass.: Eddison-Wesley, 1972.*

[15] *G Doetsch, Guide to the Application of Laplace Transforms. New York: Van Nostrand Reinhold, 1967.*

[16] *A V Efimov (editor), A Book of Problems in Mathematics. p.2 Moscow: Nauka, 1981 (in Russian).*

[17] *A V Efimov (editor), A Book of Problems in Mathematics. p.4 Moscow: Nauka, 1990 (in Russian).*

[18] *L E Elsgolts, Differential Equations and the Calculus of Variations. Moscow: Mir Publishers, 1980.*

[19] *A Erdelyi, W Magnus, F Oberhettinger and F Tricomi Tables of Integral Transforms, vol. 1. New York: McGraw-Hill, 1954.*

[20] *A F Filippov, A Book of Problems in Differential Equations. Moscow: Nauka Publishers, 1973 (in Russian).*

[21] *L R Ford, Differential Equations. New York: McGraw-Hill, 1955.*

[22] *A R Forsyth, A Treatise on Differential Equations. New York: St. Martin's Press, 1954.*

[23] *J K Hale, Ordinary Differential Equations. New York: Wiley, 1969.*

[24] *P Hartman, Ordinary Differential Equations. New York: Wiley, 1964.*

[25] *E Hille, Lectures on Ordinary Differential Equations. Reading, Mass.: Addison-Wesley, 1969.*

[26] *W Hurewicz, Lectures on Ordinary Differential Equations. Cambridge: M.I.T. Press, 1958.*

[27] *I M Gel'fand and S V Fomin, Calculus of Variations. Englewood Cliffs, N.J.: Prentice-Hall, 1963.*

[28] *H H Goldstine, A History of the Calculus of Variations. New York: Springer-Verlag, 1980.*

[29] *E Goursat, Cours d'analyses mathematique, III. Gauthier-villars, 1927.*

[30] *F S Gudymenko, I A Pavlyuk and V A Volkova, A Book of Problems in Differential Equations. Kiev, 1972 (in Ukrainian).*

[31] *E L Ince, Ordinary Differential Equation. Longmans, Green, 1927.*

[32] *E Kamke, Differentialgleichungen: Losungsmethoden und Losungen. Leipzig:Akademische Verlag, 1943; New York: Chelsea reprint, 1971.*

[33] *W Kaplan, Advanced Calculus. Reading, Mass.: Addison-Wesley, 1959.*

[34] *W Kaplan, Operational Methods for Linear Systems. Reading, Mass.: Addison-Wesley, 1962.*

[35] *W Kaplan, Ordinary Differential Equations. Reading, Mass.: Addison-Wesley, 1958.*

[36] *M L Krasnov, A I Kiselyov and G I Makarenko, A Book of Problems in Ordinary Differential Equations. Moscow: Mir Publishers, 1989.*

[37] *M L Krasnov, A I Kiselyov and G I Makarenko, Functions of Complex Variables. Operation Calculus. Theory of Stability. Moscow: Nauka, 1981.*

[38] *M L Krasnov, A I Kiselyov and G I Makarenko, Problems and Exercises in the Calculus of Variations. Moscow: Mir Publishers, 1975.*

[39] *N W Mc.Lachlan, Modern Operation Calculus. New York: Dover, 1962.*

[40] *J LaSalle and S Lefschetz, Stability by Liapunov's Direct Method, with Applications . New York: Academic Press, 1961.*

[41] *N M Matveev, A Book of Problems and Exercises in Ordinary Differential Equations. Leningrad: Leningrad Univers. Press, 1960 (in Russian).*

[42] *S G Mikhlin, Variational Methods in Mathematical Physics. New York: Macmillan, 1964.*

[43] *P J Olver, Applications of Lie groups to Differential Equations. New York: Springer-Verlag, 1986.*

[44] *H Philips, Differential Equations. New York: Wiley, 1924.*

[45] *G Sansone, Equazioni differenziali nel campo reale, P.1-2. Bologna, Zanichelli, 1948 - 1949.*

[46] *I S Sokolnikoff, Mathematical Theory of Elasticity. New York: McGraw-Hill, 1956.*

[47] *I S Sokolnikoff and R M Redheffer, Mathematics of Physics and Modern Engineering. New York: McGraw-Hill, 1966.*

[48] *M R Spiegel, Applied Differential Equations. New York: Prentice-Hall, 1967.*

[49] *J A Tierney, Differential Equations. Boston: Allyn and Bacon, 1979.*

[50] *F G Tricomi, Equazioni differenziali. Torino: Boringhieri, 1967.*

[51] *R Weinstock, Calculus of Variations, with Applications to Physics and Engineering. New York: McGraw-Hill, 1952.*

Index